# Nomenclature of Organic Compounds

## Principles and Practice

# Nomenclature of Organic Compounds

## Principles and Practice

*Edited by*

**John H. Fletcher,** The University of Connecticut
**Otis C. Dermer,** Oklahoma State University
**Robert B. Fox,** Naval Research Laboratory

for the Committee on

Nomenclature of the

Division of Organic

Chemistry of the American

Chemical Society.

ADVANCES IN CHEMISTRY SERIES **126**

AMERICAN CHEMICAL SOCIETY

WASHINGTON, D. C.     1974

ADCSAJ 126 1-337 (1974)

Library of Congress Catalog Card 73-92675

ISBN 8412-0191-9

PRINTED IN THE UNITED STATES OF AMERICA

# Advances in Chemistry Series

**Robert F. Gould,** *Editor*

# FOREWORD

ADVANCES IN CHEMISTRY SERIES was founded in 1949 by the American Chemical Society as an outlet for symposia and collections of data in special areas of topical interest that could not be accommodated in the Society's journals. It provides a medium for symposia that would otherwise be fragmented, their papers distributed among several journals or not published at all. Papers are refereed critically according to ACS editorial standards and receive the careful attention and processing characteristic of ACS publications. Papers published in ADVANCES IN CHEMISTRY SERIES are original contributions not published elsewhere in whole or major part and include reports of research as well as reviews since symposia may embrace both types of presentation.

# CONTENTS

# PREFACE

A t its initial meeting in April 1946 the newly formed Committee on Nomenclature of the Division of Organic Chemistry, ACS, took note of the obvious need among chemists for some kind of handbook to supplement the then-available published rules and methods of organic nomenclature. Except for information contained in the pamphlet "The Naming and Indexing of Chemical Compounds by Chemical Abstracts" (1945) and the 1930 Rules of the International Union of Chemistry, very little material of a definitive or official nature had yet appeared in print.

Early in 1948 the Chairman, Dr. Howard S. Nutting, distributed to the Committee a draft of a "Preamble on the Nomenclature of Organic Compounds." This document, though brief, presented the generally accepted principles then in use for forming structurally descriptive names of organic compounds. Although Dr. Nutting's draft was discussed at subsequent meetings, the proposed Preamble was never published owing to the Committee's preoccupation with other projects and its difficulty in resolving differences of opinion on questionable points. Indeed, the minutes of the meeting of September 1950 state that a majority of the Committee's members believed it useless at that time to continue work toward the publication of a nomenclature handbook.

In 1957, after a two-year hiatus following the untimely death of the Chairman, Dr. Mary Alexander, the Committee was reconstituted with one of the undersigned (JHF) as chairman and Dr. Nutting as a continuing member. With the latter's encouragement the handbook project was reactivated, and by November 1958 an outline of chapter topics had been developed and a prototype chapter written by the Chairman.

After reaching general agreement on the scope and format of the proposed handbook, the Committee began to prepare a manuscript. As work progressed each member contributed drafts of one or more chapters, which were then reviewed by the entire group; names of the individual contributors are listed, along with their affiliation at that time, on page xiv. In 1969 the completion and revision of the book were turned over to the undersigned editorial committee. The earlier chapter drafts were updated extensively and otherwise revised by this editorial committee, which therefore takes responsibility for the present form and content. In a few cases, recommendations for naming are quite different from those developed by the original contributors. For this reason the chapters do not carry the names of individual authors.

ix

The book is intended to provide the practicing chemist with a ready reference source designed to answer day-to-day questions about names and formulas of organic compounds. Comparison with other sources of information on nomenclature is natural. For various reasons the Rules issued by the IUPAC Commission on the Nomenclature of Organic Chemistry are more descriptive and less prescriptive of usage than the present book; they are also still incomplete. Both the IUPAC Rules and the practices used by *Chemical Abstracts* in compiling subject indexes have influenced us in writing this book. The major shift to systematic nomenclature adopted by *Chemical Abstracts* in 1972 is echoed and sometimes exceeded here. To be sure, indexing nomenclature, which cannot tolerate scatter of entries among alternative names, is not necessarily always the best for general use; nonetheless, we believe that many trivial (common) names should be abandoned in favor of systematic ones. The IUPAC Rules and the policies of *Chemical Abstracts* are frequently referred to (not always with specific Rules or page numbers), especially when we extend or disagree with their naming methods.

The writer of a dictionary or of a language grammar (such as is constituted by this book) must choose a course somewhere between (a) recognizing all common practices without evaluation and (b) approving some usage and condemning others. While we do not, and indeed can not, venture to set rules, we have recommended what we consider the best up-to-date practice, often with a statement of our reasons. In a few chapters the recommendations are innovative, going beyond most current usage; such are the proposals that heterocyclic systems be named with replacement names rather than the old conventional ones (Chapter 6) and that hydroxylamines and hydrazines be designated as derivatives of azane and diazane, respectively (Chapter 32). Most of the book, however, is simply an exposition of good modern usage.

The first 15 chapters are devoted to general principles; the rest cover the naming of families of compounds, mostly grouped according to the kind of functional group present. Structural formulas are used extensively both in text examples and in further illustrations listed at the ends of chapters. Reference lists of names of substituting groups (less properly called "radical" names) are provided as appendices.

The editors acknowledge with sincere appreciation the efforts of all those who have contributed their time and thought to the preparation of this book. In addition special thanks must go to Drs. Kurt Loening and Mary A. Magill of The Chemical Abstracts Service: to Dr. Loening for serving as reviewer and consultant on all the chapters and to Dr. Magill for preparing the Index.

September 1973

J. H. FLETCHER
O. C. DERMER
R. B. FOX

# ABOUT THE BOOK AND ITS USE

As indicated in the Preface, the process of assembling, reviewing, and updating subject material leading to the final version of this nomenclature manual has covered a period of about 15 years. During that time more has been accomplished in official nomenclature development and agreement at the international level (by the IUPAC Commissions) than in all previous years. The trend of the IUPAC and other nomenclature work has been unmistakably clear: systematically formed names for chemical compounds are becoming increasingly preferable to trivial (nonsystematic) names. A major step in accord with this trend has been taken recently by Chemical Abstracts Service; this is the comprehensive revision of the established methods of naming compounds for listing in the Subject Indexes of *Chemical Abstracts.* Although the changes instituted by the editors of *Chemical Abstracts* became official too late to be reflected fully in this book, many of the recommendations made here are closely consistent with the new policies adopted there.

From the beginning of the handbook project much thought has been given to the purpose of the book and the manner in which it might be used. Although designed primarily for reference use in naming individual organic molecular structures, real or hypothetical, it may also be used for instructional purposes. An attempt has been made to present basic principles in a logical sequence throughout the first 15 chapters, and numerous cross references are provided to assist the reader. However, by ordinary textbook standards the book is probably somewhat overconcise and lacking in explanatory material.

The arrangement of Chapters 16-40 is alphabetical by chapter title to enable the user to bypass the Index in most instances when looking for information on how to name a compound belonging to a given functional class—*e.g.,* an amine, an ether, or a phosphorus compound. Each chapter begins with a brief introduction indicating the scope of the chapter and defining pertinent terms used therein. Recommendations are then provided for naming compounds containing the functional group under consideration. A discussion section dealing with special problems and proscribing poor practices follows next. The chapter concludes with a list of examples illustrating acceptable usage; for these examples the preferred name is cited first in each case. Where appropriate, names that should be abandoned entirely are so marked.

In Chapters 16-40 the various functional classes are treated one at a time; compounds of mixed function are not discussed. Therefore, in naming structures containing more than one kind of functional group one should first consult Chapter 10 to determine which kind of function takes precedence and how the remaining kinds are accommodated. Having ascertained which kind of functional group ranks highest, one should next examine the chapter dealing with that class for instructions in naming the parent compound. Finally, one should consult other functional class chapters for information on naming the lower-ranking groups as substituents. Broadly applicable principles such as order of citation of prefix names, numbering, punctuation, and word separation are not dealt with in Chapters 16-40, except for those aspects specific to the functional class in question; these subjects are treated in earlier chapters as basic principles.

Although the book, as presently constituted, does not address directly the problem of deducing structures from names, it may often be helpful for this purpose. However, the most comprehensive and useful source of information for use in translating chemical names into molecular structures is provided by the Subject Indexes of *Chemical Abstracts;* these indexes contain not only systematic names but trivial names and trade names as well; they cover the published literature over a period of many years and therefore include names that are now obsolete or little used but still crop up.

In searching *Chemical Abstracts* for the meaning of an unfamiliar name of an organic compound, a good approach is to consult the Collective Subject Indexes, beginning with the most recent ones and working back timewise. If the given name is systematic, or partly so, one must usually look first for the portion of the name that designates the parent compound; having found the parent name as a main index entry, the next step is to examine the subheadings, which denote substituent groups. Even though the specific name in question is not listed, close analogs can often be found that provide guidance. For meanings of names of individual substituent groups one should consult the comprehensive alphabetical list of organic "radical" names apprearing in the Introduction to the Subject Index to Volume 66 of *Chemical Abstracts* (January-June 1967); this list provides good coverage prior to the major changes in nomenclature instituted for the Ninth Collective Period, which begins with Volume 76 (January-June 1972).

It is important to note that until November 1971 work on the nomenclature recommendations presented in this book was conducted independently without any detailed knowledge of what changes were being planned by the research and development staff of *Chemical Abstracts.* This course was deliberate; the objectives of the two groups, though closely related, were at no time identical. Chemical names best suited for indexing are not always the best choices for other purposes, and the editors of this book have given no special consideration to indexing problems. There is, however, a reassuring degree of agreement between the present recommendations and the new naming procedures adopted by *Chemical Abstracts.* This suggests that the trend toward wider use of systematic organic nomenclature will continue and very likely accelerate in formal scientific communication.

# Contributors

Leonard T. Capell, Chemical Abstracts Service, Columbus, Ohio (retired)

Otis C. Dermer, Oklahoma State University, Stillwater, Okla.

Norman G. Fisher, E. I. du Pont de Nemours & Co., Inc., Wilmington, Del.

John H. Fletcher, American Cyanamid Co., Stamford, Conn.

Robert B. Fox, Naval Research Laboratory, Washington, D.C.

Harriet Geer, Parke, Davis & Co., Detroit, Mich.

Gilbert E. Goheen, Southern Regional Research Laboratory, USDA, New Orleans, La.

Charles D. Hurd, Northwestern University, Evanston, Ill. (retired)

Floyd L. James, Miami University, Oxford, Ohio

Stanley P. Klesney, The Dow Chemical Co., Midland, Mich.

Kurt L. Loening, Chemical Abstracts Service, Columbus, Ohio

Mary Magill, Chemical Abstracts Service, Columbus, Ohio (retired)

Howard S. Nutting, The Dow Chemical Co., Midland, Mich. (retired)

Milton Orchin, University of Cincinnati, Cincinnati, Ohio

Leslie B. Poland, Ethyl Corp., Detroit, Mich.

Russell J. Rowlett, Virginia Institute for Scientific Research, Richmond, Va.

Louis Schmerling, Universal Oil Products Co., Des Plaines, Ill.

# Origin and Evolution of
# Organic Nomenclature

Alchemists frequently assigned names to substances with the intent of concealing what they were working on from their fellow workers. This attitude has gradually changed throughout the years, so that chemists now strive to reveal their latest theories or discoveries to others either orally or through the printed medium. To communicate easily with one another, it has been necessary to devise a special language which we call chemical nomenclature (*1,2,3,4,5*). We expect this nomenclature not only to define a given compound uniquely but to reveal what atoms are present, how the atoms are arranged within the molecule, and chemical relationships with other compounds as well.

Chemical nomenclature as we know it today had its beginning when Guyton de Morveau, Lavoisier, Berthollet, and Fourcroy published their "Methode de Nomenclature Chimique" in 1787. Their chief contributions were to indicate in a name the elements and, so far as possible, the relative proportions of these elements present in a given compound. Names such as alcohol, ether, and succinic acid were included in their recommendations although their primary concern was with inorganic compounds.

Organic nomenclature really got its start in the 1830's with Berzelius, Liebig, and Wöhler when names such as benzoyl chloride and ethyl iodide were proposed (*6*). The first effective consideration of organic nomenclature on an international basis came about in 1889 when an International Commission for the Reform of Chemical Nomenclature was organized. Three years later, 34 of the leading chemists from 9 European countries met and agreed upon what has become known as the **Geneva Rules** for nomenclature (*7*). The objective of this group was to provide names suitable for the systematic indexing of aliphatic organic compounds. They introduced the following principles which are still considered to be of primary importance in naming aliphatic compounds: (1) the name of the longest chain is taken as the parent name; (2) the presence of a functional or characteristic group of atoms is indicated by a suffix. They intended to complete the task by developing rules for naming cyclic compounds, but this never came to pass.

In 1911 an International Association of Chemical Societies was organized. Nomenclature reform was among the subjects considered by this group, but World War I intervened before any recommendations were made. It is interesting to note that C. I. Istrati of Bucharest, a participant in the Geneva Congress of 1892, published in 1913 what is probably the most extensive treatise (*8*) on nomenclature ever written (1210 pages). However, this treatise remained hidden from the rest of the world until long after its publication.

International cooperation was again initiated in 1920 when the **International Union of Chemistry** was organized. At its meeting in Brussels the following year, three commissions were appointed for the "reform of chemical nomenclature": one for organic, a second for inorganic, and a third for biological chemistry. Reports of these commissions were published thereafter in the *Comptes rendus de l' Union Internationale de Chimie*. The **Organic Commission** published its **Definitive Report** (*9*), consisting of 68 rules (the so-called Liege Rules), in 1930, and these were supplemented by less extensive reports of the meetings at Lucerne in 1936 and at Rome in 1938.

World War II interrupted the work of the International Union of Chemistry (IUC) and its commissions once more, and nothing further was done until a meeting was held in London in 1947. At that time, the name of the Union was changed to the **International Union of Pure and Applied Chemistry** (IUPAC), and the word "reform" was dropped from the names of its nomenclature commissions. Since 1947, the **Commission on the Nomenclature of Organic Chemistry** of IUPAC has, in the main, confined its efforts to codifying sound practices that already exist rather than working on the origination of new nomenclature. Thus, in the IUPAC Rules, acceptable alternative methods have been recognized where, for various reasons, limitation to a single method of nomenclature appears undesirable or unfeasible.

Since 1947 the Organic Commission of IUPAC has published the following sets of Definitive Rules (*10, 11, 12, 13*):

**Nomenclature of Organic Chemistry, 1949**
   Nomenclature of Organosilicon Compounds
   Changes and Additions to the Definitive Report
   Extended List of Radical Names

**Nomenclature of Organic Chemistry, 1957**
   Section A.   Hydrocarbons
   Section B.   Fundamental Heterocyclic Systems

**Nomenclature of Organic Chemistry, 1965**
   Section C.   Characteristic Groups containing Carbon,
       Hydrogen, Oxygen, Nitrogen, Halogen,
       Sulfur, Selenium, and/or Tellurium.

**Nomenclature of Organic Chemistry, 1971**
   Sections A, B, and C (slightly revised)

Today, the path of nomenclature rules from the first proposals to acceptance in definitive form by IUPAC can be a tortuous one, and it must be admitted that the elapsed time for the entire operation can sometimes be painfully long. For a proposal originating in the United States the procedure is as follows:

(1) An individual or a group of workers in a particular field of chemistry drafts new or revised rules that seem to fit its needs. Such a group may be made up of experts appointed as an *ad hoc* subcommittee of an established nomenclature committee, or it may not have any official status at all. In either case a formal report presenting the proposed rules and the reasons why they should be adopted for general use is prepared.

(2) The original report is submitted for review, revision (if necessary), and approval by an officially recognized committee on nomenclature such as that of one of the divisions of the American Chemical Society. Currently, the Organic, Inorganic, Polymer, Carbohydrate, and Fluorine Divisions have active nomenclature committees.

(3) The report is then reviewed by the American Chemical Society's Committee on Nomenclature.

(4) After approving the report, the ACS Committee submits it to the ACS Council for official acceptance on behalf of the Society.

(5) Concurrently, committees of the National Research Council's Division of Chemistry and Chemical Technology may review the report and submit it to IUPAC with their comments and recommendations. Alternatively, the report may be submitted to an IUPAC Commission directly, without review by an NRC Committee.

(6) The appropriate IUPAC Commission considers the report and its proposals along with any others on the same subject submitted from other sources. Factors such as past and current usage, adaptability to other languages, and the overall potential utility of the proposed rules are considered. Note is also taken of any relevant actions by other nomenclature committees.

(7) The IUPAC Commission then submits its recommendations to the IUPAC Council for official adoption and publication.

It has been found that the time span between Step 1 and Step 7 can be reduced greatly and the quality of new nomenclature rules improved if simultaneous attacks are made on the problem from all possible points of view. This can be accomplished by inviting experts in the field, irrespective of their nationality, and members of nomenclature committees or commissions at all levels to review progress reports and final reports of the initiating group or subcommittee at appropriate stages in the development of the new rules. On the other hand, premature publication of proposed new nomenclature rules without approval of the appropriate committees not only complicates the process leading to the official adoption of those rules but also leads to confusion among chemists and journal editors as to which kinds of names are considered acceptable for use in published reports of research.

Ideally, each IUPAC Commission should be the leader in its particular field of nomenclature and should anticipate the needs of chemists throughout the world. Practically speaking, this is quite impossible. New compounds in new fields are being reported in the literature every day. Thus, the burden of providing acceptable names for these compounds necessarily falls upon research workers and journal editors and upon abstracting services such as *Chemical Abstracts*. To maintain consistency in its indexing, the latter has published a compilation of its nomenclature rules and practices (*14, 15, 16*) for the use of its own staff and for the guidance of users. At its own discretion, *Chemical Abstracts* revises its rules in the light of new IUPAC recommendations as they become available. At the same time, IUPAC depends heavily on *Chemical Abstracts*, on Beilstein's *Handbuch*, and on journal editors for guidance when considering new nomenclature recommendations.

Although it is easy to point out many imperfections and inconsistencies in today's chemical nomenclature rules, a few words about their advantages and virtues are in order. The language of chemistry is said to involve more terms and to be better ordered than the language of any other scientific discipline. All of this has come about despite the fact that until 1949 relatively few nomenclature rules had been agreed upon internationally. The acceptance and codification of many new or revised rules by the IUPAC commissions and by other leaders in nomenclature such as *Chemical Abstracts* and the ACS nomenclature committees have done much towards maintaining clear and useful communications among chemists. Continuation of these efforts will ensure that improved means for describing and identifying chemical composition and molecular structure will be available for use in our published scientific literature and indexes, in the classroom, and on the lecture platform.

### Literature Cited

1. Patterson, A. M., Cross, E. J., *in* "Thorpe's Dictionary of Applied Chemistry," 4th

ed., I. Heilbron *et al.,* Eds., Vol. VIII, pp. 594-620, Longmans, London, 1947.
2. ADVAN. CHEM. SER. (1953) **8**.
3. Crosland, M.P., "Historical Studies in the Language of Chemistry," Harvard University Press, Cambridge, 1962.
4. Capell, L. T., Loening, K. L., *in* "Treatise on Analytical Chemistry," I. M. Kolthoff, P. J. Elving, Eds., Pt. II, Vol. 11, pp. 1-44, Interscience, New York, 1965.
5. Loening, K. L., *in* "Kirk-Othmer Encyclopedia of Chemical Technology," A. Standen, Ed., 2nd ed., Vol. 14, pp. 1-15, Interscience, New York, 1967.
6. v. Liebig, J., Wöhler, F., *Ann. Pharm.* (1932) 3, 249.
7. Tiemann, F., *Ber.* (1893) 26, 1595.
8. Istrati, C. I., "Studiu Relativ la o Nomenclatura Generala in Chimia Organica," Bucuresti Librariile Soced, 1913.
9. *Compt. rend. Conf. IUC, 10th, Liege* (1930) 57; *J. Amer. Chem. Soc.* (1933) 55, 3905.
10. *Compt. rend. Conf. IUPAC, 15th, Amsterdam* (1949) 127; *Compt. rend. Conf. IUPAC, 16th* (1951) 100 (reprints available from Chemical Abstracts Service, Columbus, Ohio 43210).
11. IUPAC, "Nomenclature of Organic Chemistry. Definitive Rules for Section A, Hydrocarbons, and Section B. Fundamental Heterocyclic Systems," Butterworths, London, 1958; *J. Amer. Chem. Soc.* (1960) 82, 5545.
12. IUPAC, "Nomenclature of Organic Chemistry. Definitive Rules for Section C. Characteristic Groups Containing Carbon, Hydrogen, Oxygen, Nitrogen, Halogen, Sulfur, Selenium, and/or Tellurium," Butterworths, London, 1965; *Pure Appl. Chem.* (1965) 11.
13. IUPAC, "Nomenclature of Organic Chemistry, Sections A, B, and C," 3rd ed., The Butterworth Group, London, 1971.
14. "The Naming and Indexing of Chemical Compounds from Chemical Abstracts," *Chem. Abstr.* (1962) 56, 1N,
15. "Selection of Index Names for Chemical Substances," *Chem. Abstr.* (1972) 76, Index Guide.
16. "Chemical Abstracts Service, Chemical Substance Name Selection Manual for the Ninth Collective Period," Chemical Abstracts, 1972.

# 2

# Acyclic and Alicyclic Hydrocarbons

On the whole, the systematic nomenclature of aliphatic hydrocarbons is so well understood and accepted that chemists generally have little difficulty in this area, and the urge to invent new trivial names is pretty well suppressed. This situation probably developed because of the inherent simplicity of the compounds themselves and the relatively early international agreement on aliphatic hydrocarbon nomenclature.

In this chapter, the following classes of compounds are discussed: the various **acyclic hydrocarbons** such as the alkanes, alkenes, alkynes, and alkadienes; the **alicyclic hydrocarbons** such as cyclopropane, cyclopropene, and their homologs; the **bridged alicyclic hydrocarbons** such as bicyclo[2.2.1]heptane; also, briefly, **bridged aromatic hydrocarbons** (*see also* Chapter 4); and, finally, **spiro alicyclic hydrocarbons** such as spiro-[3.3]heptane. Also considered are names of substituting groups that result conceptually from the removal of one or more hydrogen atoms from an aliphatic hydrocarbon molecule.

## Recommended Nomenclature Practice

**Acyclic Saturated Hydrocarbons.** Straight-chain saturated hydrocarbons have names ending in *ane,* and their class name is **alkane.** Except for the first four members of the series, which have trivial names, the first portion of the name cites the number of carbon atoms. For examples, *see* Table 2.1.

**Branched-chain hydrocarbons** have systematic names made up of two parts; the terminal portion is the name of the longest straight chain present in the compound—*i.e.,* the parent chain, and preceding this are the substituting-group names of side chains. Arabic number prefixes indicate location of the branching. Such prefixes, called locants, are listed in numerical sequence, separated from each other by commas and from the remainder of the name by a hyphen. The entire hydrocarbon name is written as one word.

$$CH_3CHCHCH_2CH_3$$
with $CH_3$ substituents on the 2nd and 3rd carbons

2,3-Dimethylpentane

## Table 2.1. Names of Straight-Chain Alkanes

| | | | |
|---|---|---|---|
| $CH_4$ | Methane | $C_{17}H_{36}$ | Heptadecane |
| $C_2H_6$ | Ethane | $C_{18}H_{38}$ | Octadecane |
| $C_3H_8$ | Propane | $C_{19}H_{40}$ | Nonadecane |
| $C_4H_{10}$ | Butane | $C_{20}H_{42}$ | Eicosane |
| $C_5H_{12}$ | Pentane | $C_{21}H_{44}$ | Heneicosane |
| $C_6H_{14}$ | Hexane | $C_{22}H_{46}$ | Docosane |
| $C_7H_{16}$ | Heptane | $C_{23}H_{48}$ | Tricosane |
| $C_8H_{18}$ | Octane | $C_{30}H_{62}$ | Triacontane |
| $C_9H_{20}$ | Nonane[a] | $C_{31}H_{64}$ | Hentriacontane |
| $C_{10}H_{22}$ | Decane | $C_{32}H_{66}$ | Dotriacontane |
| $C_{11}H_{24}$ | Undecane[b] | $C_{33}H_{68}$ | Tritriacontane |
| $C_{12}H_{26}$ | Dodecane | $C_{40}H_{82}$ | Tetracontane |
| $C_{13}H_{28}$ | Tridecane | $C_{50}H_{102}$ | Pentacontane |
| $C_{14}H_{30}$ | Tetradecane | $C_{60}H_{122}$ | Hexacontane |
| $C_{15}H_{32}$ | Pentadecane | $C_{100}H_{202}$ | Hectane |
| $C_{16}H_{34}$ | Hexadecane | $C_{132}H_{266}$ | Dotriacontahectane |

[a]Formerly called enneane.
[b]Formerly called hendecane.

Four branched alkanes have trivial names recognized in the 1957 IUPAC Rules, but these names (shown as alternates) are to be used only for the unsubstituted hydrocarbons themselves.

$$CH_3$$
$$CH_3\overset{|}{C}HCH_3$$

2-Methylpropane
Isobutane *(unsubstituted only)*

$$CH_3$$
$$CH_3\overset{|}{C}HCH_2CH_3$$

2-Methylbutane
Isopentane *(unsubstituted only)*

$$CH_3$$
$$CH_3-\overset{|}{\underset{|}{C}}-CH_3$$
$$CH_3$$

2,2-Dimethylpropane
Neopentane *(unsubstituted only)*

$$CH_3$$
$$CH_3\overset{|}{C}HCH_2CH_2CH_3$$

2-Methylpentane
Isohexane *(unsubstituted only)*

In branched hydrocarbons the longest straight chain is numbered from one end to the other with Arabic numerals, the direction being chosen so as to give the lowest numbers possible to the positions of side chains. When alternative series of locants containing the same number of terms are compared term by term, that series is "lowest" which contains the lowest number at the first point of difference.

$$CH_3CHCCH_2CH_2CH_2CCH_3$$

with substituents:
$CH_3$, $CH_3$, $H_3C$, $CH_3$, $CH_3$

2,2,6,6,7-Pentamethyloctane
*not* 2,3,3,7,7-Pentamethyloctane

If two or more different species of side chains are present, their names are cited in alphabetical order. In applying alphabetical order, multiplying prefixes such as *di* and *tri* are disregarded. Thus, ethyl precedes dimethyl. Structure-defining prefixes, when separated from the name by a hyphen and italicized, are also disregarded; thus, neopentyl and isopropyl are considered to start with *n* and *i*, respectively, but in *sec*-butyl, *tert*-butyl, and *tert*-pentyl, the first letter is considered to be *b*, *b*, and *p*, respectively.

$$CH_3CH_2CHCHCH_2CH_2CHCH_3$$

with substituents:
$CH_3$, $CH_3$, $CH_2CH_3$

5-Ethyl-2,6-dimethyloctane

$$CH_3CH_2CHCH_2CHCH_2CH_2CHCH_3$$

with substituents:
$CH_3$, $CH_3$, $CH_3CHCH_2CH_3$

5-*sec*-Butyl-2,7-dimethylnonane

If chains of equal length are competing for selection as the parent chain in a branched alkane, the choice goes to the chain carrying the greatest number of side chains.

$$CH_3CHCHCH_2CH_2CHCH_3$$

with substituents:
$CH_3$, $CH_3$, $CH_2CH_3$

3-Ethyl-2,6-dimethylheptane
*not* 5-Isopropyl-2-methylheptane

Note that the alternative name 5-ethyl-2,6-dimethylheptane would not satisfy the earlier stated requirement for lowest possible locant numbers.

**Substituting univalent groups** derived from acyclic saturated hydrocarbons by conceptual removal of one hydrogen atom are named by replacing the ending *ane* of the hydrocarbon name with *yl*. Numbering begins with the carbon atom having the free bond. The class name for such groups is **alkyl**. Eight trivial names for branched alkyl groups are recognized in the 1957 IUPAC Rules. These, together with the corresponding systematic names, are included in Table 2.2.

**Substituting bivalent groups** derived from acyclic saturated hydrocarbons by conceptual removal of two hydrogen atoms are named in either of two ways. If both free bonds are on the same carbon atom, the ending *ane* of the hydrocarbon name is replaced with *ylidene*. The class name for such groups is **alkylidene**. Trivial names for branched alkylidene

## Table 2.2. Names of Straight-Chain and Branched Alkyl Groups

| | | | |
|---|---|---|---|
| $CH_3-$ | Methyl | $CH_3CH_2CH_2CH_2CH_2-$ | Pentyl |
| $CH_3CH_2-$ | Ethyl | | |
| $CH_3CH_2CH_2-$ | Propyl | $CH_3\overset{\underset{\mid}{CH_3}}{C}HCH_2CH_2-$ | 3-Methylbutyl *Isopentyl |
| $CH_3\overset{\underset{\mid}{CH_3}}{C}H-$ | 1-Methylethyl *Isopropyl | $CH_3CH_2CH_2\overset{\underset{\mid}{CH_3}}{C}H-$ | 1-Methylbutyl |
| $CH_3CH_2CH_2CH_2-$ | Butyl | $CH_3CH_2\overset{\underset{\mid}{CH_3CH_2}}{C}H-$ | 1-Ethylpropyl |
| $CH_3\overset{\underset{\mid}{CH_3}}{C}HCH_2-$ | 2-Methylpropyl *Isobutyl | $CH_3-\overset{\underset{\mid}{CH_3}}{\overset{\mid}{C}}-CH_2-$ | 2,2-Dimethyl-propyl *Neopentyl |
| $CH_3CH_2\overset{\underset{\mid}{CH_3}}{C}H-$ | 1-Methylpropyl *sec-Butyl | $CH_3CH_2-\overset{\underset{\mid}{CH_3}}{\overset{\mid}{C}}-$ | 1,1-Dimethyl-propyl *tert-Pentyl |
| $CH_3-\overset{\underset{\mid}{CH_3}}{\overset{\mid}{C}}-$ | 1,1-Dimethylethyl *tert-Butyl | $CH_3CH_2CH_2CH_2CH_2CH_2-$ | Hexyl |
| | | $CH_3\overset{\underset{\mid}{CH_3}}{C}HCH_2CH_2CH_2-$ | 4-Methylpentyl *Isohexyl |

* For the unsubstituted group only

groups corresponding to those approved for alkyl groups are recognized in the 1957 IUPAC Rules (*see* Table 2.2). Usage is heavily in favor of methylene over methylidene for the first member of the series.

| | | | |
|---|---|---|---|
| $CH_2{<}$ | Methylene Methylidene | $CH_3CH_2\overset{\underset{\mid}{CH_3}}{C}{<}$ | 1-Methylpropylidene *sec-Butylidene |
| $CH_3CH{<}$ | Ethylidene | | |
| $CH_3CH_2CH{<}$ | Propylidene | $CH_3-\overset{\underset{\mid}{CH_3}}{\overset{\mid}{C}}-CH{<}$ | 2,2-Dimethylpropylidene *Neopentylidene |
| $CH_3\overset{\underset{\mid}{CH_3}}{C}{<}$ | 1-Methylethylidene *Isopropylidene | | |

* For the unsubstituted group only

If the two free bonds are on different carbon atoms, the straigh?-chain group terminating in these two carbon atoms is named systemati ally by citing the number of methylene ($CH_2$) groups making up the chain. Side chains are named as substituting groups. The name ethylene has been used much more widely than dimethylene for the first member of the series.

$-CH_2CH_2-$      Ethylene
                     Dimethylene

$-CH_2CH_2CH_2-$   Trimethylene

$$\overset{\displaystyle CH_3}{\underset{\displaystyle |}{-CH_2CH-}}$$
Methylethylene
*Propylene

*For the unsubstituted group only

$-CH_2CH_2CH_2CH_2-$   Tetramethylene

$$\overset{\displaystyle CH_3CH_2}{\underset{\displaystyle |}{-CH_2CH-}}$$
Ethylethylene
*not* Butylene

$$-CH_2\overset{\displaystyle CH_3}{\underset{\displaystyle |}{CH}}CH_2-$$
2-Methyltrimethylene

$$-CH_2-\overset{\displaystyle CH_3}{\underset{\displaystyle \underset{\displaystyle CH_3}{|}}{\overset{\displaystyle |}{C}}}-CH_2-$$
2,2-Dimethyltrimethylene
*not* Neopentylene

**Substituting trivalent groups** derived from acyclic saturated hydrocarbons by conceptual removal of three hydrogen atoms from the same carbon atom are named by replacing the ending *ane* of the hydrocarbon name with *ylidyne*. The class name for such groups is alkylidyne.

$-CH{<}$     Methylidyne
                   $CH_3CH_2C{\displaystyle \stackrel{\diagup}{\diagdown}}$     Propylidyne

If the free bonds are on each end of a chain and also on an intermediate carbon atom, the ending *triyl* is affixed to the full name of the hydrocarbon; however, the 1957 IUPAC Rules also recognize endings combining *yl* and *ylidene* as shown.

$CH{\displaystyle \stackrel{\diagup}{\diagdown}}$     Methylidyne
            Methanylylidene
          *not* Methine

$-CH_2CH{<}$    1,1-Ethanetriyl
                Ethanylylidene

$$-\overset{\displaystyle CH_3}{\underset{\displaystyle |}{CHCH}}{<}$$
2-Methyl-1,1,2-ethanetriyl
1-Methylethan-1-yl-2-ylidene

$$-CH_2\underset{\displaystyle |}{CH}CH_2CH_2-$$
1,2,4-Butanetriyl

**Higher-valent groups** are named using the endings *tetrayl, pentayl, hexayl,* etc., employing lowest possible numbers to indicate positions of the free bonds; however, the 1957 IUPAC Rules also recognize endings combining *yl, ylidene,* and *ylidyne.*

$-\overset{|}{\underset{|}{C}}-$　　Methanetetrayl　　　　$\overset{\diagdown}{\diagup}CCH_2C\overset{\diagup}{\diagdown}$　　1,1,1,3,3,3-Propanehexayl

　　　　　　　　　　　　　　　　　　　　　　　　　　Propanediylidyne

$-CH_2CHCHCH_2CH_2CHCH_2-$　　　　1,2,3,6,7-Heptanepentayl
　　　　　$|\quad|$　　　　　$|$

$\overset{\diagup}{\diagdown}CCH_2CH_2CH\overset{\diagdown}{\diagup}$　　　　　　1,1,1,4,4-Butanepentayl

　　　　　　　　　　　　　　　　1-Butanylidene-4-ylidyne

**Acyclic Unsaturated Hydrocarbons.** Straight-chain unsaturated hydrocarbons containing one double bond are named by replacing the ending *ane* of the corresponding saturated hydrocarbon name with *ene*. If the chain contains two double bonds, the name is given the ending *adiene;* if three double bonds, *atriene;* etc. The corresponding class names are **alkene, alkadiene, alkatriene,** etc. If the chain contains double bonds connecting at least three consecutive carbon atoms, the double bonds are called cumulative double bonds, and the class name **cumulene** is given to such compounds. Thus, 1,2,3-butatriene, $CH_2{=}C{=}C{=}CH_2$, is a cumulene. In all unsaturated hydrocarbons the chain is numbered so as to give the lowest possible numbers to carbon atoms bearing double bonds. Although the trivial names **ethylene** and **allene** are recognized in the 1957 IUPAC Rules, the systematic names ethene and propadiene are much to be preferred.

| | | | |
|---|---|---|---|
| $CH_2{=}CH_2$ | Ethene | $CH_3CH_2CH_2CH{=}CH_2$ | 1-Pentene |
| | *not* Ethylene | | |
| | | $CH_2{=}C{=}CH_2$ | Propadiene |
| | | | *not* Allene |
| $CH_3CH{=}CH_2$ | Propene | | |
| | *not* Propylene | | |
| | | $CH_2{=}CHCH{=}CH_2$ | 1,3-Butadiene |
| $CH_3CH{=}CHCH_3$ | 2-Butene | | |

If a straight-chain hydrocarbon contains one or more triple bonds, the endings *yne, adiyne, atriyne,* etc. are used, and the corresponding class names are **alkyne, alkadiyne, alkatriyne,** etc. The chain is numbered so as to give the lowest possible numbers to carbon atoms bearing triple bonds. The trivial name **acetylene** is recognized in the 1957 IUPAC Rules, but the systematic name ethyne is shorter and much to be preferred.

| | | | |
|---|---|---|---|
| $HC{\equiv}CH$ | Ethyne | $HC{\equiv}C{-}C{\equiv}CH$ | Butadiyne |
| | Acetylene | | |
| $CH_3C{\equiv}CH$ | Propyne | $CH_3C{\equiv}CCH_2C{\equiv}CH$ | 1,4-Hexadiyne |

If a straight-chain hydrocarbon contains both double and triple bonds, the endings *enyne, adienyne, enediyne,* etc. are used. Numbers as low as possible are given to carbon atoms bearing double and triple bonds (considered together), but if there is a further choice, those bearing double bonds are assigned the lowest numbers. Class names are **alkenyne, alkadienyne, alkenediyne,** etc.

| | |
|---|---|
| $HC{\equiv}CCH{=}CH_2$ | Butenyne |
| $CH_3C{=}CHC{\equiv}CH$ | 3-Penten-1-yne |
| $HC{\equiv}CCH_2CH{=}CH_2$ | 1-Penten-4-yne |
| $HC{\equiv}CCH{=}CHCH{=}CH_2$ | 1,3-Hexadien-5-yne |

**Branched-chain unsaturated hydrocarbons** are named as derivatives of the straight-chain component—*i.e.*, the parent chain—containing the maximum number of double and triple bonds. If there is a choice among such parent chains, the longest is selected; if there is a further choice, the chain containing more double bonds is selected. The trivial name isoprene is recognized in the 1957 IUPAC Rules.

$$CH_3C{=}CH_2$$ with $CH_3$

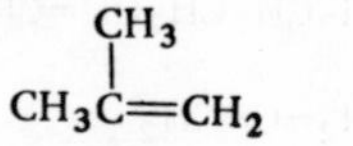

2-Methylpropene
*not* Isobutene

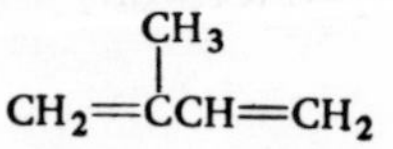

$$CH_2{=}CCH{=}CH_2$$ with $CH_3$

2-Methyl-1,3-butadiene
Isoprene (*unsubstituted only*)

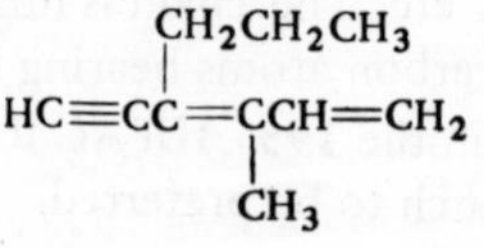

$$HC{\equiv}CC{=}CCH{=}CH_2$$ with $CH_2CH_2CH_3$ and $CH_3$

3-Methyl-4-propyl-1,3-hexadien-5-yne

**Substituting unsaturated groups** are named analogously to their saturated counterparts by describing and indicating location of the unsaturation. Class names are **alkenyl, alkynyl, alkadienyl, alkadiynyl, alkadienynyl, alkenediynyl,** etc. Several trivial names for unsaturated hydrocarbon groups are recognized in the 1957 IUPAC Rules.

$CH_2=CH-$      Ethenyl
                  *Vinyl

$HC\equiv CCH_2-$    2-Propynyl
             *not* Propargyl

$CH_3CH=CH-$   1-Propenyl

$CH_2=CHCH_2-$
          2-Propenyl
          *Allyl

$-CH=CH-$       Ethenylene
                  *Vinylene

$CH_2=CHCH=$   2-Propenylidene

$$CH_3=\overset{\overset{\displaystyle CH_3}{|}}{C}-$$
          1-Methylethenyl
          *Isopropenyl

$-CH_2CH=CH-$   Propenylene

$CH_2=C=$       Ethenylidene
                  *Vinylidene

$$CH_3=\overset{\overset{\displaystyle CH_3}{|}}{C}CH_2-$$
          2-Methyl-2-propenyl
        *not* Isobutenyl

$$-CH=\overset{\overset{\displaystyle |}{}}{C}CH=CH-\text{ 1,3-Butadiene-1,2,4-triyl}$$

$HC\equiv C-$      Ethynyl

* For the unsubstituted group only

**Alicyclic Hydrocarbons.** **Saturated monocyclic hydrocarbons** are named by attaching the prefix *cyclo* to the name of the acyclic straight-chain hydrocarbon possessing the same number of carbon atoms. Their class name is **cycloalkane**.

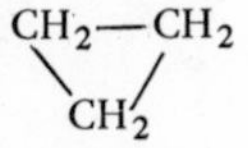

$$\underset{\underset{\displaystyle CH_2}{\diagdown\diagup}}{CH_2-CH_2}$$     Cyclopropane

Hydrocarbon side chains are named as substituting groups except when a single ring system is attached to a single chain containing a greater number of carbon atoms or when more than one ring system is attached to the same chain.

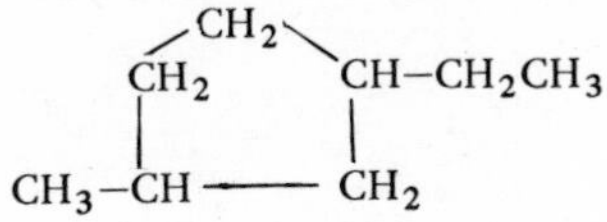

$$\begin{array}{c} \overset{\displaystyle CH_2}{\diagup \quad \diagdown} \\ CH_2 \qquad\quad CH-CH_2CH_3 \\ |\qquad\qquad\qquad | \\ CH_3-CH \overline{\qquad\qquad} CH_2 \end{array}$$     1-Ethyl-3-methylcyclopentane

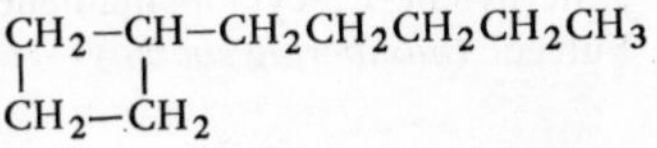

$$\begin{array}{l} CH_2-CH-CH_2CH_2CH_2CH_2CH_3 \\ |\qquad\quad | \\ CH_2-CH_2 \end{array}$$     1-Cyclobutylpentane

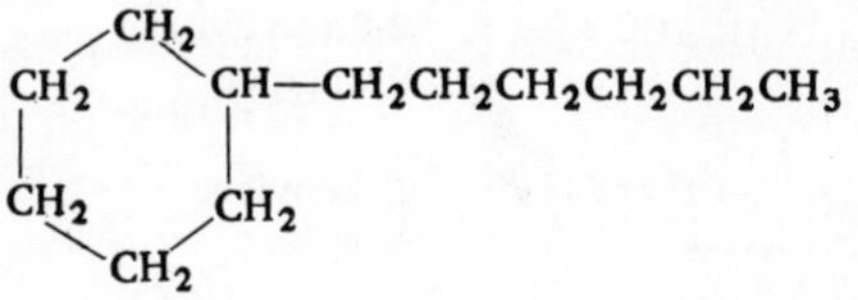

Hexylcyclohexane
1-Cyclohexylhexane

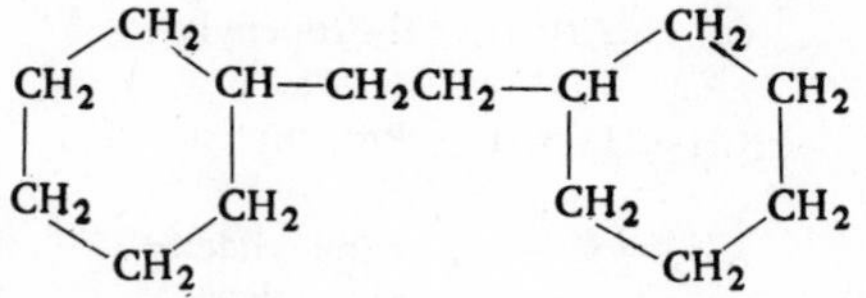

1,2-Dicyclohexylethane

**Unsaturated monocyclic hydrocarbons** are named by replacing the ending *ane* of the corresponding saturated alicyclic hydrocarbon name with the appropriate suffix, *ene,, yne, adiene, adiyne, enyne,* etc. As with acyclic compounds, lowest possible numbers are given to carbon atoms bearing double and triple bonds (considered together). The trivial name **fulvene** is recognized in the 1957 IUPAC Rules.

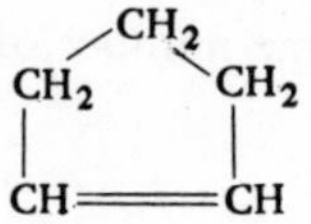

Cyclopentene

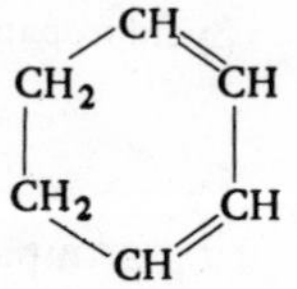

1,3-Cyclohexadiene

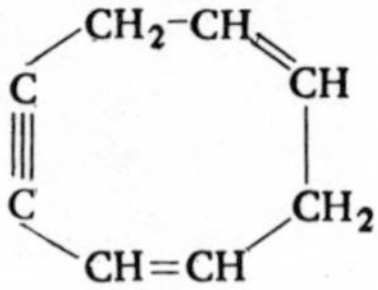

1,6-Cyclooctadien-3-yne

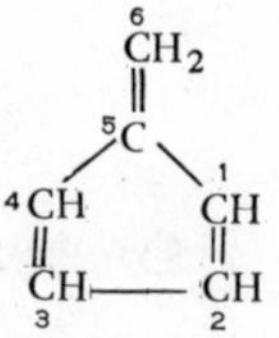

5-Methylene-1,3-cyclopentadiene
Fulvene (*numbering shown*)

Substituting monocyclic groups are named by the principles presented earlier in this chapter for substituting acyclic groups. Positions of double and triple bonds are indicated by lowest possible numbers, once the carbon atoms bearing free bonds have been considered.

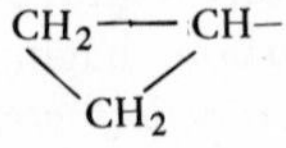

$$CH_2 \underset{CH_2}{\overset{CH-}{\diagdown}}$$

Cyclopropyl

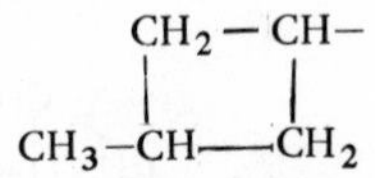

$$\begin{array}{c} CH_2 - CH- \\ | \quad\quad | \\ CH_3-CH---CH_2 \end{array}$$

3-Methylcyclobutyl

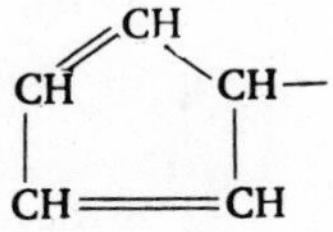

$$\begin{array}{c} CH \\ CH \quad\quad CH- \\ | \quad\quad\quad \\ CH===CH \end{array}$$

2,4-Cyclopentadienyl

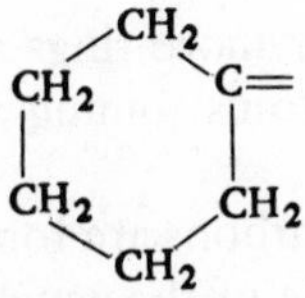

$$\begin{array}{c} CH_2 \\ CH_2 \quad C= \\ | \quad\quad\quad | \\ CH_2 \quad CH_2 \\ CH_2 \end{array}$$

Cyclohexylidene

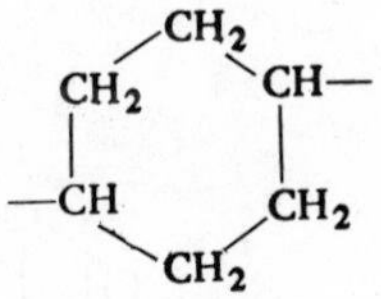

$$\begin{array}{c} CH_2 \\ CH_2 \quad CH- \\ -CH \quad\quad CH_2 \\ CH_2 \end{array}$$

1,4-Cyclohexylene

$$\begin{array}{c} CH \\ CH \quad\quad CH \\ | \quad\quad\quad | \\ CH \quad\quad CH \\ CH \end{array}$$

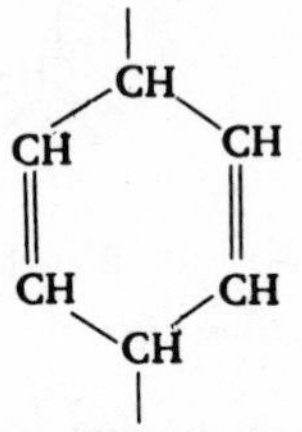

2,5-Cyclohexadien-1,4-ylene

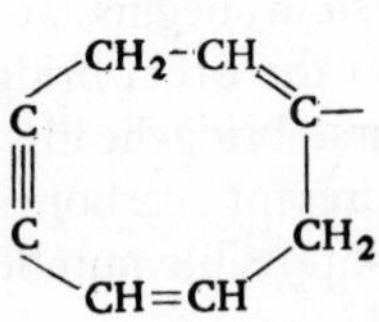

$$\begin{array}{c} CH_2\text{-}CH \\ C \quad\quad C- \\ | \quad\quad\quad \\ C \quad\quad CH_2 \\ CH=CH \end{array}$$

1,6-Cyclooctadien-4-yn-1-yl

**Bridged Alicyclic Hydrocarbons.** Cyclic hydrocarbons containing one or more pairs of carbon atoms common to two or more rings are called bridged hydrocarbons.

Bridged alicyclic hydrocarbons are named by the von Baeyer method. In applying this method, the number of rings is taken as being equal to the minimum number of bond scissions required to convert the bridged ring system into an acyclic hydrocarbon having the same number of carbon atoms. The prefixes *bicyclo, tricyclo, tetracyclo*, etc. are used to denote the number of rings so determined.

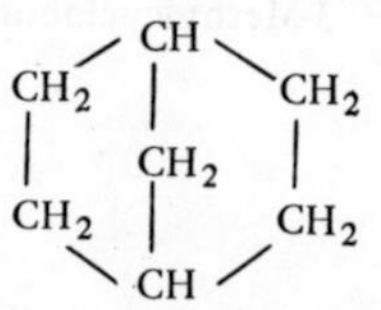

A bicyclo system 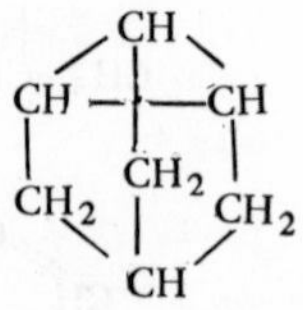                    A tricyclo system

The carbon atoms common to two or more rings are called bridgeheads, and each bond, atom, or chain of atoms joining these carbon atoms is called a bridge.

Names of bridged alicyclic hydrocarbons are formed by combining the appropriate *cyclo* prefix with the name of the acyclic hydrocarbon (Table 2.1) containing the same total number of carbon atoms and interposing an expression within brackets denoting the number of atoms in each bridge cited in descending order.

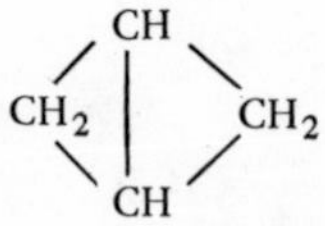

Bicyclo[1.1.0]butane 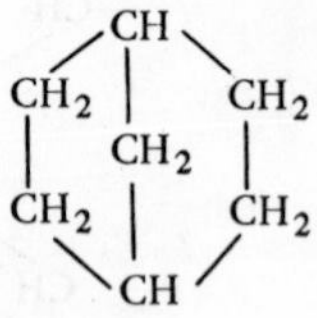                Bicyclo[2.2.1]heptane

Numbering of a bridged ring system begins at one bridgehead and proceeds along the longest bridge to the other bridgehead, then along the next longest bridge back to the first bridgehead; the shortest bridge is numbered last. If unsaturation is present, carbon atoms bearing double and triple bonds are given the lowest possible numbers consistent with the ring-numbering principles just stated.

$$
\begin{array}{c}
\overset{7}{CH_2} \!-\!-\!- \overset{1}{CH} \!-\!-\!- \overset{2}{CH_2} \\
|\qquad\qquad |^{8}\qquad\quad | \\
H_3C\!-\!C\!-\!CH_3 \quad \overset{}{CH_2} \\
|\qquad\qquad\qquad |^{3} \\
\underset{6}{CH_2} \!-\!-\!- \underset{5}{CH} \!-\!-\!- \underset{4}{CH_2}
\end{array}
\qquad
\begin{array}{c}
\overset{9}{CH_2} \!-\!-\!- \overset{1}{CH} \!-\!-\!- \overset{2}{CH_2} \\
|\qquad\qquad |\qquad\qquad |^{3} \\
\overset{8}{CH} \quad \overset{10}{CH_2} \quad \overset{}{CH_2} \\
\|\qquad\qquad |\qquad\qquad |^{4} \\
\phantom{x}\quad \overset{11}{CH_2} \quad \overset{}{CH_2} \\
\|\qquad\qquad |\qquad\qquad | \\
\underset{7}{CH} \!-\!-\!- \underset{6}{CH} \!-\!-\!- \underset{5}{CH_2}
\end{array}
$$

8,8-Dimethylbicyclo[3.2.1] octane    Bicyclo[4.3.2] undec-7-ene

In naming systems containing more than two rings, superscript numerals are used to specify location of additional bridges.

$$
\begin{array}{c}
CH \\
/ \quad \ \backslash \\
CH \!-\!|\!-\! CH \qquad \text{Tricyclo}[1.1.0.0^{2:4}]\text{butane} \\
\backslash \quad \ / \\
CH
\end{array}
$$

For further examples illustrating naming of more complex bridged alicyclic hydrocarbons, *see* the 1957 IUPAC Rules (Rule A-32).

**Substituting groups** derived from bridged alicyclic hydrocarbons by conceptual removal of one or more hydrogen atoms are named by the principles presented earlier in this chapter for acyclic and cyclic groups. Lowest possible numbers consistent with requirements for von Baeyer ring numbering are given first to carbon atoms having free bonds, then to carbon atoms bearing double and triple bonds.

$$
\begin{array}{c}
\overset{6}{CH} \!-\!-\!- \overset{1}{CH} \!-\!-\!- \overset{2}{CH}\!- \\
\|\qquad\qquad |\qquad\qquad | \\
\phantom{x}\quad CH_2 \\
\|\qquad\qquad |\qquad\qquad | \\
\underset{5}{CH} \!-\!-\!- \underset{4}{CH} \!-\!-\!- \underset{3}{CH_2}
\end{array}
\qquad \text{Bicyclo}[2.2.1]\text{ hept-5-en-2-yl}
$$

**Bridged aromatic hydrocarbons** are not named by the von Baeyer method but rather by treating each bridge as a bivalent substituting group attached to an aromatic ring system. Either of two types of name may be used: (1) for bridged structures containing only one aromatic ring system, the well established names provided by the 1957 IUPAC Rules are recommended; (2) for more complex bridged aromatic structures, the recently developed phane nomenclature offers advantages. Both methods are described in Chapter 4 since the polycyclic hydrocarbons involved are usually classified as aromatic rather than alicyclic.

**Spiro Alicyclic Hydrocarbons.** Cyclic hydrocarbons containing one or more single carbon atoms common to two or more rings are called spiro hydrocarbons. The common atom is called a spiro atom, and the linkage between the rings is called a spiro union.

**Spiro hydrocarbons** are named by combining the prefix *spiro* with the name of the acyclic hydrocarbon (Table 2.1) containing the same total number of carbon atoms and interposing an expression within brackets denoting the number of atoms (other than the spiro atom) in each ring.

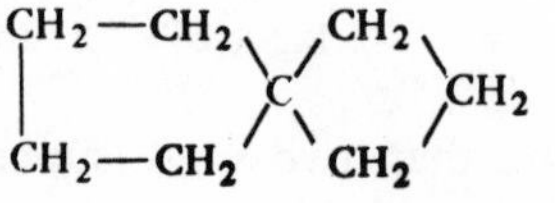

$$\begin{array}{ccc} CH_2-CH_2 & CH_2 & \\ | & \diagdown C \diagup & \diagdown CH_2 \\ CH_2-CH_2 & CH_2 & \end{array} \qquad \text{Spiro[3.4]octane}$$

Numbering of a spiro ring system begins at a ring carbon atom adjacent to the spiro atom (if the rings are of unequal size, the smaller ring is numbered first) and proceeds around one ring, through the spiro atom, and thence around the other ring. If unsaturation is present, carbon atoms bearing double and triple bonds are given the lowest possible numbers consistent with the ring-numbering principles just stated.

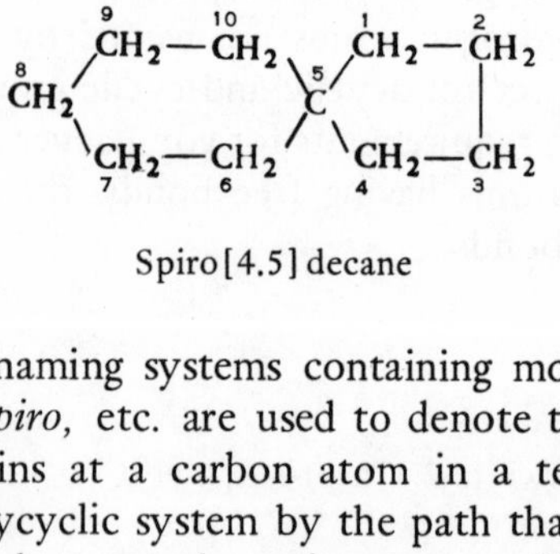

Spiro[4.5]decane 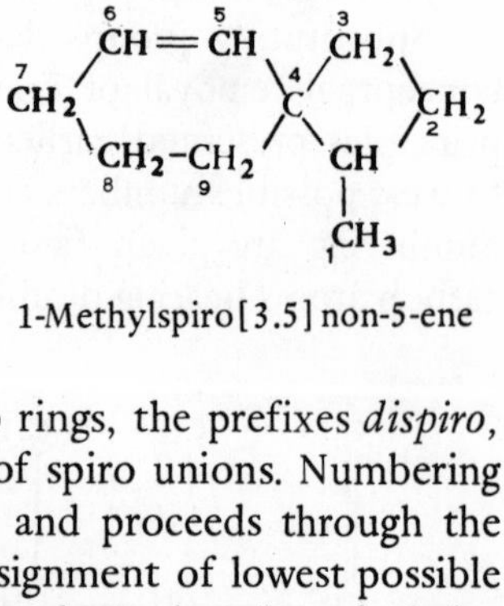            1-Methylspiro[3.5]non-5-ene

In naming systems containing more than two rings, the prefixes *dispiro,* *trispiro,* etc. are used to denote the number of spiro unions. Numbering begins at a carbon atom in a terminal ring and proceeds through the polycyclic system by the path that enables assignment of lowest possible numbers to the spiro atoms. Within the brackets, lengths of chains between spiro atoms are cited in the order which corresponds to the sequence of numbering of the total ring system.

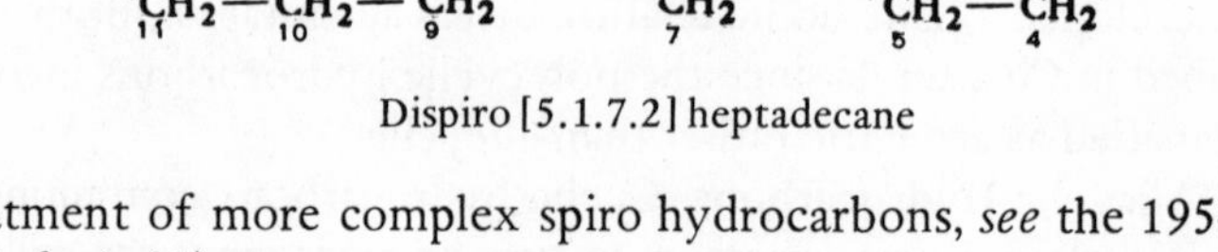

Dispiro[5.1.7.2]heptadecane

For treatment of more complex spiro hydrocarbons, *see* the 1957 IUPAC Rules (Rule A-41).

## Discussion

In naming saturated hydrocarbons the prefix $n$ (meaning normal) has often been used to denote a straight chain—*e.g.*, *n*-hexane. This practice, although permissible, is unnecessary since hexane is now by definition the name of the hydrocarbon having six carbon atoms arranged in a straight chain.

There has also been some use in the literature of a "derivative" system of hydrocarbon nomenclature, sometimes called the "simple-nucleus" system. According to this system, the compound $(CH_3)_3CH$ has been called trimethylmethane, the compound $(CH_3)_2C=CHCH_3$, trimethyl-ethylene, etc. In this system the first member of a homologous series is considered the parent, the names of the substituting groups that replace the hydrogen atoms of the parent are placed before the parent name, and the whole name is written as one word. Other examples are methyl-acetylene for $CH_3C\equiv CH$ and divinylacetylene for $CH_2=CHC\equiv CCH=CH_2$. Such names have the virtue of simplicity, but they are not recommended because a nomenclature scheme based on this system would become cumbersome on even limited extension.

As indicated above under *Recommended Nomenclature Practice*, a number of trivial names for branched-chain hydrocarbons and branched-chain substituting hydrocarbon groups are still recognized in the 1957 IUPAC Rules but only when these hydrocarbons and groups are not substituted. Thus, isobutane is acceptable, but chloroisobutane is not. The simplest and best procedure is to use the recommended systematic names in all instances, even though these are usually somewhat longer than the corresponding trivial names.

Names for branched-chain substituting groups such as 2-propyl (for 1-methylethyl) and 3-pentyl (for 1-ethylpropyl) are obviously attractive but have not received official recognition and cannot, therefore, be recommended.

Since ethylene and propylene are officially recognized as names of bivalent substituting hydrocarbon groups, these terms should be abandoned as names for the unsaturated hydrocarbons $C_2H_4$ and $n$-$C_3H_6$, which are correctly named ethene and propene, respectively. The trivial name "allene" for $CH_2=C=CH_2$ should also be abandoned in favor of propadiene.

The trivial name "propargyl" for the substituting group $HC\equiv CCH_2$ should no longer be used. The correct name is 2-propynyl. Ethyne is preferred over acetylene even though the latter is widely known and used.

For a few acyclic hydrocarbons of symmetrical structure, names formed by doubling names of component substituting groups have found some usage—*e.g.*, biisopropyl for $(CH_3)_2CHCH(CH_3)_2$ and biallyl for $CH_2=CHCH_2CH_2CH=CH_2$. Such names should be abandoned.

# 3

# Monocyclic Aromatic Hydrocarbons

Unsaturated hydrocarbon ring systems containing one or more six-membered rings in each of which the unsaturation may be represented formally by three alternating (conjugated) double bonds have long been known as aromatic hydrocarbons. The class name for these compounds is **arene**.

This chapter deals only with monocyclic arenes—*i.e.*, with benzene and substituted benzenes. Polycyclic arenes are the subject of Chapters 4 and 5, and unsaturated alicyclic (nonaromatic) hydrocarbons are treated in Chapter 2.

## Recommended Nomenclature Practice

**Simple monocyclic arenes** are best named systematically as derivatives of the parent compound benzene, although trivial names for a few substituted benzenes are recognized in the 1957 IUPAC Rules.

Benzene

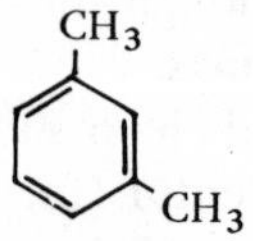

1,3-Dimethylbenzene
*m*-Xylene

Methylbenzene
Toluene

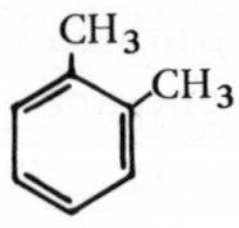

1,2-Dimethylbenzene
*o*-Xylene

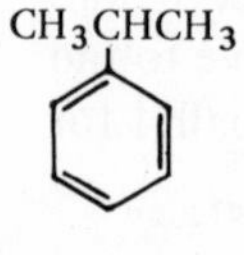

Isopropylbenzene
Cumene

1,4-Dimethylbenzene
*p*-Xylene

20

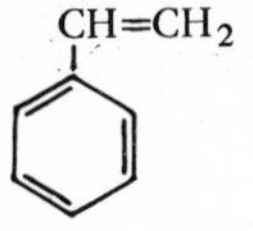

CH=CH$_2$

Ethenylbenzene
Styrene

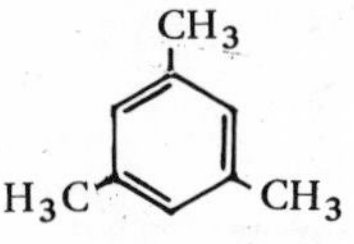

1,3,5-Trimethylbenzene
Mesitylene

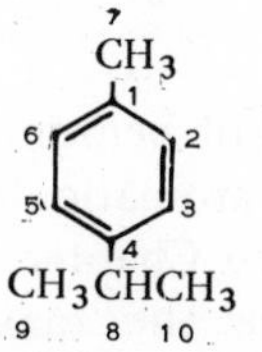

1-Isopropyl-4-methylbenzene
*p*-Cymene (*fixed numbering shown*)

All other simple monocyclic aromatic hydrocarbons are best named systematically as derivatives of benzene, particularly when an added substituent is identical with one already present.

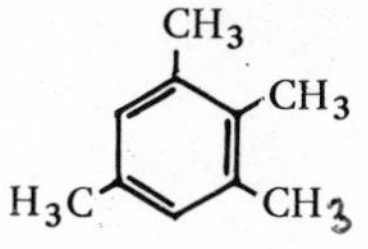

1,2,3,5-Tetramethylbenzene
*not* 2-Methylmesitylene

Positions of substituents are indicated by Arabic numerals except that *o-* (ortho), *m-* (meta), and *p-* (para) may be used instead of 1,2-, 1,3-, and 1,4-, respectively, for disubstituted benzene derivatives only. Lowest possible numbers are given to the carbon atoms carrying substituents, choice between alternatives being made as with the aliphatic hydrocarbons (Chapter 2); however, when a name is based on a recognized trivial name other than benzene, priority for lowest numbers is given to substituents implied by the parent trivial name. The isomeric cymenes have fixed numbering, as in the example (*p*-cymene) shown above.

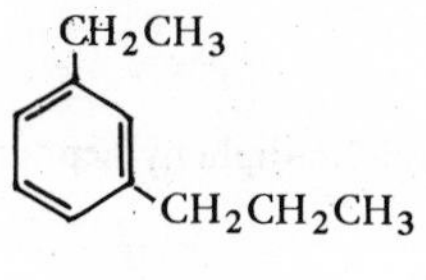

1-Ethyl-3-propylbenzene
*m*-Ethylpropylbenzene

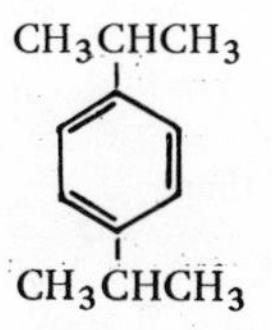

1,4-Diisopropylbenzene
*p*-Diisopropylbenzene
*not* *p*-Isopropylcumene

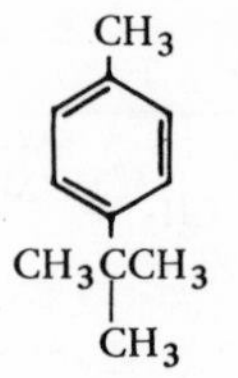

$$CH_3$$

1-*tert*-Butyl-4-methylbenzene
4-*tert*-Butyltoluene
*p-tert*-Butyltoluene

$$CH_3\overset{|}{\underset{CH_3}{C}}CH_3$$

**Complex monocyclic arenes** composed of benzene rings and aliphatic chains are named either as substituted aromatic hydrocarbons or as substituted aliphatic hydrocarbons (*see also* Chapter 2). Choice between these methods is made (1) to produce the maximum number of substitutions in the parent compound, and (2) so that smaller structural units are named as substituents on a larger parent compound. Thus, a hydrocarbon containing several aliphatic groups attached to benzene as the nucleus is named as a derivative of benzene; similarly, a hydrocarbon containing two or more benzene rings attached to one straight chain is named as a derivative of the straight-chain hydrocarbon.

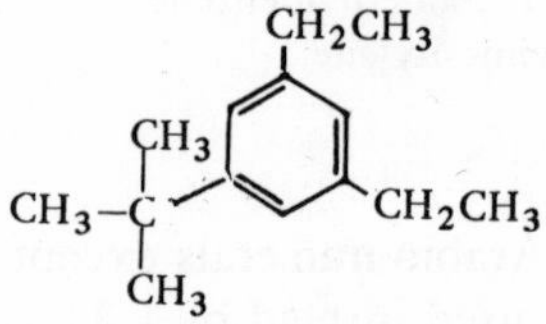

$$CH_2CH_3$$
$$CH_3$$
$$CH_3-\overset{|}{\underset{CH_3}{C}} \quad CH_2CH_3$$

1-*tert*-Butyl-3,5-diethylbenzene

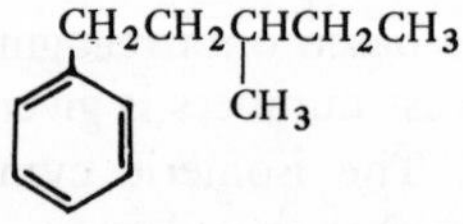

$$CH_2CH_2\overset{|}{\underset{CH_3}{C}HCH_2CH_3}$$

3-Methyl-1-phenylpentane

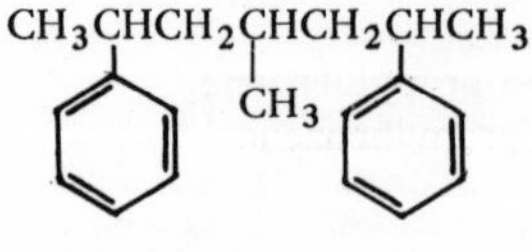

$$CH_3CHCH_2\overset{|}{\underset{CH_3}{C}HCH_2CHCH_3}$$

4-Methyl-2,6-diphenylheptane

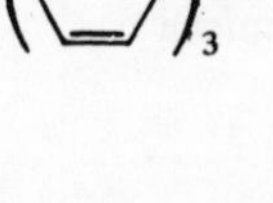

$$\left(\bigcirc\right)_3 CH$$

Triphenylmethane

Finally, a hydrocarbon composed of one aliphatic chain and one benzene ring is usually named as a derivative of the larger structural unit. If the aliphatic chain is unsaturated, the compound may be named as a derivative of that chain, regardless of chain size; however, when the unsaturated chain contains no more than three carbon atoms, the compound is usually named as a derivative of benzene.

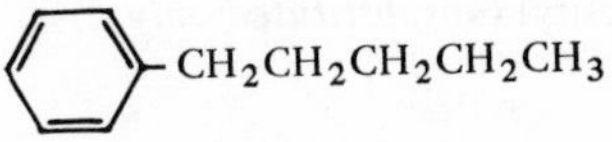

$CH_2CH_2CH_2CH_2CH_3$      Pentylbenzene

$CH_2CH_2CH_2CH_2CH_2CH_2CH_3$      1-Phenylheptane

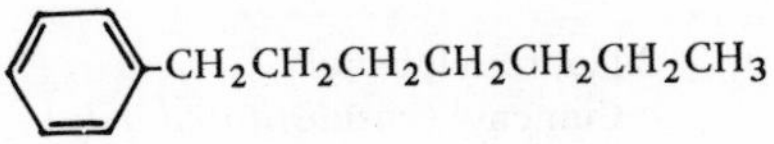

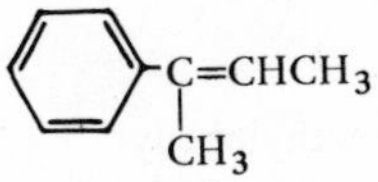

$C=CHCH_3$      2-Phenyl-2-butene
$CH_3$

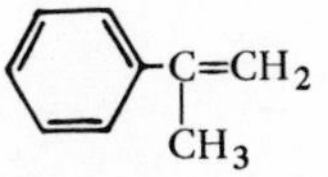

$C=CH_2$      Isopropenylbenzene
$CH_3$              2-Phenylpropene

**Substituting univalent groups** derived from monocyclic arenes by conceptual removal of one hydrogen atom from a carbon atom of the ring are best named as derivatives of the parent substituting group *phenyl,* although a few other trivial names are recognized in the 1957 IUPAC Rules as shown below. The class name for these groups is *aryl.* Numbering begins with the carbon atom having the free bond, except when substituents are implied rather than expressed as in a recognized trivial name other than phenyl.

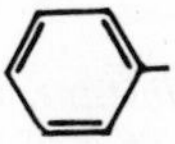

Phenyl

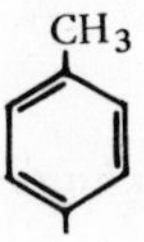

$CH_3$

4-Methylphenyl
*p*-Tolyl (*unsubstituted only*)

2,6-Dimethylphenyl
2,6-Xylyl (*unsubstituted only*)

2,4,6-Trimethylphenyl
Mesityl (*unsubstituted only*)

2-Isopropylphenyl
*o*-Cumenyl (*unsubstituted only*)

5-Isopropyl-2-methylphenyl
*p*-Cym-2-yl (*unsubstituted only*)

2-Isopropyl-5-methylphenyl
*p*-Cym-3-yl (*unsubstituted only*)

**Substituting bivalent groups** are named as derivatives of the parent substituting group *phenylene*, the carbon atoms having free bonds being assigned lowest possible numbers. For the phenylene group itself, use of *o-*, *m-*, or *p-* is an acceptable alternative to numerals. The class name for these groups is *arylene.*

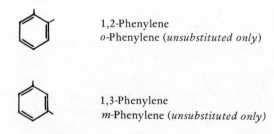

1,2-Phenylene
*o*-Phenylene (*unsubstituted only*)

1,3-Phenylene
*m*-Phenylene (*unsubstituted only*)

1,4-Phenylene
*p*-Phenylene *(unsubstituted only)*

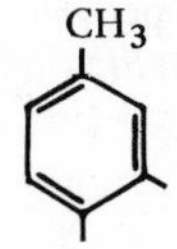

4-Methyl-1,2-phenylene
*not* 3,4-Tolylene

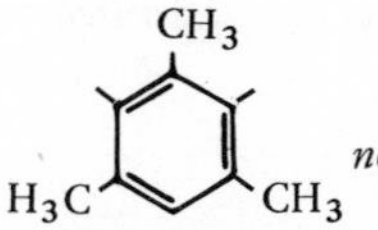

2,4,6-Trimethyl-1,3-phenylene
*not* 2,4-Mesitylene

Substituting groups having three or more free bonds are named by adding suffixes *triyl, tetrayl,* etc. to the systematic name of the corresponding hydrocarbon.

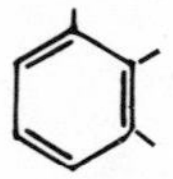

1,2,3-Benzenetriyl

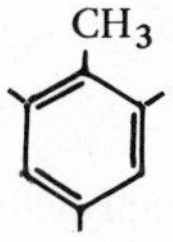

2-Methyl-1,3,5-benzenetriyl
*not* 2,4,6-Toluenetriyl

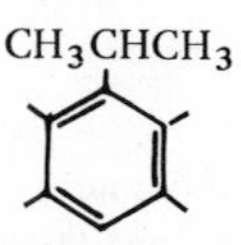

3-Isopropyl-1,2,4,5-benzenetetrayl
*not* 2,3,5,6-Cumenetetrayl

**Substituting groups** derived from monocyclic arenes by conceptual removal of hydrogen from a side chain are best named systematically as aryl derivatives of the appropriate aliphatic substituting group *(see* Chapter 2), although a few trivial names are recognized in the 1957 IUPAC Rules as shown on the next page.

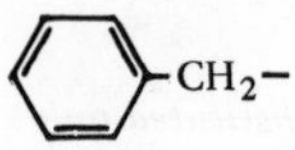

| | |
|---|---|
| $C_6H_5-CH_2-$ | Phenylmethyl<br>Benzyl (*unsubstituted only*) |
| $C_6H_5-CH=CHCH_2-$ | 3-Phenyl-2-propenyl<br>Cinnamyl (*unsubstituted only*) |
| $C_6H_5-CH_2CH_2-$ | 2-Phenylethyl<br>Phenethyl (*unsubstituted only*) |
| $C_6H_5-CH=CH-$ | 2-Phenylethenyl<br>Styryl (*unsubstituted only*) |
| 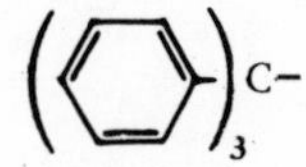 $(C_6H_5-)_3C-$ | Triphenylmethyl<br>Trityl (*unsubstituted only*) |
| $C_6H_5-CH\!<$ | Phenylmethylene<br>Benzylidene (*unsubstituted only*) |
| $C_6H_5-CH=CHCH\!<$ | 3-Phenyl-2-propenylidene<br>Cinnamylidene (*unsubstituted only*) |
| $C_6H_5-C\!\lll$ | Phenylmethylidyne<br>Benzylidyne (*unsubstituted only*) |

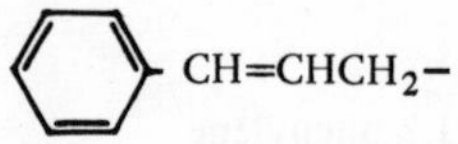

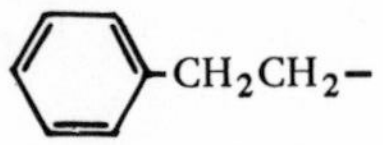

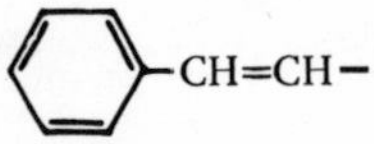

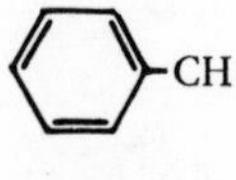

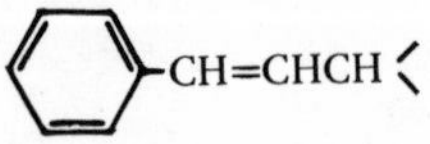

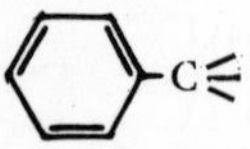

As indicated, these trivial names should be used only for the unsubstituted groups themselves.

## Discussion

Broadly speaking, the class of aromatic hydrocarbons includes some compounds that do not contain a benzene or "benzenoid" ring—*i.e.*, an isolated or fused six-membered ring having three alternating (conjugated) double bonds, but the benzene ring is by far the most frequently occurring structural unit within the large group of compounds considered to exhibit aromatic properties.

The unsubstituted six-carbon monocyclic aromatic hydrocarbon has almost always been called benzene despite the fact that the 1930 IUC Rules ( Liége Rules) permitted the used of "phene", a name which is not mentioned in the 1957 IUPAC Rules.

The name benzyne has been given to the unisolated compound 1,3-cyclohexadien-5-yne, which has been postulated as an intermediate in nucleophilic substitution reactions of many benzene derivatives.

The name "benzol" for benzene, borrowed from the German language, has appeared sometimes in semitechnical articles and in advertising, but it should never be used in scientific writing as an English word denoting the hydrocarbon $C_6H_6$.

Systematic names are recommended for all substituted benzenes, but it is recognized that the common trivial names toluene, *o*-xylene, *m*-xylene, *p*-xylene, and styrene will find continued use. The compounds mesitylene, cumene, and cymene are not encountered as frequently, and their trivial names can be abandoned more conveniently.

The 1957 IUPAC Rules indicate that *o*-, *m*-, and *p*- may always be used in place of the corresponding Arabic numerals in naming disubstituted benzene derivatives. On the other hand, until recently, *Chemical Abstracts* limited the use of *o*-, *m*-, and *p*- to compounds in which both substituents are alike (as in *p*-diethylbenzene) or the substituents are different but one is expressed as part of a trivial name (as in *m-tert*-butyltoluene); now it has stopped using these prefixes altogether. Thus the present trend is toward exclusive use of Arabic numerals, and abandonment of *o*-, *m*-, and *p*- is recommended (*see also* Chapter 12).

The prefix *ar* (aromatic) may be used to indicate that a substituent is attached to a carbon atom in an aromatic ring rather than in a side chain. For example, *ar,ar*-dimethylstyrene indicates that the two methyl groups are both attached to ring carbon atoms rather than to those of the ethenyl group and that the exact positions of the methyl groups are unknown.

It is recommended that trivial names of substituting groups be used only for the unsubstituted groups. One reason is obvious from the following example of the difficulty that results when a trivial name is used as the parent name of a substituted group.

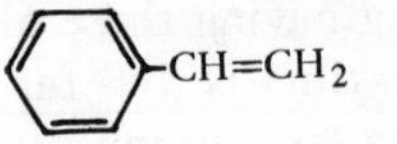

$\bigcirc$—CH=CH$_2$    Styrene

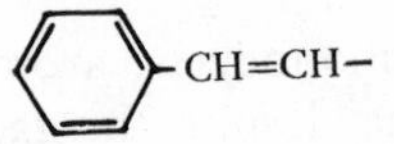

$\bigcirc$—CH=CH—    Styryl

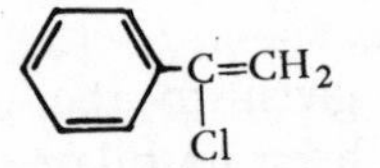

$\bigcirc$—$\underset{\underset{Cl}{|}}{C}$=CH$_2$    α-Chlorostyrene

$\bigcirc$—$\underset{\underset{Cl}{|}}{C}$=CH—    β-Chlorostyryl

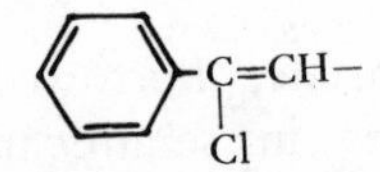

In this case a different sequence of Greek lettering unfortunately has been established for the styryl group as compared with the corresponding hydrocarbon, styrene. This confusing situation can easily be avoided by using systematic names—*i.e.*, (1-chloroethenyl)benzene instead of α-chlorostyrene and 2-chloro-2-phenylethenyl instead of β-chlorostyryl.

# Polycyclic Aromatic Hydrocarbons

This chapter deals with the nomenclature of aromatic hydrocarbons composed of two or more rings joined in such a manner that each component ring shares two or more carbon atoms with at least one other component ring. Such a combination of rings is called an **aromatic ring system**, and the component rings are said to be fused to one another. Depending upon the number of rings present, a ring system may be designated **bicyclic, tricyclic,** etc.; thus, a **polycyclic** ring system is one containing two or more component rings. Monocyclic aromatic hydrocarbons are treated in Chapter 3.

The class of **polycyclic aromatic hydrocarbons** includes ring systems containing the maximum number of alternating (conjugated) double bonds; although the component rings may vary in size, the benzene ring is by far the most common.

Polycyclic ring systems in which the joining of the component rings is direct but does not involve sharing of carbon atoms are discussed in Chapter 5.

*Recommended Nomenclature Practice*

The **35 fused polycyclic aromatic hydrocarbons** listed in Table 4.1 have names recognized by the 1957 IUPAC Rules. Some of these names are formed systematically, but many are trivial. Numbering of each ring system is fixed as shown.

Table 4.1.  Polycyclic Aromatic Hydrocarbons

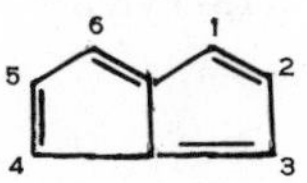

(1) Pentalene

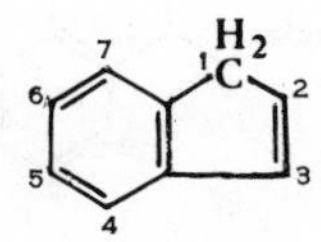

(2) Indene
(1*H* isomer shown)

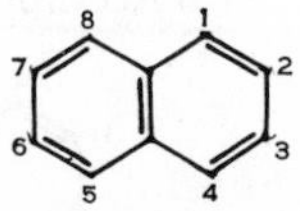

(3) Naphthalene

## Table 4.1 Polycyclic Aromatic Hydrocarbons (Continued)

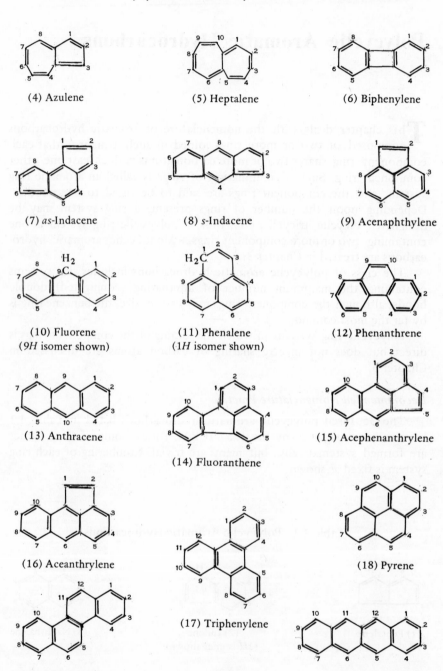

(4) Azulene

(5) Heptalene

(6) Biphenylene

(7) as-Indacene

(8) s-Indacene

(9) Acenaphthylene

(10) Fluorene
(9H isomer shown)

(11) Phenalene
(1H isomer shown)

(12) Phenanthrene

(13) Anthracene

(14) Fluoranthene

(15) Acephenanthrylene

(16) Aceanthrylene

(17) Triphenylene

(18) Pyrene

(19) Chrysene

(20) Naphthacene

## Table 4.1 Polycyclic Aromatic Hydrocarbons (Continued)

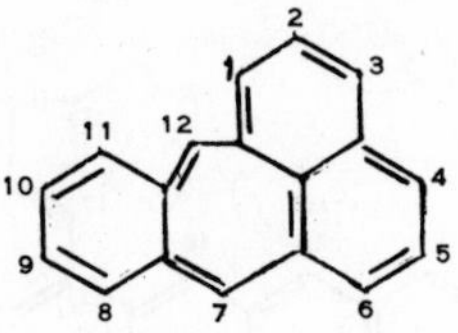

(21) Pleiadene

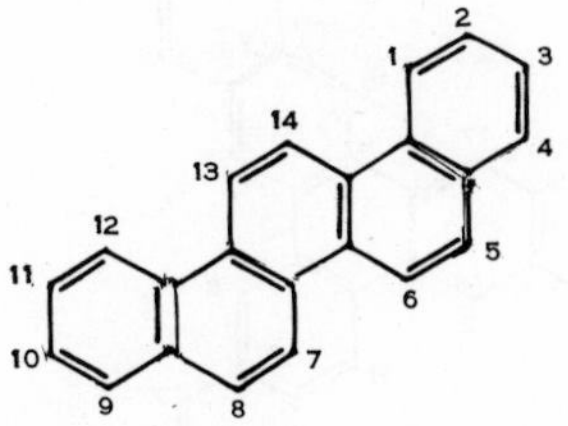

(22) Picene

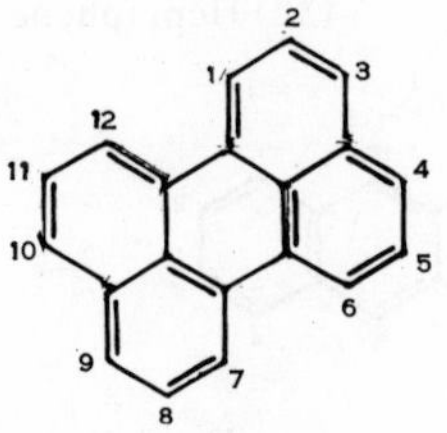

(23) Perylene

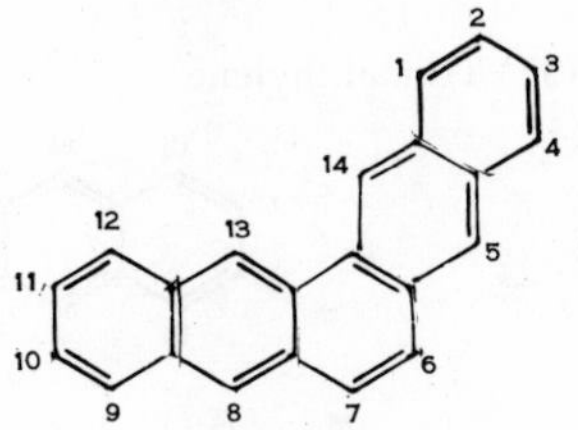

(24) Pentaphene

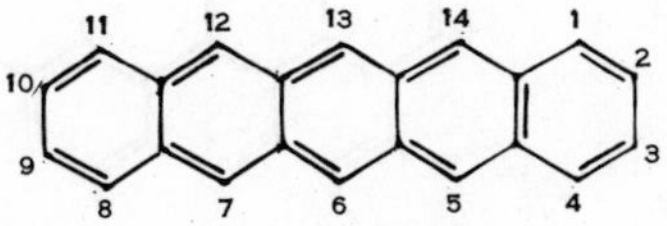

(25) Pentacene

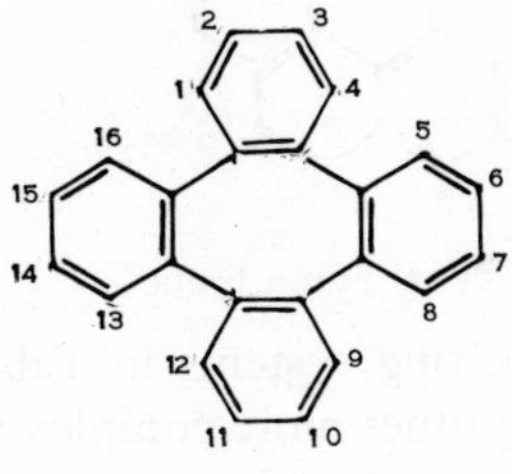

(26) Tetraphenylene

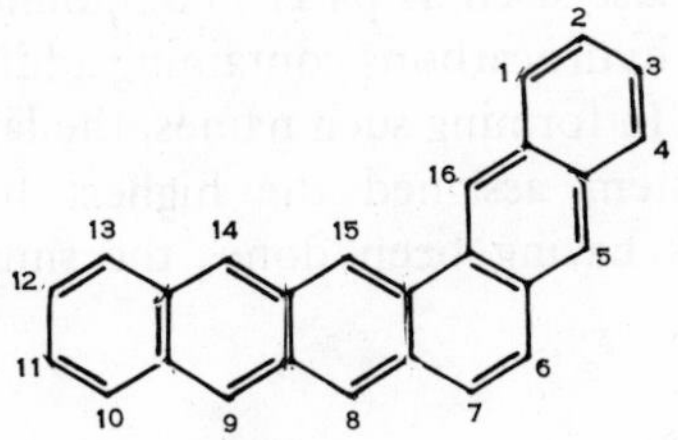

(27) Hexaphene

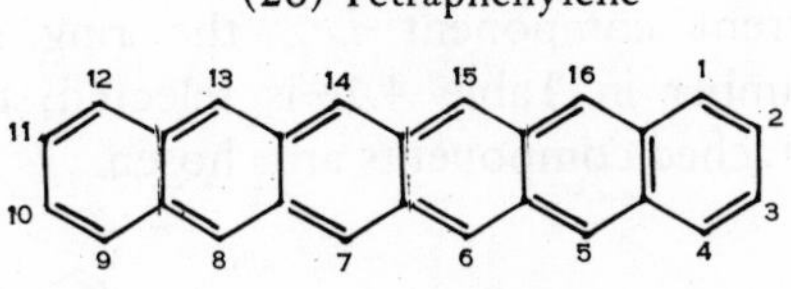

(28) Hexacene

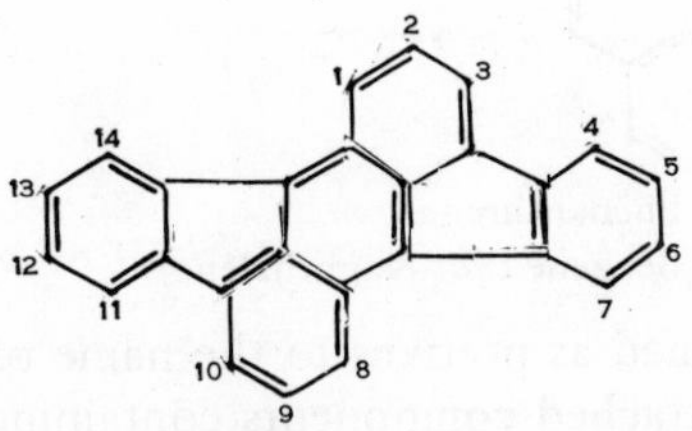

(29) Rubicene

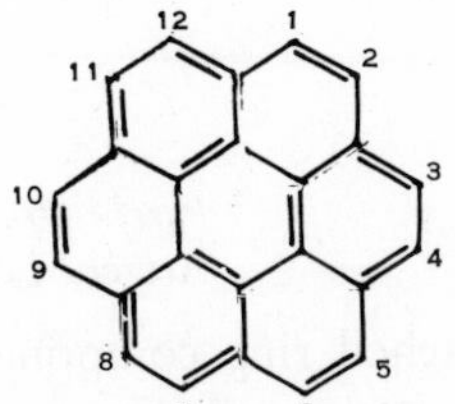

(30) Coronene

## Table 4.1  Polycyclic Aromatic Hydrocarbons  (Continued)

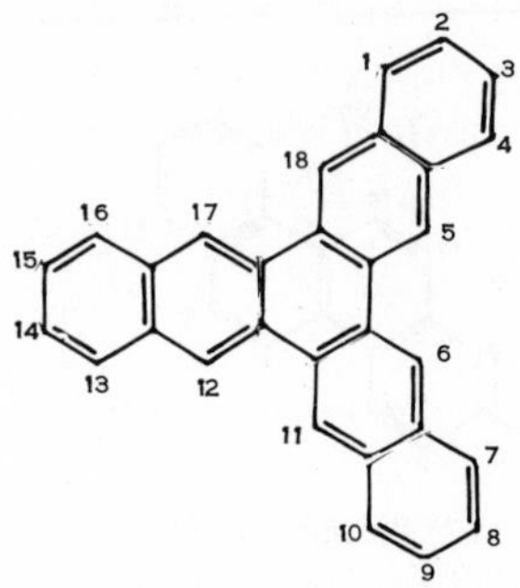

(31) Trinaphthylene

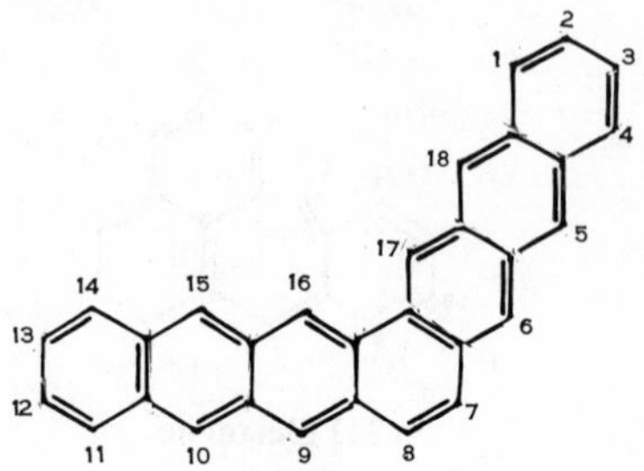

(32) Heptaphene

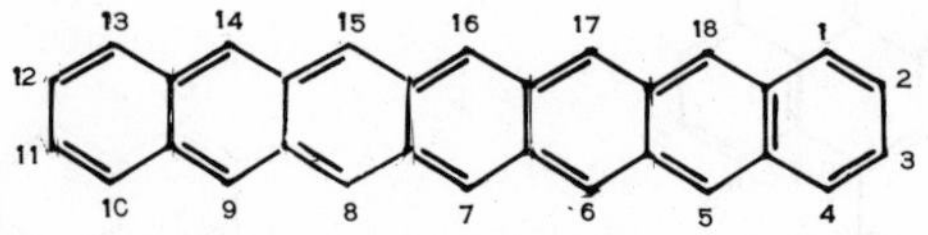

(33) Heptacene

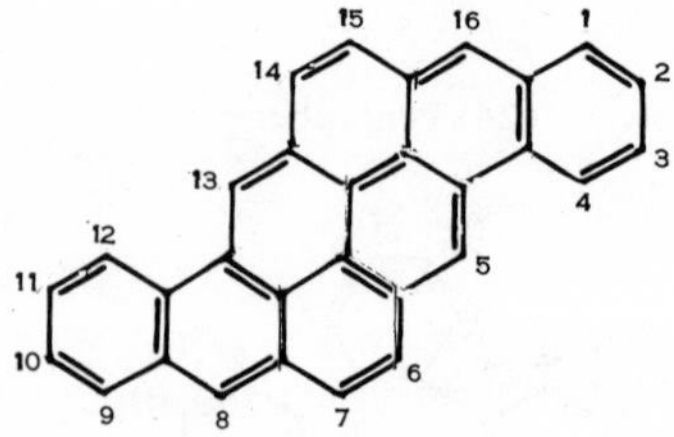

(34) Pyranthrene

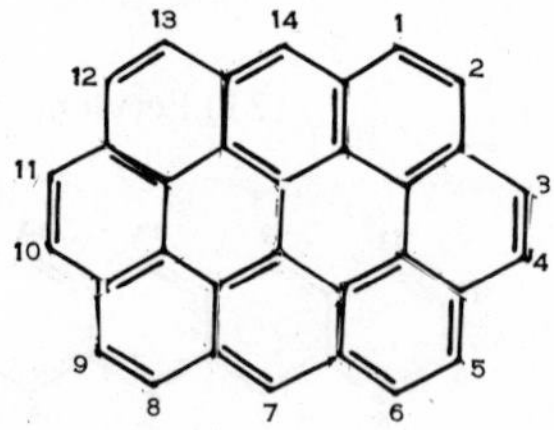

(35) Ovalene

The ring systems in Table 4.1 are also used as parent compounds in naming other more complex polycyclic hydrocarbons containing additional rings fused to the parent component. In forming such names, the largest parent component—*i.e.,* the ring system assigned the highest listing number in Table 4.1—is selected; this having been done, the simplest attached components are chosen.

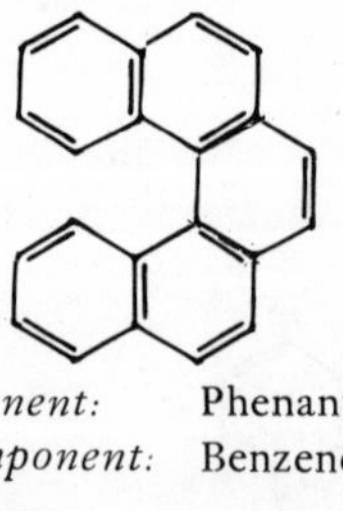

*Parent component:*    Phenanthrene
*Attached component:*   Benzene (*not* Naphthalene)

Attached ring components are named as prefixes to the name of the parent component. For **monocyclic attached components** containing the maximum number of conjugated double bonds, prefixes indicating the number of carbon atoms in the attached ring are used,—*e.g., cyclopenta, cyclohepta, cycloocta*—except that *benzo* is customarily used in place of *cyclohexa.* Prefix names for **polycyclic attached components** are formed by changing the ending *ene* of the aromatic hydrocarbon name to *eno,* but five shortened forms are recognized in the 1957 IUPAC Rules:

Acenaphtho for Acenaphthyleno
Anthra for Anthraceno
Naphtho for Naphthaleno
Perylo for Peryleno
Phenanthro for Phenanthreno

Location of an attached component is expressed by a lower-case letter in italics that denotes the side (pictorially speaking) of the parent ring system to which the component is fused; in addition, a number pair is used to denote the side of the attached component involved in the fusion. The sides of the parent component are lettered *a, b, c,* etc., Side *a* lying between Positions 1 and 2, Side *b* between Positions 2 and 3 (or in certain cases, 2 and 2a), Side *c* between Positions 3 and 4, etc. Where a choice is available, letters occurring as early in the alphabet as possible are selected. The number pair (if needed) indicating the fused side of the attached ring is then prefixed to the letter indicating the fused side of the parent component; numbers as low as possible are used, and their order of citation conforms to the direction of lettering of the parent component. (Number pairs are obviously unnecessary with attached monocyclic components.) When two or more different components are fused to equivalent sides of the parent ring system, their prefix names are cited alphabetically, and the location of the first component cited is given the earliest letter of the alphabet possible. Finally, the combination of numbers and letters is enclosed in brackets immediately following the prefix designating the attached component.

Thus, for the above example, the conceptual process of fusion may be represented schematically as follows:

Dibenzo[c,g] phenanthrene

After the appropriate letters (and numbers when necessary) within the brackets defining the fusion process have been established, the complete ring system is numbered as a unit without regard to the original numbering patterns of the parent and attached components.

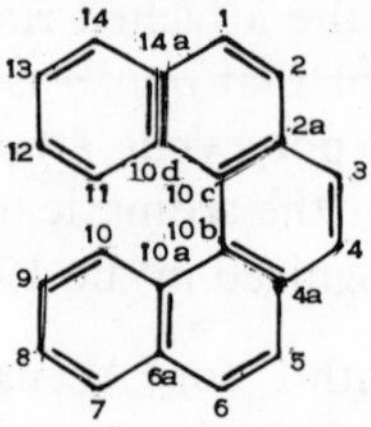

For numbering purposes the polycyclic structure is depicted so that the largest possible number of rings lie in a horizontal row, the correct orientation being (by convention) that in which the maximum number of rings is above and to the right of the horizontal row. The structure so oriented is numbered beginning with the carbon atom which is in the most counterclockwise unfused position of the uppermost ring lying farthest to the right. Enumeration is completed by numbering the carbon atoms in clockwise direction, omitting atoms which are common to two or more rings. Shared atoms are then numbered by adding the letters *a, b,* etc. (not italicized) to the number of the position immediately preceding; they are given the lowest numbers possible when there is a choice. (For further details concerning ring numbering, *see* the 1957 IUPAC Rules and "The Ring Index.") In the following examples, numerals outside the rings indicate numbering of the complete ring system; letters and numbers inside the rings apply to the separate components prior to fusion.

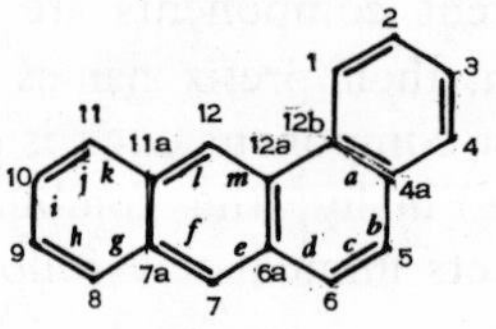

Benz[*a*]anthracene

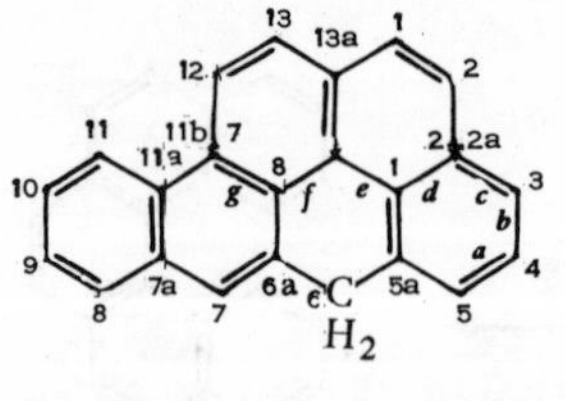

6*H*-Naphtho[2,1,8,7-*defg*]naphthacene

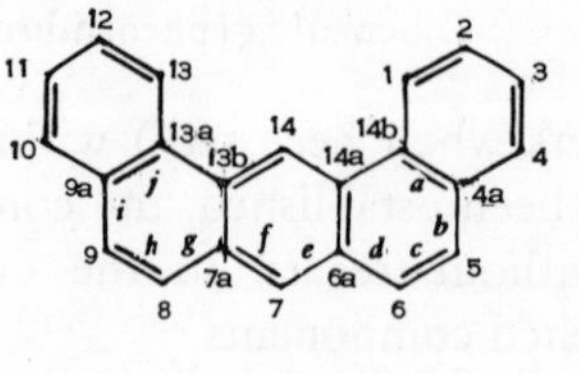

Dibenz[*a,j*]anthracene
*not* Naphtho[2,1-*b*]phenanthrene

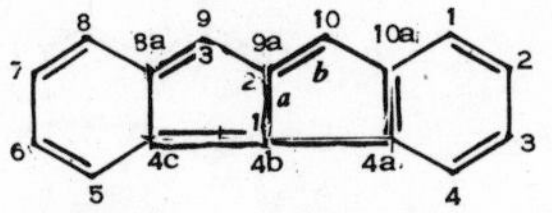

Indeno[1,2-*a*]indene

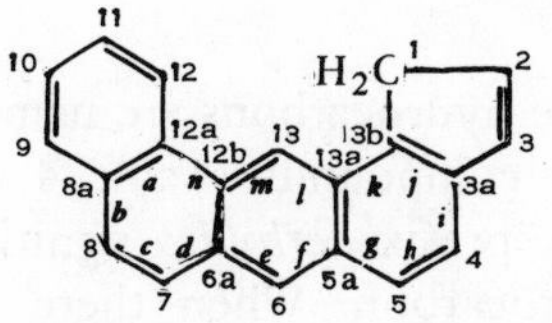

1*H*-Benzo[*a*]cyclopent[*j*]anthracene

Note elision of the *o* of *benzo* and the *a* of *cyclopenta* preceding the *a* of anthracene among the names shown for the above examples (*see also* Discussion section).

Three time-honored exceptions to systematic ring numbering are recognized in the 1957 IUPAC Rules:

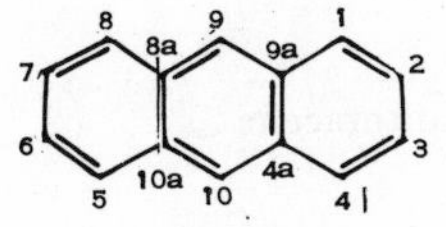

Anthracene

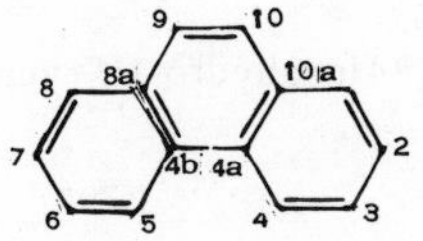

Phenanthrene

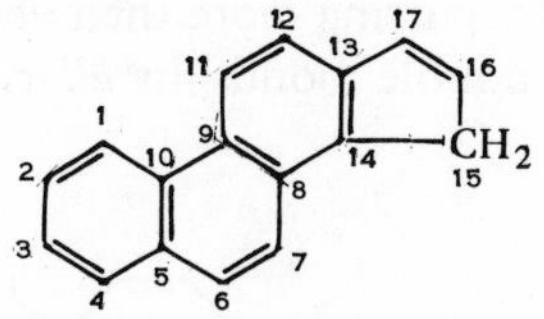

Cyclopenta[*a*]phenanthrene
(15*H* isomer shown)

As illustrated by some of the above examples, a ring system with the maximum number of conjugated double bonds can exist in two or more forms differing only in the position of an "extra" hydrogen atom, usually appearing in a $CH_2$ group. Location of this hydrogen atom is indicated by a prefix consisting of an italic capital *H* immediately preceded by the appropriate number locant. Hydrogen so occurring is properly called **indicated hydrogen**.

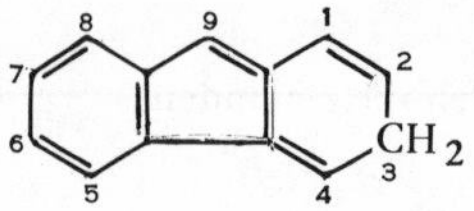

3*H*-Fluorene 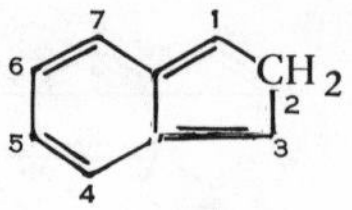                                    2*H*-Indene

**Hydrogenated polycyclic aromatic hydrocarbons** are named as derivatives of the corresponding aromatic compounds (Table 4.1) using the prefixes *dihydro, tetrahydro,* etc.; the prefix *perhydro* signifies complete hydrogenation of an aromatic hydrocarbon. When there is a choice, numbers as low as possible are used for positions of the added hydrogen atoms. However, if the choice involves indicated hydrogen, the latter is given preference for assigning the lowest available number.

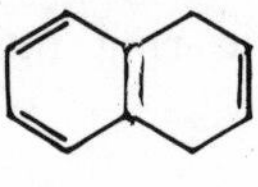

1,4-Dihydronaphthalene

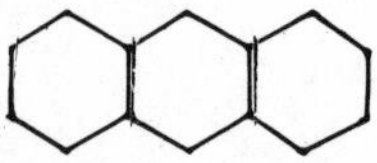

Perhydroanthracene

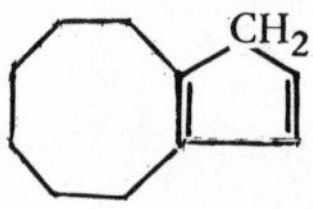

4,5,6,7,8,9-Hexahydro-1*H*-cyclopentacyclooctene

As illustrated by the last example above, the ending *ene* used in conjunction with the prefix *cyclo* (occurring more than once) implies the maximum number of conjugated double bonds in *all* rings of a fused polycyclic hydrocarbon.

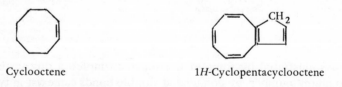

Cyclooctene                            1*H*-Cyclopentacyclooctene

Seven trivial names for partially hydrogenated polycyclic aromatic hydrocarbons are recognized in the 1957 IUPAC Rules. These are shown in Table 4.2, with the corresponding systematic names listed second in each case.

## Table 4.2. Hydrogenated Polycyclic Aromatic Hydrocarbons

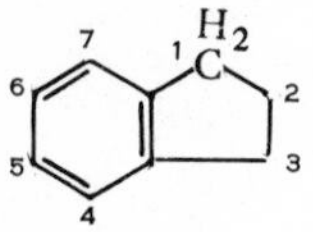

Indan
2,3-Dihydro-1*H*-indene

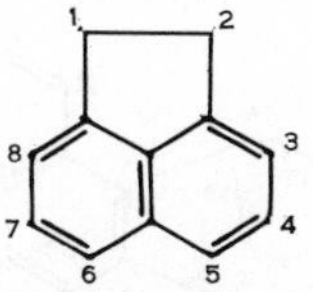

Acenaphthene
1,2-Dihydroacenaphthylene

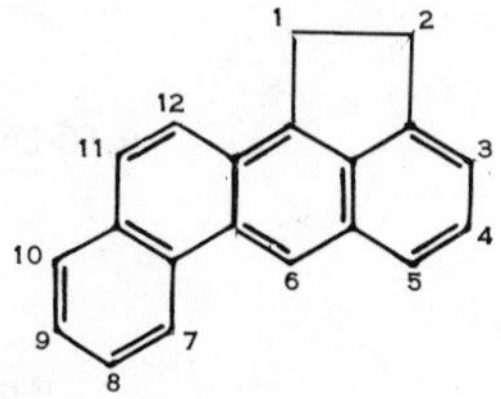

Cholanthrene
1,2-Dihydrobenz[*j*]aceanthrylene

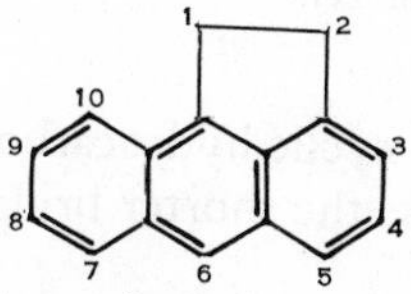

Aceanthrene
1,2-Dihydroaceanthrylene

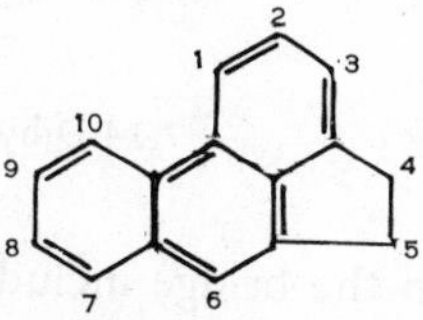

Acephenanthrene
4,5-Dihydroacephenanthrylene

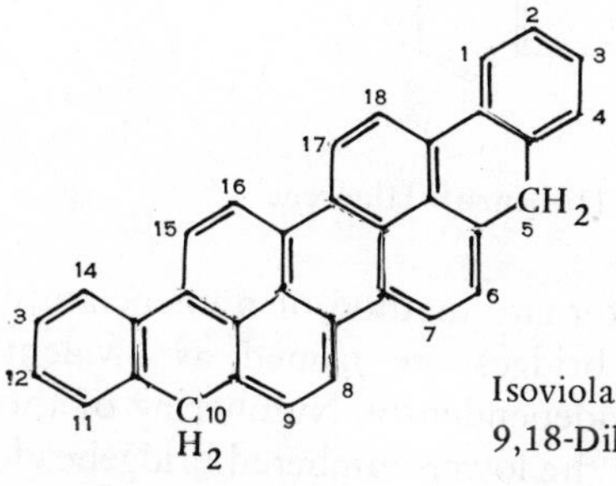

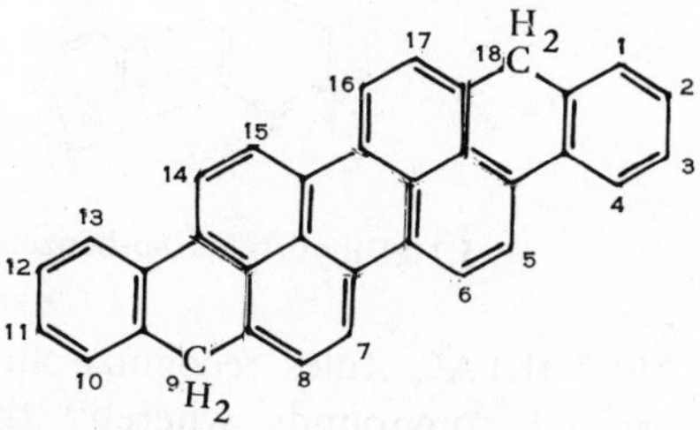

Isoviolanthrene
9,18-Dihydrodinaphtho[1,2,3-*cd*:1',2',3'-*lm*]perylene

Violanthrene
5,10-Dihydrodinaphtho[1,2,3-*cd*:3',2',1'-*lm*]perylene

**Bridged polycyclic aromatic hydrocarbons** (*see also* Chapter 2) are named by attaching a prefix name for each bridge to the name of the parent unbridged ring system. Prefix names of acyclic bridges are derived from those of the corresponding hydrocarbon chains by replacing the final *ane* or *ene* of the hydrocarbon name with *ano* or *eno*. Positions of bridges are indicated by number pairs locating the points of attachment of each bridge to the parent ring system. When there is a choice, the points of attachment (called bridgeheads) are given the lowest numbers possible. The carbon atoms in the bridges are then numbered in turn, starting each time with the bridge atom next to the bridgehead possessing the higher number.

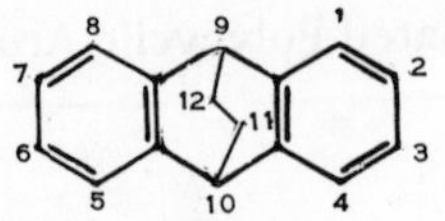

9,10-Dihydro-9,10-ethanoanthracene

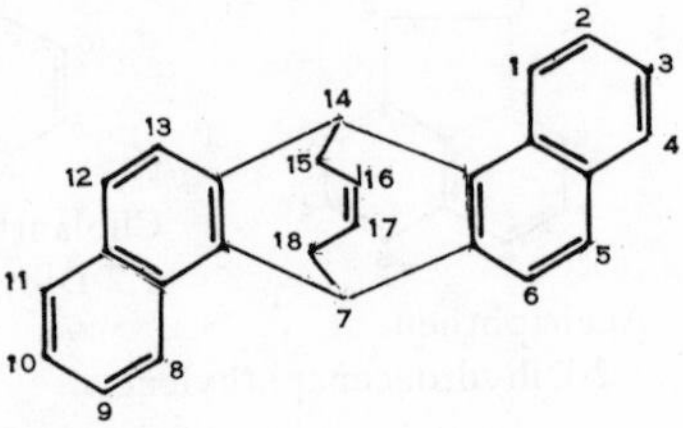

7,14-Dihydro-7,14-(2-buteno)dibenz[a,b]anthracene

When the bridge includes carbon atoms contained in a cyclic hydrocarbon group—i.e., when there are, in effect, two bridge paths—the shorter bridge path is numbered first.

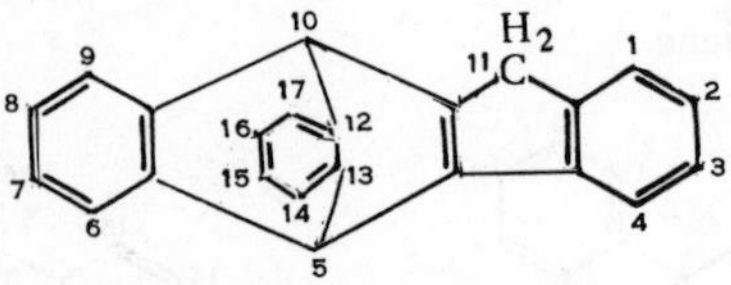

5,10-Dihydro-5,10-o-benzeno-11H-benzo[b]fluorene

The 1957 IUPAC Rules recognize an alternate method of nomenclature for bridged compounds whereby the bridges are named as bivalent substituting groups and are numbered independently. Numbering of the bridge in each case is started adjacent to the lower-numbered bridgehead, and numbers assigned to the bridge carbon atoms are primed.

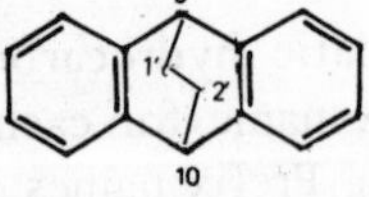

9,10-Dihydro-9,10-ethyleneanthracene

The first-described method, used by *Chemical Abstracts*, is the recommended one.

  **Substituting univalent groups** derived from fused polycyclic hydrocarbons are named by changing the final *e* of the hydrocarbon name to *yl*. The carbon atoms having free bonds are given numbers as low as possible consistent with the fixed numbering of the hydrocarbon.

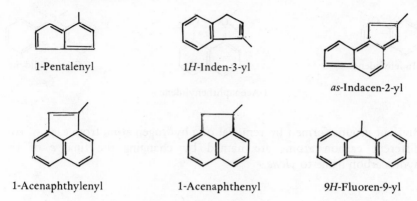

1-Pentalenyl          1*H*-Inden-3-yl

*as*-Indacen-2-yl

1-Acenaphthylenyl        1-Acenaphthenyl        9*H*-Fluoren-9-yl

Three abbreviated forms, naphthyl, anthryl, and phenanthryl, in which the *yl* ending replaces more than the final *e* of the hydrocarbon name are recognized in the 1957 IUPAC Rules.

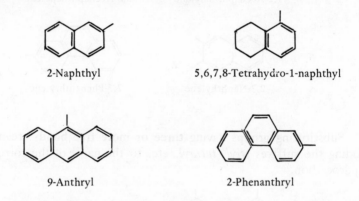

2-Naphthyl          5,6,7,8-Tetrahydro-1-naphthyl

9-Anthryl          2-Phenanthryl

However, these abbreviated forms are used only for the simple ring systems as such. Substituting groups derived from fused derivatives of these ring systems are named systematically.

Benz[*a*]anthracen-12-yl
*not* 12-Benz[*a*]anthryl

**Substituting bivalent groups** derived from fused polycyclic hydrocarbon by conceptual removal of two hydrogen atoms from a single carbon atom are named by changing the final *e* of the hydrocarbon name to *ylidene*.

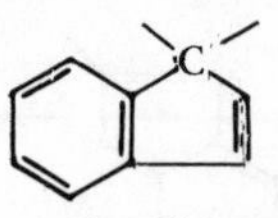

1-Indenylidene

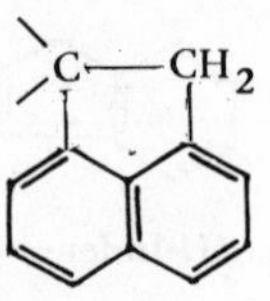

1-Acenaphthenylidene

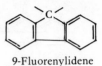

9-Fluorenylidene

Bivalent groups formed by removal of a hydrogen atom from each of two different carbon atoms are named by changing the final *e* of the hydrocarbon name to *ylene*.

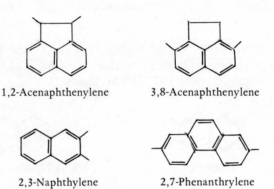

1,2-Acenaphthenylene        3,8-Acenaphthenylene

2,3-Naphthylene        2,7-Phenanthrylene

**Substituting groups** having **three or more free bonds** are named by adding the suffixes *triyl*, *tetrayl*, etc. to the name of the corresponding hydrocarbon.

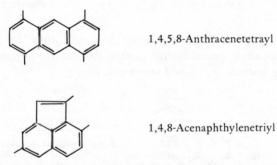

1,4,5,8-Anthracenetetrayl

1,4,8-Acenaphthylenetriyl

**Substituting groups** derived from fused polycyclic hydrocarbons by conceptual removal of hydrogen from a **side chain** are named systematically as aryl derivatives of the appropriate acyclic substituting group (*see* Chapter 2).

$CH_2CH_2-$

2-(1-Naphthyl)ethyl

## Discussion

As yet, no generally applicable systematic nomenclature for fused polycyclic hydrocarbons based on a single repeating structural unit (most often, the benzene ring) has been adopted, except for those cases in which several benzene rings (five or more) occur in a linear arrangement— *e.g.,* pentacene and hexacene (*see* Table 4.1). Since the frequency of occurrence in the published literature of compounds involving polycyclic ring systems containing more than three rings drops off rapidly as the number of component rings increases, the codified usage-based approach to naming such compounds presented in the 1957 IUPAC Rules has proved adequate to date, even though it relies heavily on trivial names.

The 1957 IUPAC Rules recognize trivial names for 35 fused polycyclic aromatic hydrocarbons (Table 4.1) and name all others (except those in Table 4.2) as substituted derivatives of these. When the substituents are side chains, it is relatively simple to designate their identity and locations. When, on the other hand, the substituents are fused rings, this cannot be done as readily because the preferred parent ring system may not be immediately obvious. The 1957 IUPAC Rules do make it possible to choose a single, preferred parent component, but for most chemists a simpler procedure is to consult "The Ring Index" (*1*), which lists and names all rings and ring systems appearing in the published literature through 1956; Supplements I, II, and III extend this coverage through 1963. If the ring system in question has not been reported and listed, one should consult the "Introduction to The Ring Index" in conjunction with the 1957 IUPAC Rules and the *Chemical Abstracts Index Guide (2)* for cases where the information here proves inadequate.

It is customary to elide the final *o* or *a* of prefixes in fusion names when the name of the parent ring system begins with a vowel—*e.g.,* benz[*a*]anthracene for benzo[*a*]anthracene, except that such elision is not practiced with the prefixes anthra and phenanthro nor with those ending in *eno*. Spelling conventions of this kind tax the memory, and it is difficult to advocate strict adherence to them. The corresponding fully spelled forms are clear and unambiguous, and, therefore, can not be termed unacceptable.

The nonsystematic numbering allowed for cyclopenta[*a*]phenanthrene does not apply to the other isomers of cyclopentaphenanthrene. The exception made for the [*a*] isomer is in deference to the importance of this ring system in the chemistry of steroids.

A specialized nomenclature for bridged aromatic hydrocarbons designed to simplify the naming of structures more complex than the examples shown above was proposed by Brandes H. Smith in his book "Bridged Aromatic Compounds" (3). This proposal modified and extended the "cyclophane" type of name that first appeared in the literature in the early 1950's for structures in which two benzene rings are doubly connected by polymethylene bridges. Other authors have suggested and described similar but still more general naming systems for phanes (4, 5, 6); the proposals are under review by nomenclature committees.

## Literature Cited

1. Patterson, A. M., Capell, L. T. and Walker, D. F. "The Ring Index" 2nd ed., American Chemical Society, Washington, D. C., 1960; also, "Supplements I-III," 1963-65.
2. *Chemical Abstracts Index Guide,* published annually beginning with Vol. 69 (July-Dec. 1968). Also available separately from Subscription Fullfillment, Chemical Abstracts Service, Columbus, Ohio 43210.
3. Smith, B. H., "Bridged Aromatic Compounds," Academic Press, New York, 1964.
4. Vögtle, F., and Neumann, P. *Tetrahedron Lett.* (1969) 5329; *Tetrahedron* (1970) 26, 5847.
5. Hirayama, K., *Tetrahedron Lett.* (1972) 2109.
6. Kauffmann, Th., *Tetrahedron* (1972) 28, 5183.

# Hydrocarbon Ring Assemblies

The term **ring assembly** denotes a polycyclic hydrocarbon in which two or more rings (or fused ring systems) are joined directly by single or double bonds and in which the number of such direct ring junctions is one less than the total number of rings and ring systems. For example, two benzene rings joined by a single bond, or two cyclohexane rings joined by a single or double bond, are classed as ring assemblies. On the other hand two benzene rings joined by two single bonds form a fused polycyclic system (*see* Chapter 4).

Ring assemblies

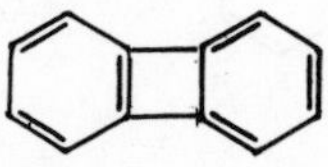

A fused polycyclic system

## Recommended Nomenclature Practice

Hydrocarbon **ring assemblies** consisting of **two identical rings or ring systems** joined by a single or double bond are named by placing the prefix *bi* before the name of the component ring system, except that the name biphenyl is commonly used for the assembly consisting of two benzene rings. Alternative names based on the substituting group name of the component hydrocarbon are also recognized in the 1957 IUPAC Rules.

Numbering of a ring assembly is based on that of the component ring or ring system, unprimed numbers being assigned to one ring or ring system and primed numbers to the other. When necessary, to avoid ambiguity  the points at which the component rings are joined are indicated by placing appropriate locants before or within the name. If the two rings are joined by a double bond, a capital Greek delta ($\Delta$) with superscript number locants is required in  the hydrocarbon-type name to denote such joining.

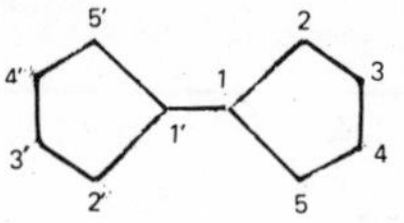

Bicyclopentane
Bicyclopentyl

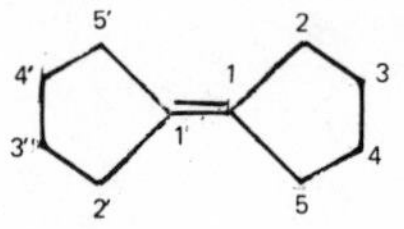

$\Delta^{1,1'}$-Bicyclopentane
Bicyclopentylidene

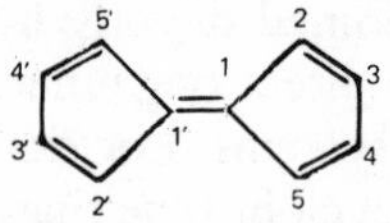

$\Delta^{1,1'}$-Bi-2,4-cyclopentadiene
Bi-2,4-cyclopentadien-1-ylidene

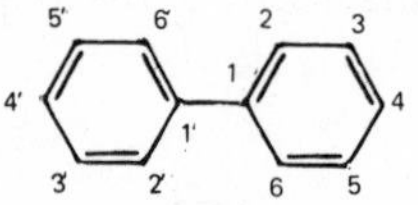

Biphenyl (Bibenzene, though correctly
    formed, is hardly ever used).
    *not* Diphenyl

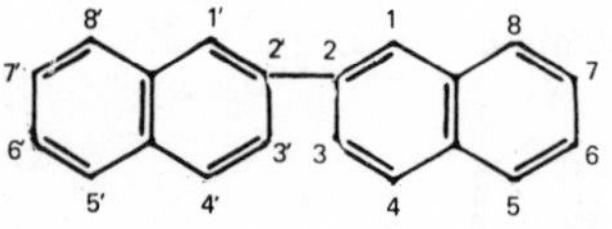

2,2'-Binaphthalene
2,2'-Binaphthyl

When there is a choice in numbering, unprimed numbers are given to the ring or ring system which has the lower-numbered point of joining.

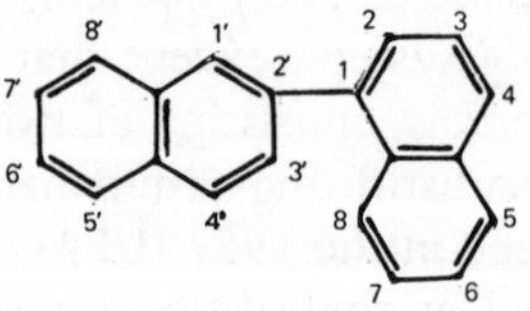

1,2'-Binaphthalene
1,2'-Binaphthyl

When component ring systems having the same point of attachment contain substituents at various positions, lowest possible numbers are given to the substituents, an unprimed number being considered to be lower than the same number when primed. Primed and unprimed numbers are arranged in a single series in ascending numerical order.

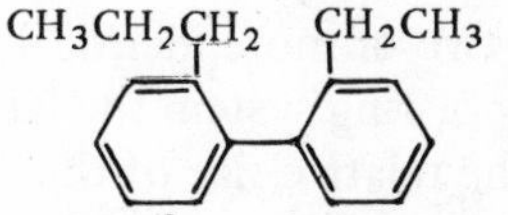

CH₃CH₂CH₂  CH₂CH₃

2-Ethyl-2′-propylbiphenyl

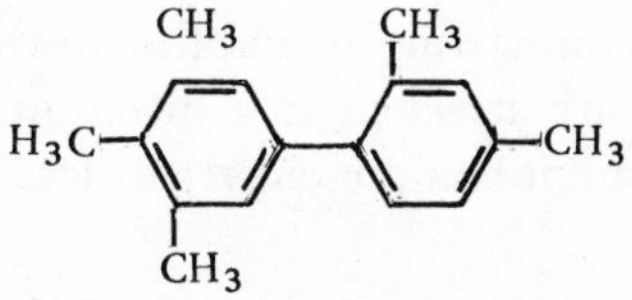

H₃C—⟨⟩—⟨⟩—CH₃

CH₃  CH₃

CH₃

2,3′,4′,4′,5′-Pentamethylbiphenyl

*not* 2′,3,4,4′,5-Pentamethylbiphenyl

When a **hydrocarbon ring assembly** is composed of **unlike rings** or ring systems, the name is formed by selecting one ring or ring system as the parent component and treating the other components as substituents. The parent component is chosen so as to contain the maximum number of rings; if there is a further choice, the component containing the larger ring is selected. Ordinary numbering principles are followed—*i.e.,* primed numbers are not required for this type of ring assembly.

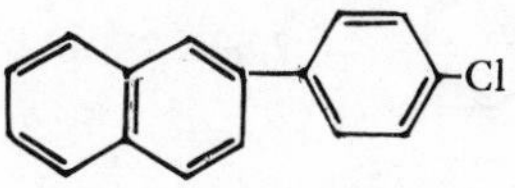

2-(4-Chlorophenyl)naphthalene

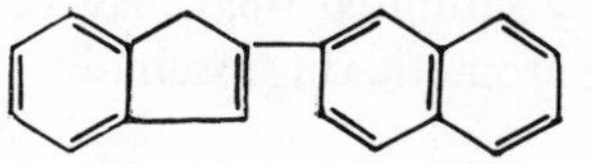

2-(1*H*-Inden-2-yl)naphthalene

However, the parent component may itself be an assembly of identical rings that utilizes primed numbers.

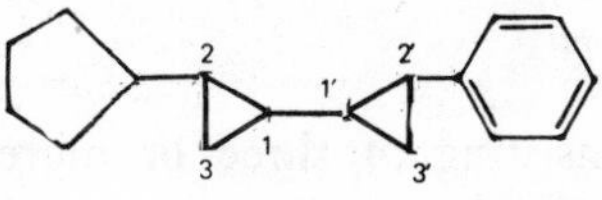

2-Cyclopentyl-2′-phenylbicyclopropane

If there is a choice, the component ring in the lower state of hydrogenation is selected as the parent.

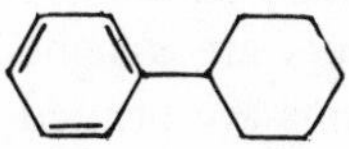

Cyclohexylbenzene

In choosing between two polycyclic aromatic hydrocarbons of Table 4.1, the ring system assigned the higher listing number is selected as the parent.

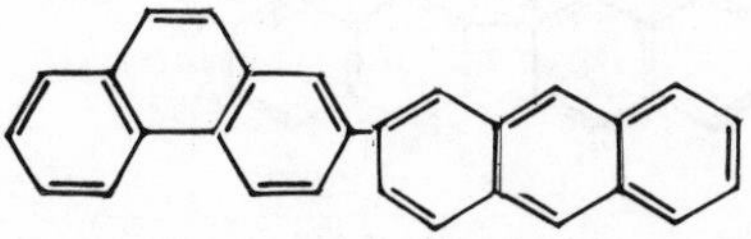

2-(2-Phenanthryl)anthracene

In naming ring assemblies in which two or more identical rings or ring systems are attached to a different ring or ring system the latter is chosen as the parent component regardless of the relative size of the rings.

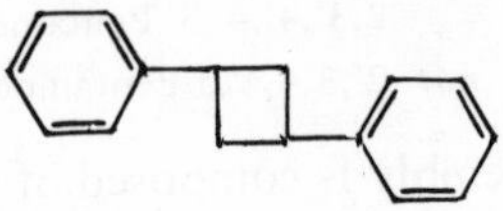

1,3-Diphenylcyclobutane

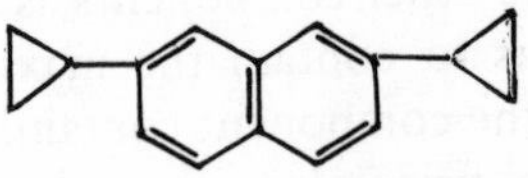

2,7-Dicyclopropylnaphthalene

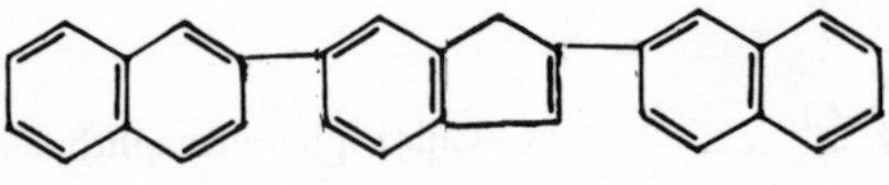

2,6-Di-2-naphthyl-1H-indene

Ring assemblies containing partially saturated rings may be named as hydro derivatives of the corresponding aromatic ring assembly.

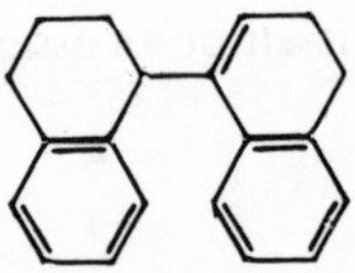

1,2,3,3'4,4'-Hexahydro-1,1'-binaphthalene

Hydrocarbon ring assemblies consisting of three or more identical rings or ring systems are named by prefixing a term indicating the number of rings before the name of the hydrocarbon corresponding to the repetitive unit. The following multiplying prefixes are used: *ter, quater, quinque, sexi, septi, octi, novi, deci,* etc. Designation of points of attachment of component rings is unnecessary in the cyclopropane series. With other ring systems, unprimed numbers are assigned to one of the terminal ring systems, and the other systems are primed serially. Lowest possible numbers are given to the points of attachment.

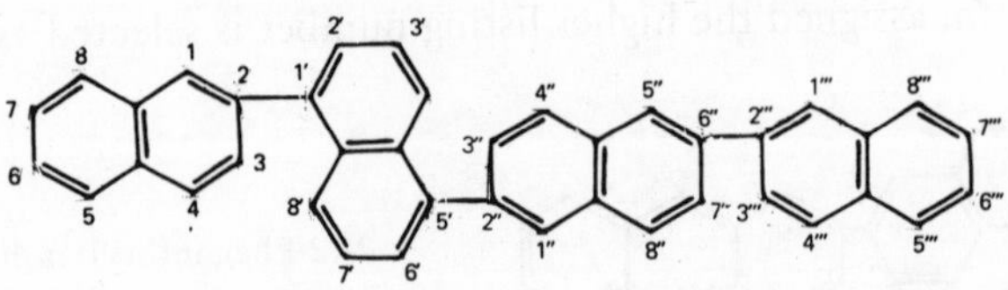

2,1',5',2'',6'',2'''-Quaternaphthalene

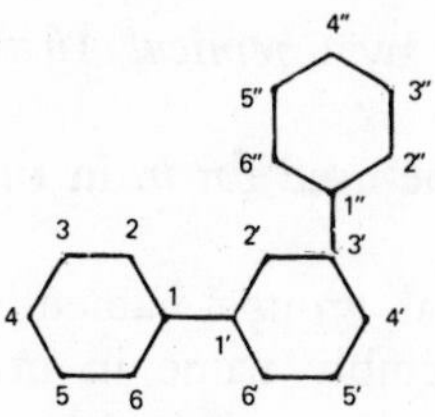

1,1′:3′,1″-Tercyclohexane

Analogously to biphenyl, assemblies of benzene rings are commonly named by using the appropriate multiplying prefix with the substituting group name, phenyl, rather than with the hydrocarbon name, benzene.

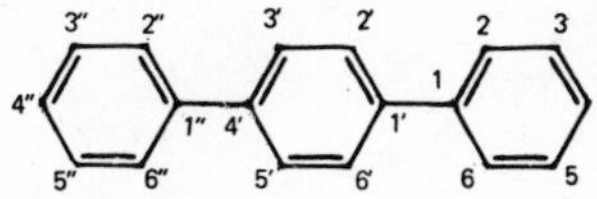

*p*-Terphenyl

or

1,1′:4′,1″-Terphenyl

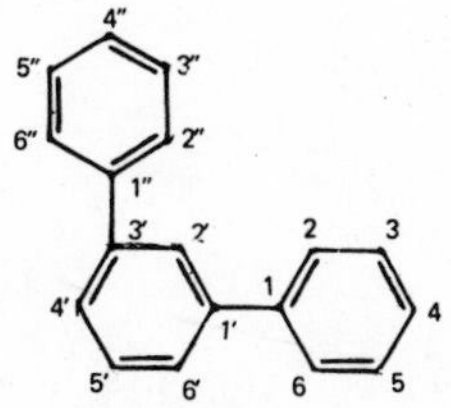

*m*-Terphenyl

or

1,1′:3′,1″-Terphenyl

When a ring assembly consists of three or more phenyl groups each attached to the same benzene ring, the compound is named as a substituted benzene—*e.g.,* 1,3,5-triphenylbenzene and hexaphenylbenzene.

## Discussion

Two fundamentally different methods are currently used for naming ring assemblies, both employing the same series of multiplying prefixes. One is based on hydrocarbon names—*e.g.,* bicyclohexane—and the other on substituting group names—*e.g.,* bicyclohexyl. Both methods are used by *Chemical Abstracts,* selection in individual cases being made on the basis of the particular compound or type of compound involved.

In this book, except for the well-established polyphenyl series, the hydrocarbon type of naming is recommended, chiefly for consistency with the remainder of systematic organic nomenclature, so as to reserve the *yl* ending for structures which are substituting groups in the usual sense.

Although the 1957 IUPAC Rules recognize as ring assemblies polycyclic systems containing non-identical components—*e.g.,* 2-phenyl-naphthalene—this practice is not followed by *Chemical Abstracts* and is not recommended here.

The multiplying prefix *di* should not be used for *bi* in ring assembly names, as in "diphenyl" for biphenyl.

When ring assemblies carry functional groups named as suffixes, *Chemical Abstracts* encloses the ring assembly name in brackets—e.g., [1,1'-binaphthalene]-2-carboxylic acid.

# Heterocyclic Systems

**H**eterocyclic systems are ring compounds containing atoms of at least two different elements as ring members. **Organic heterocyclic systems** contain one or more "foreign" elements such as oxygen, sulfur, or nitrogen in addition to carbon; atoms of such elements conceptually replacing carbon in a ring system have long been called **hetero atoms.** In recent years, however, the meaning of the term hetero atom has been broadened to include atoms other than carbon occurring in chains as well as in rings (*see* Chapter 7). The general method of naming organic ring and chain systems based on the hetero atom concept is known as **replacement nomenclature.**

*Recommended Nomenclature Practice*

**Organic heterocyclic systems** are preferably named by relating them to the corresponding hydrocarbon ring systems (carbocyclic systems) by using replacement nomenclature. In replacement names, hetero atoms are denoted by prefixes ending in *a,* as shown in Table 6.1; if two or more replacement prefixes are required in a single name, they are cited in the

Table 6.1.  Hetero Atom Prefix Names[a]

| Element | Valence | Prefix Name |
|---------|---------|-------------|
| Oxygen | II | Oxa |
| Sulfur | II | Thia |
| Selenium | II | Selena |
| Tellurium | II | Tellura |
| Nitrogen | III | Aza |
| Phosphorus | III | Phospha |
| Arsenic | III | Arsa |
| Antimony | III | Stiba |
| Bismuth | III | Bismutha |
| Silicon | IV | Sila |
| Germanium | IV | Germa |
| Tin | IV | Stanna |
| Lead | IV | Plumba |
| Boron | III | Bora |
| Mercury | II | Mercura |

[a] Listing is in descending order of precedence.

order of their listing in the table. Lowest possible numbers consistent with the numbering of the corresponding carbocyclic system are assigned to the hetero atoms, then to carbon atoms bearing double or triple bonds. Locants are cited immediately preceding the prefixes or suffixes to which they refer.

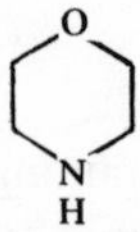

1-Oxa-4-azacyclohexane

Thiacyclopent-2-ene

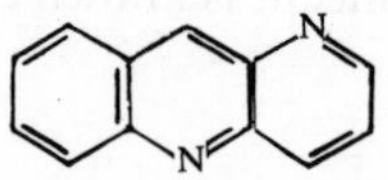

1,10-Diazaanthracene

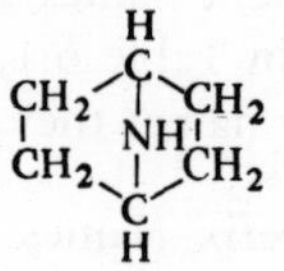

7-Azabicyclo[2.2.1]heptane

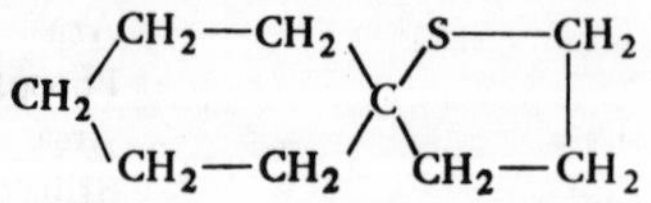

1-Thiaspiro[4.5]decane

If the corresponding carbocyclic system is partially or completely hydrogenated and is named as a *hydro* derivative of a polycyclic aromatic hydrocarbon (*see* Chapter 4), then the parent heterocyclic system is considered to be that which contains the maximum number of conjugated or isolated double bonds compatible with valency restrictions imposed by the hetero atoms; any additional hydrogen, if present, is cited using the appropriate *H* and/or *hydro* prefixes.

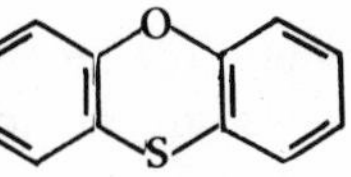

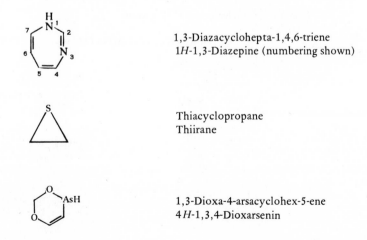

9-Oxa-10-thiaanthracene

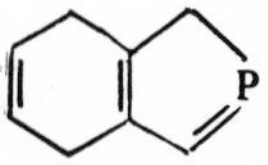

4,7-Dihydro-1*H*-2-phosphaindene

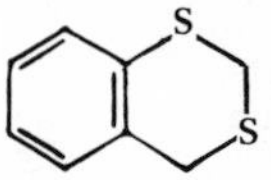

2*H*,4*H*-1,3-Dithianaphthalene

As an alternative to replacement nomenclature, an extension of the **Hantzsch-Widman** proposal (*1,2*) may be used for naming hetero rings containing three to ten atoms. (This method does not involve reference to the corresponding carbocyclic system.) Hetero atoms are denoted by the same prefixes used in replacement nomenclature (Table 6.1), but the final *a* is elided immediately preceding another vowel. Numbering of the ring begins with the atom of the element standing highest in Table 6.1 and proceeds in the direction that gives lowest possible numbers to the other hetero atoms in the ring. The degree of saturation and the size of the ring are denoted by the suffixes listed in Table 6.2. In the following examples preferred replacement names are shown (first) for comparison.

1,3-Diazacyclohepta-1,4,6-triene
1*H*-1,3-Diazepine (numbering shown)

Thiacyclopropane
Thiirane

1,3-Dioxa-4-arsacyclohex-5-ene
4*H*-1,3,4-Dioxarsenin

As in replacement nomenclature, Hantzsch-Widman names for **partially saturated hetero rings** are formed by using *hydro* prefixes, except for four- and five-membered rings, for which the special name endings shown in Table 6.3 are used, provided, however, that the hetero ring in question can accommodate conjugated unsaturation.

A number of well-established **trivial names** for heterocyclic systems are recognized in the 1957 IUPAC Rules. These are included in Table 6.4.

### Table 6.2. Hantzsch-Widman Suffixes

| Ring Members | Rings containing Nitrogen | | Rings without Nitrogen | |
|---|---|---|---|---|
| | Unsaturated[a] | Saturated | Unsaturated[a] | Saturated |
| 3 | -irine | -iridine | -irene | -irane |
| 4 | -ete | -etidine | -ete | -etane |
| 5 | -ole | -olidine | -ole | -olane |
| 6 | -ine[b] | [c] | -in[b] | -ane[d] |
| 7 | -epine | [c] | -epin | -epane |
| 8 | -ocine | [c] | -ocin | -ocane |
| 9 | -onine | [c] | -onin | -onane |
| 10 | -ecine | [c] | -ecin | -ecane |

[a] Having the maximum number of conjugated double bonds.
[b] Immediately preceding -ine or -in the special prefix names phosphor, arsen, and antimon are used rather than phospha, arsa, and stiba.
[c] Expressed by prefixing perhydro to the name of the corresponding unsaturated ring.
[d] For Si, Ge, Sn, and Pb perhydro is prefixed to the name of the corresponding unsaturated ring.

### Table 6.3. Special Hantzsch-Widman Suffixes

| Ring Members | Rings containing Nitrogen | Rings without Nitrogen |
|---|---|---|
| 4 | -etine | -etene |
| 5 | -oline | -olene |

These suffixes indicate partial saturation.

In the following examples, preferred replacement names are shown (first) for comparison.

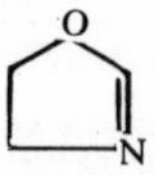

1-Oxa-3-azacyclopent-2-ene
1,3-Oxazolin-2-ene

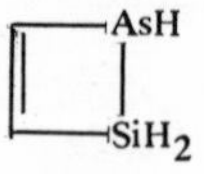

1-Arsa-2-silacyclobut-3-ene
1,2-Arsasilet-3-ene

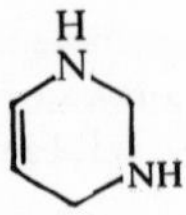

1,3-Diazacyclohex-4-ene
1,2,3,4-Tetrahydro-1,3-diazine

Both Hantzsch-Widman names and certain trivial names may be used in forming names for **fused heterocyclic systems** according to the principles given for polycyclic hydrocarbons in Chapter 4; however, replacement names for these ring systems are preferred. For details concerning the application of the **fusion method** of nomenclature to heterocyclic systems *see* 1957 IUPAC Rule B-3 or the introductory section of "The Ring Index" (*3*). In the following examples, preferred replacement names are shown (first) for comparison.

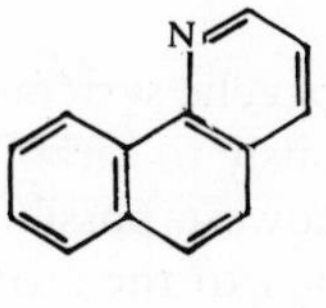

4-Azaphenanthrene
Benzo[*b*]quinoline

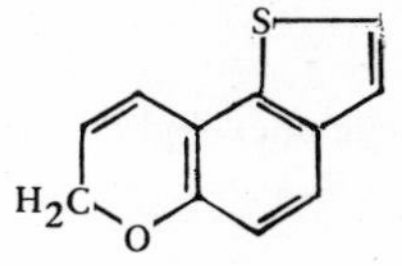

7*H*-6-Oxa-1-thiacyclopenta[*a*]naphthalene
7*H*-Thieno[2,3-*f*]chromene

**Heterocyclic ring assemblies** are preferably named by applying the replacement principle, but alternative names based on the other methods described above are acceptable.

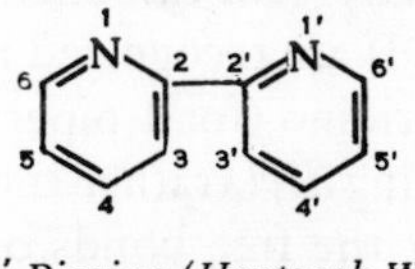

2,2'-Biazine (*Hantzsch-Widman*)
2,2'-Bipyridine (*trivial*)

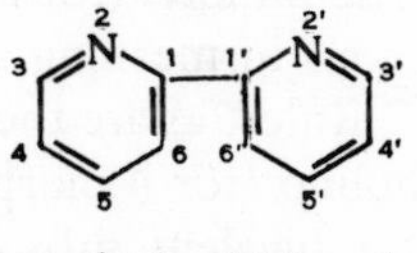

2,2'-Diazabiphenyl

Naming of hydrocarbon ring assemblies is described in Chapter 5.

When a hetero atom carries a **positive charge**, the final *a* of the replacement prefix is changed to *onia*. Trivial names for cationic heterocyclic systems are formed by attaching the suffixes *ium, diium,* etc. (with appropriate elision) to the name of the corresponding nonionic ring system.

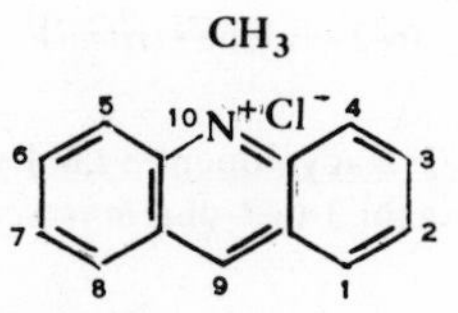

9-Methyl-9-azoniaanthracene chloride
10-Methylacridinium chloride
    (numbering shown)

10,10-Dimethyl-9-oxa-10-azoniaanthracene
  bromide
10,10-Dimethylphenoxazin-10-ium bromide

1,8-Dimethyl-1,8-diazonianaphthalene
  sulfate
1,8-Dimethyl-1,8-naphthyridinediium
  sulfate

**Substituting groups** derived from heterocyclic systems are preferably named by the replacement principle, analogously to carbocyclic substituting groups (*see* Chapters 2, 3, and 4). Lowest possible numbers are assigned first to the hetero atoms rather than to the atoms bearing free bonds.

1,3-Dioxacyclohex-4-yl

When Hantzsch-Widman or trivial nomenclature is used, certain exceptions (as compared with carbocyclic systems) prevail in forming names of substituting groups: (a) the shortened forms furyl (for furanyl), pyridyl (for pyridinyl), piperidyl (for piperidinyl), quinolyl (for quinolinyl), isoquinolyl (for isoquinolinyl), and thienyl (for thiophenyl which is ambiguous in meaning and should not be used) are recognized in the 1957 IUPAC Rules, as are the modified forms piperidino (for 1-piperidinyl) and morpholino (for 4-morpholinyl); (b) the ending *diyl* (rather than *ylene*) is used for bivalent substituting groups having the free bonds on different atoms. In the following examples, the preferred replacement names are shown first.

Azacyclopenta-2,4-dien-3-yl
1*H*-Azol-3-yl  *(Hantzsch-Widman)*
1*H*-Pyrrol-3-yl  *(trivial)*

Oxacyclohexa-3,5-dien-2-ylidene
2*H*-Oxin-2-ylidene  *(Hantzsch-Widman)*
2*H*-Pyran-2-ylidene  *(trivial)*

1-Oxa-3-azacyclopent-3-en-4-yl
1,3-Oxazol-3-in-4-yl  *(Hantzsch-Widman)*

1-Thia-2-azacyclopenta-2,4-diene-3,4-diyl
1,2-Thiazole-3,4-diyl (*Hantzsch-Widman*)
3,4-Isothiazolediyl (*trivial*)

1-Oxa-4-azacyclohex-4-yl
Perhydro-4*H*-1,4-oxazin-4-yl (*Hantzsch-Widman*)
Morpholino (*trivial*)

1-Oxa-4-azacyclohexane-2,4,6-triyl
Perhydro-4*H*-1,4-oxazine-2,4,6-triyl
  (*Hantzsch-Widman*)
2,4,6-Morpholinetriyl (*trivial*)

## Discussion

The general trend in recent years toward systematic organic nomenclature has resulted in an increasing preference for heterocyclic names
over other alternatives even though the hetero ring atoms, owing to their
functional characteristics, might not be treated analogously in open-chain
structures. Examples may be found among compounds such as lactones,
lactams, and sultams as well as cyclic anhydrides, imides, ethers, sulfides,
and amines; in the latter three cases, heterocyclic names are now used
almost exclusively.

Further systematization is evident within heterocyclic nomenclature
itself although many trivial names have appeared in the literature over the
years. Several systems of naming have evolved and are now officially
recognized. The Hantzsch-Widman method (*1,2*), proposed more than 80
years ago, provides a systematic approach to naming simple hetero rings of
up to 10 atoms and produces compact names that have found increasing
application in published articles and indexes. Using Hantzsch-Widman
names, as well as certain approved trivial names, the fusion method of
naming polycyclic hydrocarbon systems has been extended to multiple-
ring heterocyclic systems (*3*). However, owing to the increased number of
possible variations in structure introduced by the presence of the hetero
atoms, as compared with carbocyclic systems, the rules for naming
heterocyclic systems are necessarily more complex. This had led to the
extensive use of trivial names, both individually and as names of
components in quasi-systematic names for fused polycyclic hetero ring
systems.

A much simpler and more straightforward approach has been found
in replacement nomenclature, better known, perhaps, to many chemists as
"oxa-aza" nomenclature or simply as "a" nomenclature. As indicated
above under Recommended Nomenclature Practice, this method involves

the use of simple prefixes to denote replacement of one or more carbon atoms in a cyclic hydrocarbon by hetero atoms. Thus, with only a few additional rules, the principles used in naming carbocyclic systems are sufficient for naming heterocyclic systems as well. For these reasons, replacement nomenclature is much to be preferred to the Hantzsch-Widman and fusion methods although the latter are well established.

Two closely similar versions of replacement nomenclature have been proposed and used. The version recommended in this chapter is that of *Chemical Abstracts*. The other, called the Stelzner method, differs only in the manner of choosing the hydrocarbon system considered to correspond to the heterocyclic system being named (*see* 1957 IUPAC Rule B-4). In many instances the *Chemical Abstracts* and Stelzner names for heterocycles are identical.

Replacement prefixes are non-detachable—*i.e.*, they are considered to be part of the parent name of the heterocyclic system and are not separated when one forms inverted names for indexing. Thus, in the subject indexes of *Chemical Abstracts* one finds the entry 1-Oxa-3-sila-cyclobutane rather than Cyclobutane, 1-oxa-3-sila-.

Replacement names are based on names of hydrocarbons, not heterocyclic systems—*e.g.*, 1,4-diazanaphthalene, not 4-azaquinoline.

The Smith proposal (4) for naming bridged aromatic structures, described briefly in Chapter 4, is applicable to bridged heterocycles and provides names such as furanophane, pyridinophane, etc.

"The Ring Index" (3) lists and names all heterocyclic systems appearing in the published literature through 1956, and the Supplements I, II, and III extend this coverage through 1963. However, since the names listed are mainly those noted in the literature, replacement names are not included in every instance. A second good source is the *Chemical Abstracts Index Guide* (5).

Additive names for heterocyclic systems, such as tetramethylene oxide (for tetrahydrofuran), should be abandoned. However, ethylene oxide and propylene oxide were used until 1972 as entries in subject indexes of *Chemical Abstracts* and are considered acceptable for the unsubstituted compounds; the same is true of the trivial name ethylenimine. This does not hold for "styrene oxide," although the latter has received considerable usage.

## Table 6.4. Examples of Acceptable Usage

Examples 7-53 correspond to Compounds 1-47 as listed in 1957 IUPAC Rule B-2.11; the trivial names shown for these structures may be used in forming fusion names for more complex heterocyclic systems. Examples 54-67 correspond to Compounds 1-14 as listed in 1957 IUPAC Rule B-2.12; the trivial names shown for these structures are not recommended for use in fusion names. Numbering is shown only for those structures for which the numbering corresponding to the trivial name differs from that corresponding to the replacement name.

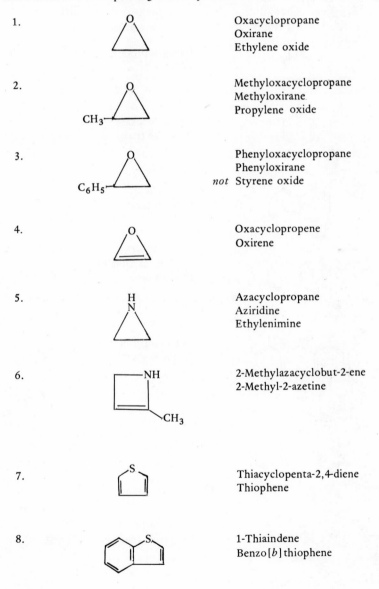

1. Oxacyclopropane
   Oxirane
   Ethylene oxide

2. Methyloxacyclopropane
   Methyloxirane
   Propylene oxide

3. Phenyloxacyclopropane
   Phenyloxirane
   *not* Styrene oxide

4. Oxacyclopropene
   Oxirene

5. Azacyclopropane
   Aziridine
   Ethylenimine

6. 2-Methylazacyclobut-2-ene
   2-Methyl-2-azetine

7. Thiacyclopenta-2,4-diene
   Thiophene

8. 1-Thiaindene
   Benzo[*b*]thiophene

9.

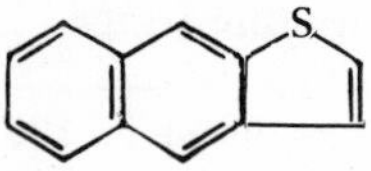

1-Thiacyclopental[b]naphthalene
Naphtho[2,3-b] thiophene

10.

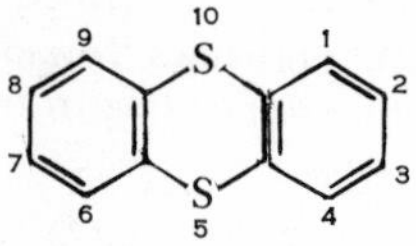

9,10-Dithiaanthracene
Thianthrene (numbering shown)

11.

Oxacyclopenta-2,4-diene
Furan

12.

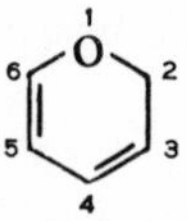

Oxacyclohexa-2,4-diene
2H-Pyran (numbering shown)

13.

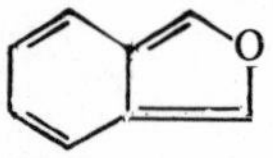

2-Oxaindene
Isobenzofuran

14.

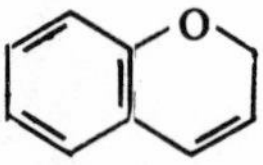

2H-1-Oxanaphthalene
2H-Chromene

15.

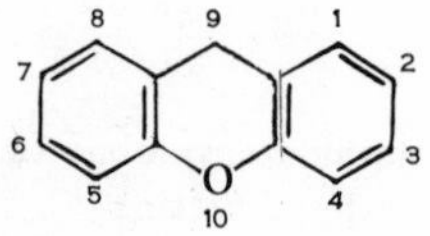

10H-9-Oxaanthracene
Xanthene (numbering shown)

16.

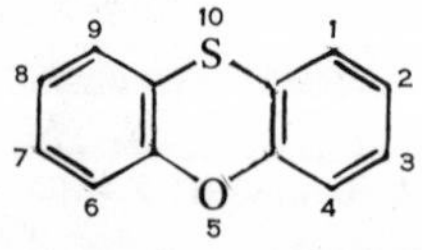

9-Oxa-10-thiaanthracene
Phenoxathiin (numbering shown)

17.

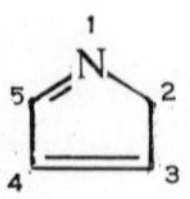

Azacyclopenta-1,3-diene
2H-Pyrrole (numbering shown)

18.   Azacyclopenta-2,4-diene
1*H*-Pyrrole

19. 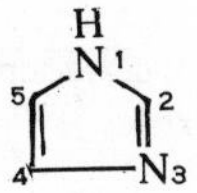  1,3-Diazacyclopenta-1,4-diene
Imidazole (numbering shown)

20. 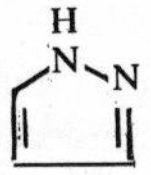  1,2-Diazacyclopenta-2,4-diene
Pyrazole

21. 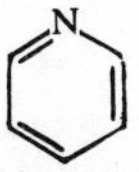  Azabenzene
Pyridine

22. 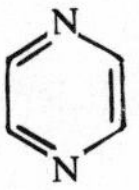  1,4-Diazabenzene
Pyrazine

23. 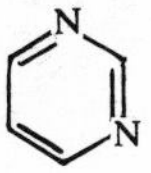  1,3-Diazabenzene
Pyrimidine

24. 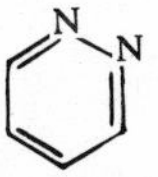 1,2-Diazabenzene
Pyridazine

25. 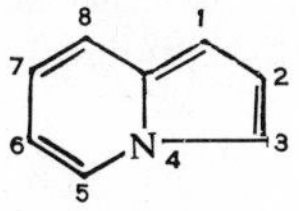 3a-Azaindene
Indolizine (numbering shown)

26. 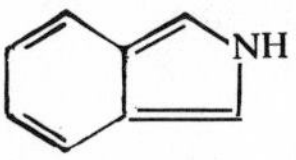  2*H*-2-Azaindene
Isoindole

27. 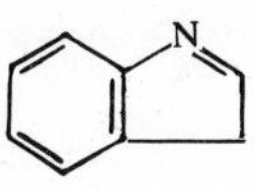  3*H*-1-Azaindene
3*H*-Indole

28.

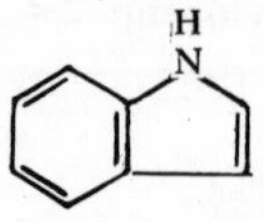

1*H*-1-Azaindene
Indole

29.

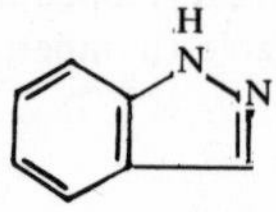

1*H*-1,2-Diazaindene
1*H*-Indazole

30.

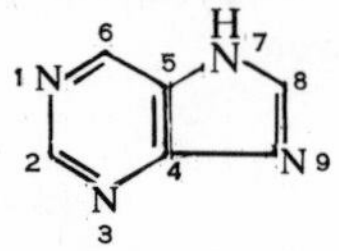

1*H*-1,3,4,6-Tetraazaindene
Purine (numbering shown)

31.

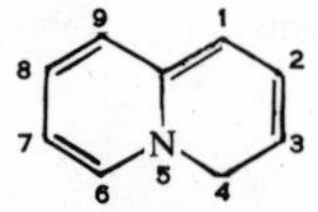

4*H*-4a-Azanaphthalene
4*H*-Quinolizine (numbering shown)

32.

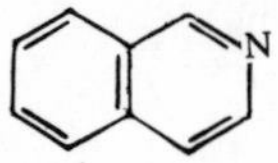

2-Azanaphthalene
Isoquinoline

33.

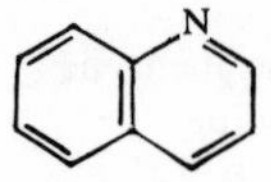

1-Azanaphthalene
Quinoline

34.

2,3-Diazanaphthalene
Phthalazine

35.

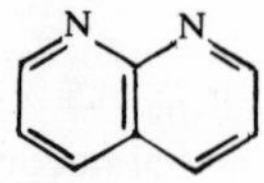

1,8-Diazanaphthalene
1,8-Naphthyridine

36.

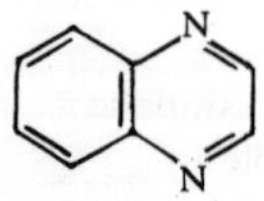

1,4-Diazanaphthalene
Quinoxaline

37.

1,3-Diazanaphthalene
Quinazoline

38. 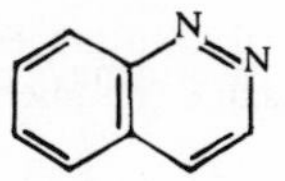                                        1,2-Diazanaphthalene
                                           Cinnoline

39. 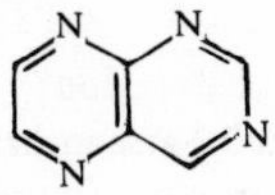                                        1,3,5,8-Tetraazanaphthalene
                                           Pteridine

40. 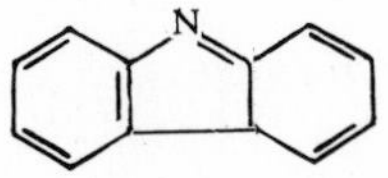                                        4a*H*-9-Azafluorene
                                           4a*H*-Carbazole

41. 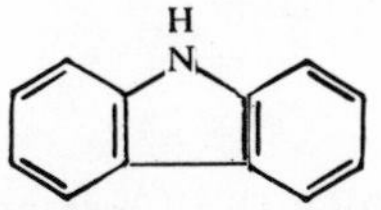                                        9*H*-9-Azafluorene
                                           Carbazole

42. 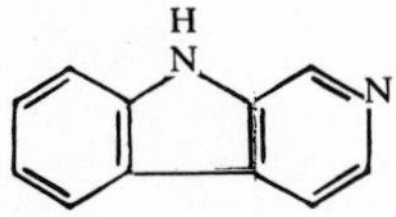                                        9*H*-2,9-Diazafluorene
                                           β-Carboline

43. 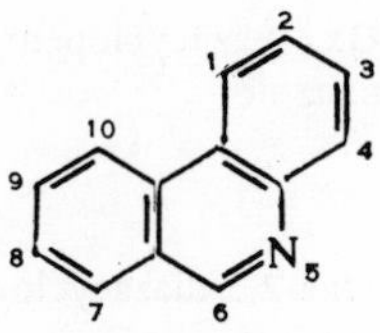                                        9-Azaphenanthrene
                                           Phenanthridine (numbering shown)

44. 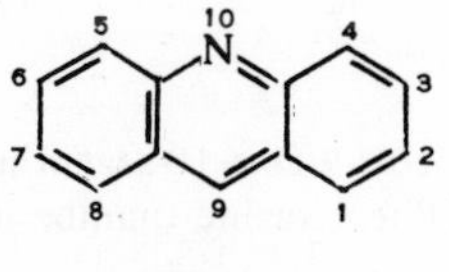                                        9-Azaanthracene
                                           Acridine (numbering shown)

45. 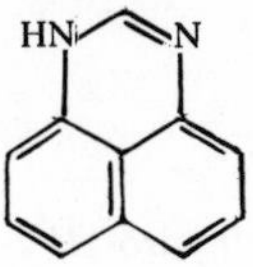                                        1*H*-1,3-Diazaphenalene
                                           Perimidine

46. 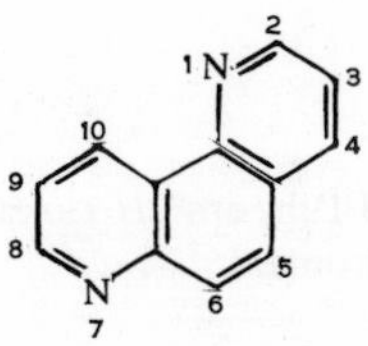                                        1,5-Diazaphenanthrene
                                           1,7-Phenanthroline (numbering shown)

47.

9,10-Diazaanthracene
Phenazine (numbering shown)

48.

9-Aza-10-arsaanthracene
Phenarsazine (numbering shown)

49.

1-Thia-2-azacyclopenta-2,4-diene
Isothiazole

50.

10H-9-Thia-10-azaanthracene
Phenothiazine (numbering shown)

51.

1-Oxa-2-azacyclopenta-2,4-diene
Isoxazole

52.

1-Oxa-2,5-diazacyclopenta-2,4-diene
Furazan

53.

10H-9-Oxa-10-azaanthracene
Phenoxazine (numbering shown)

54.

3,4-Dihydro-1H-2-oxanaphthalene
Isochroman

55.

3,4-Dihydro-2H-1-oxanaphthalene
Chroman

56.

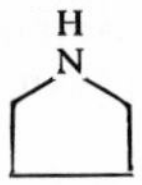

Azacyclopentane
Pyrrolidine

57.

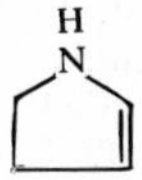

Azacyclopent-2-ene
2-Pyrroline

58.

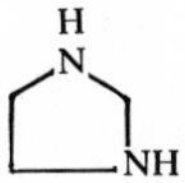

1,3-Diazacyclopentane
Imidazolidine

59.

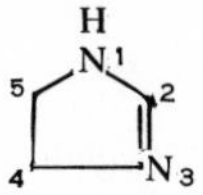

1,3-Diazacyclopent-1-ene
2-Imidazoline (numbering shown)

60.

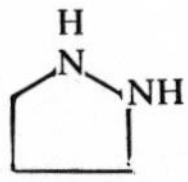

1,2-Diazacyclopentane
Pyrazolidine

61.

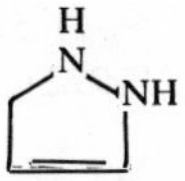

1,2-Diazacyclopent-3-ene
3-Pyrazoline

62.

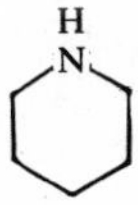

Azacyclohexane
Piperidine

63.

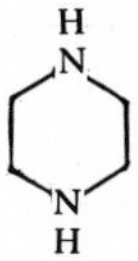

1,4-Diazacyclohexane
Piperazine

64.

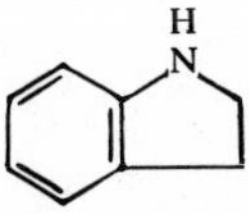

1-Azaindan
Indoline

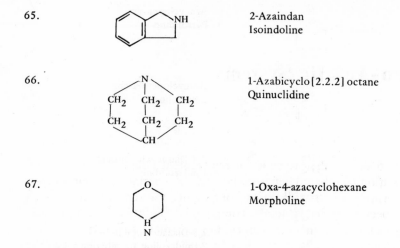

65.                                               2-Azaindan
                                                  Isoindoline

66.                                               1-Azabicyclo[2.2.2]octane
                                                  Quinuclidine

67.                                               1-Oxa-4-azacyclohexane
                                                  Morpholine

## Literature Cited

1. Hantzsch, A., Weber, K., *Ber.* (1887) **20**, 3119.
2. Widman, O., *J. Prakt. Chem.* (1888) **38**, 185.
3. Patterson, A. P., Capell, L. T., Walker, D. F., "The Ring Index," 2nd ed., American Chemical Society, Washington, D. C., 1960; also Supplements I-III, 1963-65.
4. Smith, B. H., "Bridged Aromatic Compounds," Academic, New York, 1964; *cf.* Voegtle, F., Neumann, P., *Tetrahedron Lett.* (1969) 5329 and *Tetrahedron* (1971) **27**, 5847; Hirayama, K., *Tetrahedron Lett.* (1972) 2109; Kauffmann, Th., *Tetrahedron* (1972) **28**, 5183.
5. *Chemical Abstracts Index Guide.* Published annually beginning with Vol. 69 (July-Dec. 1968). Available from Subscription Fulfillment, Chemical Abstracts Service, Columbus, Ohio 43210.

# Hetero-Acyclic Systems

Chapter 6 details the principles of replacement nomenclature and their application to organic heterocyclic systems. Originally developed and used exclusively for naming cyclic compounds, the replacement concept has since been extended to acyclic compounds, chiefly because it provides relatively simple systematic names for complex structures—*i.e.*, for straight chains containing more than two hetero atoms. However, in the broadest sense a **hetero-acyclic system** is any chain, branched or unbranched, containing atoms of at least two different elements. Treatment of **organic hetero chains** by replacement nomenclature is the subject of this chapter. Application of the replacement method to inorganic chains is illustrated for silicon compounds in Chapter 38.

*Recommended Nomenclature Practice*

In replacement nomenclature, **straight-chain organic hetero-acyclic systems** are named analogously to heterocyclic systems (*see* Chapter 6) using the prefixes shown in Table 6.1. The hetero chain always terminates with carbon.

$CH_3OCH_2CH_2OCH_2OCH_2CH_3$        2,5,7-Trioxanonane

$CH_3SCH_2OCH_2CH_2NHCH_3$        4-Oxa-2-thia-7-azaoctane

Lowest possible numbers are assigned first to the hetero atoms and then to atoms bearing double or triple bonds as described in Chapter 2.

$CH_3SCH_2CH_2OCH_2NHCH{=}CH_2$        5-Oxa-2-thia-7-azanon-8-ene

In naming **branched-chain organic hetero-acyclic systems** the parent chain is selected by applying the following criteria in the order of listing:

(1) The longest straight chain containing the maximum number of hetero atoms.

(2) The longest straight chain containing the maximum number of the species of hetero atom standing highest in Table 6.1.

(3) The longest straight chain carrying the maximum number of double and triple bonds.

$$CH_3OCH_2CH_2OCH\begin{cases}CH_2OCH_3\\CH_2CH_2CH_2CH_3\end{cases}$$   4-Butyl-2,5,8-trioxanonane

$$CH_3SCH_2CH_2SCH\begin{cases}CH_2SCH_3\\CH_2NHCH_3\end{cases}$$   4-(2-Azapropyl)-2,5,8-trithianonane

$$CH_3CH_2OCH_2CH_2OCH_2CH\begin{cases}CH=CHOC\equiv COCH_3\\CH_2CH_2OC\equiv COCH_3\end{cases}$$

8-(2,5-Dioxaheptyl)-2,5,11,14-tetraoxapentadec-6-ene-3,12-diyne

As with the parent chain, hetero chains named as **substituting groups** are chosen so as to terminate with carbon.

$$CH_3CH_2OCH_2CH_2OCH\begin{cases}OCH_2CH_2OCH_3\\CH_2OCH_2CH_2OCH_2CH_2OCH_3\end{cases}$$

10-(3-Oxabutyloxy)-2,5,8,11,14-pentaoxahexadecane
*not* 10-(1,4-Dioxapentyl)-2,5,8,11,14-pentaoxahexadecane

If a hetero chain carries one or more functional groups (Table 8.1), these are treated by substitutive nomenclature as described in Chapters 8 and 10. Groups representing the principal function are assigned lowest possible numbers *after* the hetero atoms have been considered and *before* considering unsaturation. As with hydrocarbon derivatives, the hetero chain carrying the maximum number of principal functional groups is selected as the parent chain.

$$CH_3NHCH_2CH_2NHCH_2CH_2NHCH_2CH_2OH$$
2,5,8-Triazadecan-10-ol

$$HC\equiv COCH_2CH_2OCH_2CH_2OCH_2CH_2OH$$
3,6,9-Trioxaundec-10-yn-1-ol

$$CH_3OCH_2CH_2OCH_2CH\begin{cases}CH_2COOH\\OCH_2COOH\end{cases}$$

4-(2,5-Dioxahexyl)-3-oxahexanedioic acid
*not* 7-(Carboxymethyl)-2,5,8-trioxadecan-10-oic acid

When a hetero atom occurs in a chain as part of a functional group containing other atoms—*e.g.*, in —COO— or —CONH—, it is not treated by replacement nomenclature unless several such functional groups can be named by including them in a single hetero chain.

$$CH_3OCH_2CH_2OCH_2CH_2COOCH_2CH_2OCH_3$$

3-Oxabutyl 2,5-dioxaoctan-8-oate

$$CH_3OCH_2CH_2CONHCH_2CONHCH_2CH_2CH_2NHCOCH_2CH_2SH$$

16-Mercapto-2-oxa-6,9,13-triazahexadecane-5,8,14-trione

## Discussion

In applying replacement nomenclature to acyclic compounds two fundamental points have generated considerable debate among chemists serving on official nomenclature committees: (1) the degree of structural complexity required to justify the use of replacement names, and (2) the manner of treating functional groups contained wholly or partially in hetero chains. The 1965 IUPAC Rules provide specific recommendations concerning functional groups, but they comment only in general terms on the scope of application of replacement nomenclature.

Concerning the question of complexity of structure, it is easily seen that when only one or two hetero atoms occur in a chain system, replacement names offer little advantage. Thus, for $CH_3CH_2OCH_2CH_2-$ $OCH_2CH_2OCH_2CH_2OCH_2CH_2OH$ the replacement name 3,6,9,12-tetra-oxatetradecan-1-ol is simpler than 2-[2-[(2-ethoxyethoxy)ethoxy]-ethoxy]ethanol, whereas for $CH_3OCH_2CH_2OCH_3$ the names 2,5-dioxa-hexane and 1,2-dimethoxyethane are of about equal complexity.

Concerning the treatment of functional groups, one objection to replacement naming is the resulting scission of complex groups such as $-CONH-$ into their components with loss of the group's functional identity—*e.g.,* the naming of $-CONH-$ piecewise as $-CO-$ and $-NH-$ (*see* last example above under Recommended Nomenclature Practice). Although this objection is valid on the basis of the general importance of functionality in systematic organic nomenclature, the same kind of scission has received widespread acceptance for naming ring compounds—*e.g.,* 2-pyridinone, wherein the amide function is treated as a ketone group connected to a heterocyclic amino group; hence, logically, the objection applies in both instances. Thus, until a better solution is proposed and adopted, replacement nomenclature is recommended as the most practical means of forming systematic names for highly complex open-chain structures containing several functional groups such as $-CONH-$, $-COO-$, $-CO-O-CO$ and $-C(=NH)NH-$.

Another objection to the application of replacement nomenclature to acyclic compounds, also based on considerations of functionality, concerns the treatment of near-terminal nitrogen atoms in an open-chain system as members of the hetero chain rather than as substituted amine functions. For example, it has been argued that the compound $CH_3NHCH_2CH_2NHCH_2CH_2OCH_2CH_2NHCH_2CH_2N(CH_3)_2$ should be named as a substituted oxa diaza diamine rather than as an oxa tetraaza alkane. Obviously, this preference is a matter of taste since the diamine name violates the generally accepted principle of treating like groups alike. Nevertheless, either type of name, properly formed, would be clear and unambiguous.

As with any other method, replacement nomenclature has both advantages and disadvantages. To attempt to place arbitrary limitations on its application would be shortsighted because the real value of any nomenclature system can be determined only by usage. This book, therefore, simply provides recommendations for forming replacement names of hetero-acyclic systems based on both the 1965 IUPAC Rules and *Chemical Abstracts* practice; it does, however, state a clear preference for replacement nomenclature in naming heterocyclic systems (Chapter 6).

Finally, a word about numbering of hetero chains is needed. Unfortunately, the 1965 IUPAC rules and *Chemical Abstracts* practice do not agree on this point. In this book, the latter method is recommended, not only because of *Chemical Abstracts'* world-wide leadership in the use of systematic nomenclature but also because the numbering method chosen is closely parallel to that of both IUPAC and *Chemical Abstracts* for naming hetero rings. As with the objections discussed above, the cause of disagreement on numbering of hetero-acyclic systems lies in the degree of importance given to functional groups in forming systematic names; the 1965 IUPAC Rules recommend that in chain compounds functional groups be given consideration for lowest numbers *before* hetero atoms are considered, yet they do not prescribe this treatment for ring compounds.

In summary, replacement nomenclature provides a simple and broadly useful method for naming both heterocyclic and hetero-acyclic systems. To date, in neither class has it been applied extensively to compounds of relatively simple structure for which other names, trivial or systematic, are readily available. However, the ultimate scope of its application may well be much broader than that of current usage because of the general trend toward more systematic names, particularly for use in indexes such as those of *Chemical Abstracts.*

## Table 7.1. Examples of Acceptable Usage

1. $CH_3SCH_2CH_2OCH_2CH_2CH_2CH_2CH_2OCH_2CH_2SCH_3$
   5,11-Dioxa-2,14-dithiapentadecane

$$SCH_3$$
$$|$$

2. $CH_3SCH_2CH_2NHCH_2CH_2NHCH_2CHCH_2CH_2CH_2CH_3$
   3-Butyl-2,11-dithia-5,8-diazadodecane

3. $CH_3CH_2CH_2CH_2SCH_2CH_2\overset{\displaystyle CH_2CH_2OCH_2OCH_3}{\underset{|}{N}}CH_2CH_2OCH_2NHCH_2CH_3$

7-(3-Thiaheptyl)-2,4,10-trioxa-7,12-diazatetradecane

4. $CH_3CH_2OCH_2CH_2OCH_2CH_2OCH_2CH_2OCH_2CH_2S-$

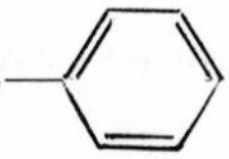

1-(Phenylthio)-3,6,9,12-tetraoxatetradecane

5. $CH_3SiH_2CH_2HgCH_2SiH_2CH_3$

2,6-Disila-4-mercuraheptane

6. $CH_3\overset{\displaystyle CH_3}{\underset{|}{B}}NHCH_2CH_2OCH_2CH_2NH\overset{\displaystyle CH_3}{\underset{|}{B}}CH_3$

2,10-Dimethyl-6-oxa-3,9-diaza-2,10-diboraundecane

7. $BrCH_2CH_2OCH_2CH_2SCH_2CH_2NO_2$

1-Bromo-8-nitro-3-oxa-6-thiaoctane

8. $CH_3CH_2NHCH_2CH_2OCH_2CH_2OCH_2CH_2NHCH_2COOH$

6,9-Dioxa-3,12-diazatetradecan-1-oic acid

9. $CH_3\overset{\displaystyle CH_3}{\underset{\displaystyle \underset{|}{CH_3}}{\underset{|}{Si}}}\,O\overset{\displaystyle CH_3}{\underset{\displaystyle \underset{|}{CH_3}}{\underset{|}{Si}}}CH_2\overset{}{\underset{\displaystyle \underset{|}{COOH}}{CH}}CH_2\overset{\displaystyle CH_3}{\underset{\displaystyle \underset{|}{CH_3}}{\underset{|}{Si}}}\,O\overset{\displaystyle CH_3}{\underset{\displaystyle \underset{|}{CH_3}}{\underset{|}{Si}}}CH_3$

2,2,4,4,8,8,10,10-Octamethyl-3,9-dioxa-2,4,8,10-tetrasilaundecane-6-carboxylic acid

10.

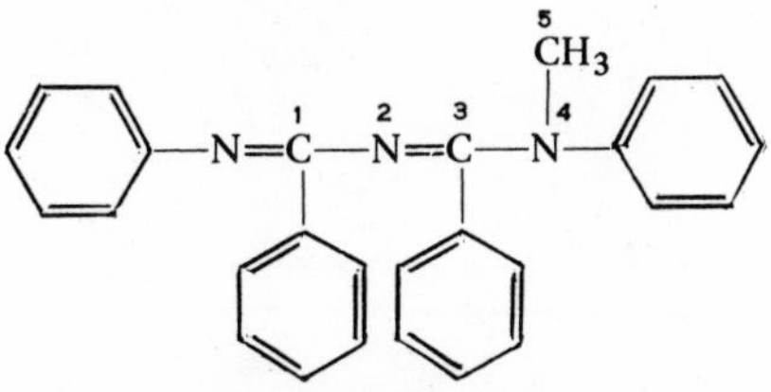

*N*, 1,3,4-Tetraphenyl-2,4-diazapent-2-en-1-imine

11. $CH_3OCH_2CH_2OCH_2CH_2OCH_2CH_2NHCH_2CH_2OH$

2,5,8-Trioxa-11-azatridecan-13-ol

12. $CH_3NHCH_2CH_2\overset{\displaystyle CH_3}{\underset{|}{N}}CH_2OCH_2CH_2NH_2$

5-Methyl-7-oxa-2,5-diazanonan-9-amine

13.

$CH_3CH_2OCOCH_2CH_2OCH_2CH_2OCH_2CH_2OCH_2CH_2OCH_2CH_2COOCH_2CH_3$

Diethyl 4,7,10,13-tetraoxahexadecanedioate

14.

$HOOC(CH_2)_8CONH(CH_2)_6NHCO(CH_2)_8CONH(CH_2)_6NHCO(CH_2)_8COOH$

10,19,28,37-Tetraoxo-11,18,29,36-tetraazahexatetracontanedioic acid

$$CH_2OCH_3$$
$$|$$
15.      $HOCH_2CH_2SCH_2CHNHCHCH_2OCH_2CH_2SCH_2CH_2OH$
$$|$$
$$CH_2OCH_2OCH_3$$

8-(2,4-Dioxapentyl)-10-(2-oxapropyl)-6-oxa-3,12-dithia-9-azatetradecane-1,14-diol

16.      $CH_3CH_2NHCH_2CH_2OCH_2CH_2SCH_2CH_2NHCH_2CH_2CONHCH_2CH_3$

N-Ethyl-6-oxa-9-thia-3,12-diazapentadecan-15-amide

# Functional Compounds:

# Substitutive Nomenclature

Naming of chain and ring systems has been covered in Chapters 2 through 7. With the principles of these chapters as a basis it is possible by invoking the principle of substitution to name any functional compound systematically no matter how many or what kinds of functional groups it contains. Although other methods (*see* Chapter 9) for naming functional compounds have found wide use over the years, none is so broadly applicable as substitutive nomenclature.

To organic chemists the word **function** has come to denote an atom or a recognized combination of atoms that confers characteristic chemical properties upon the molecule in which it occurs. The term **functional group** is also used commonly in the same sense since more often than not functions comprise groups of atoms rather than just single atoms. Examples of common functions are Cl, OH, $NH_2$, =O, and COOH. Carbon-to-carbon unsaturation and hetero atoms in rings are considered nonfunctional for nomenclature purposes. [The term "characteristic group," as introduced and used in the 1965 IUPAC Rules, has essentially the same meaning as function but will not be used in this book.]

Compounds of **simple function** are defined as those containing a function of one kind only. This function may be repeated several times in a molecule, in which case the compound is said to be one of **multiple function**. Compounds of **mixed function** are those which contain more than one kind of function. Thus, a polyfunctional compound may be one either of multiple function or of mixed function.

**Substitution** means the replacement of one or more hydrogen atoms in a given compound by some other kind of atom or group of atoms, functional or nonfunctional. In substitutive nomenclature each substituting atom or group (commonly called a substituent) is cited as either a prefix or suffix to the name of the compound or substituting group to which it is attached; the latter is called the **parent compound** or **parent group**. For nomenclature purposes all substituting atoms and groups have been assigned prefix names; a majority of functional groups have also been given other names for use as suffixes.

This chapter deals with basic principles of substitutive nomenclature that can be applied to all classes of functional compounds. Recommendations for specific treatment of each functional class are provided in later chapters.

### Recommended Nomenclature Practice

If the compound being named is of **mixed function**—*i.e.,* contains more than one kind of functional group—one of these kinds is selected for citation as a suffix to the name of the parent compound, and all other substituting groups are cited as prefixes (*see* Chapter 10).

Prefix and suffix names for commonly occurring functional groups are listed in Table 8.1. A more comprehensive compilation of prefix names for both functional and nonfunctional (hydrocarbon) groups will be found in the Appendix.

If the compound being named is of **simple** or **multiple function**—*i.e.,* contains only one kind of functional group — as many of these as possible are cited in the suffix.

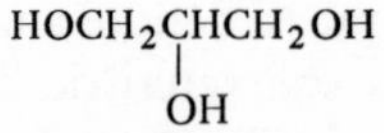

$$HOCH_2CHCH_2OH$$
$$|$$
$$OH$$

1,2,3-Propanetriol
*not* 2-Hydroxy-1,3-propanediol
*not* 2,3-Dihydroxy-1-propanol

Remaining groups are cited as prefixes.

$$CH_2OH$$
$$|$$
$$HOCH_2CCH_2OH$$
$$|$$
$$CH_2OH$$

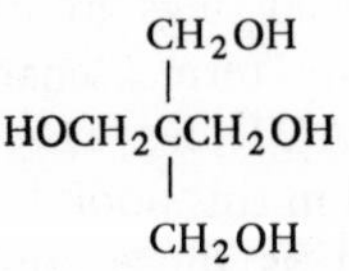

2,2-Bis(hydroxymethyl)-1,3-propanediol

However, some functional groups have not been assigned suffix names and therefore are always cited as prefixes (*see* Table 8.1).

$$CH_3CH_2Cl$$                          Chloroethane

$$C_6H_5NO_2$$                          Nitrobenzene

If more than one simple (unsubstituted) prefix name is required, these are arranged alphabetically.

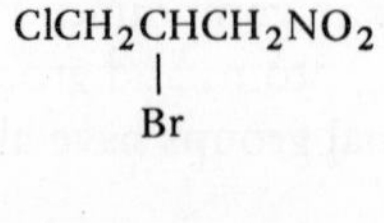

$$ClCH_2CHCH_2NO_2$$
$$|$$
$$Br$$

2-Bromo-1-chloro-3-nitropropane

Multiplying prefixes such as *di, tri,* and *tetra* do not affect the alphabetical arrangement of simple prefix names.

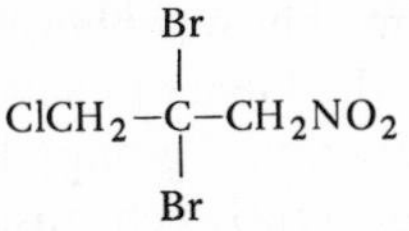

$$\underset{\underset{Br}{|}}{\overset{\overset{Br}{|}}{ClCH_2-C-CH_2NO_2}}$$

2,2-Dibromo-1-chloro-3-nitropropane

However, the prefix name for a substituted substituent—*i.e.*, a complex prefix name—is considered to begin with the first letter of its complete name including any multiplying prefix present. Complex prefix names are enclosed in parentheses to avoid ambiguity, whether locants are cited or not.

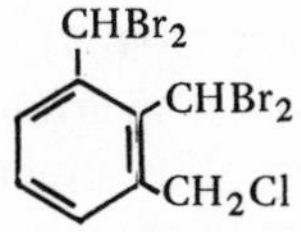

1-(Chloromethyl)-2,3-bis(dibromomethyl)benzene

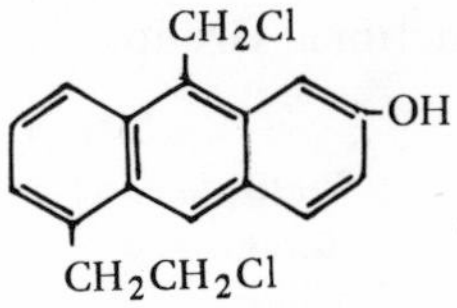

5-(2-Chloroethyl)-9-(chloromethyl)-2-anthracenol

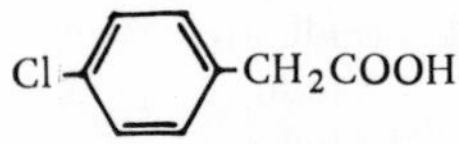

(4-Chlorophenyl)ethanoic acid

## Discussion

Substitutive nomenclature as described in this chapter and in Chapter 10 is the most generally useful method yet developed for naming organic compounds. Other methods (*see* Chapter 9) have been proposed and are still used to a limited extent for various reasons, but these could be abandoned entirely without any real loss to chemical recording and communication. Exclusive use of substitutive names is a worthwhile future objective for authors and indexers alike.

In this book the term prefix name is used rather than "radical name" to denote the name of a substituting atom or group cited as a prefix, even though the latter terminology has been employed by *Chemical Abstracts* and IUPAC. For nomenclature purposes the word "radical" is best reserved for designation of a chemical species in which one atom or more carries an unpaired electron (*see* Chapter 28).

One might well ask why, in substitutive nomenclature, all functional groups have not been assigned suffix names. The answer is that the substitutive system was not designed all at one time but has evolved from incomplete rules and principles agreed upon by the Geneva Congress in 1892. By far the most frequently occurring functional groups lacking suffix names are the halogens, $NO_2$, and OR. Apparently, chemists have not felt a strong need for suffixes to designate these groups, perhaps because of their relatively low order of reactivity compared with other functions.

In naming a complex compound substitutively, alphabetical order is by far the most widely used method for citing two or more prefix names. (An alternative method stemming from Beilstein's *Handbuch der organischen Chemie* and based on complexity of substituting groups, although described in the 1965 IUPAC Rules (Section C-16.3), is not recommended.) Although deviation from alphabetical order in arranging a series of prefix names does not result in ambiguity, this practice is undesirable in publications because it results in names that must be reconstructed for indexing or index searching.

### Table 8.1. Substitutive Names of Functional Groups
*A. Groups Cited Only as Prefixes[a]*

| Formula | Prefix Name |
|---|---|
| $-Br$ | Bromo |
| $-Cl$ | Chloro |
| $-ClO$ | Chlorosyl |
| $-ClO_2$ | Chloryl |
| $-ClO_3$ | Perchloryl |
| $-F$ | Fluoro |
| $-I$ | Iodo |
| $-IO$ | Iodosyl |
| $-IO_2$ | Iodyl |
| $-I(OH)_2$ | Dihydroxyiodo |
| $-N=C=O$ | Carbonylamino |
| $-N=C=S$ | Thiocarbonylamino |
| $=N_2$ | Diazo |
| $-N_3$ | Azido |
| $-N=C$ | Carbonylamino |
| $-NO$ | Nitroso |
| $-NO_2$ | Nitro |
| $=N(O)OH$ | *aci*-Nitro |
| $-OC\equiv N$ | Cyanato |
| $-OOH$ | Hydroperoxy |
| $-OR$ | R-oxy |
| $-OOR$ | R-dioxy |
| $-SC\equiv N$ | Thiocyanato |
| $-SR$ | R-thio |
| $-S(O)R$ | R-sulfinyl |
| $-SO_2R$ | R-sulfonyl |
| $-SSR$ | R-dithio |

[a] The functional groups in part A have no suffix names, and in substitutive nomenclature they are always cited as prefixes. The order of this listing has no significance for nomenclature purposes.

## Table 8.1 Continued

### B. *Groups Cited as Either Prefixes or Suffixes*[b]

| *Formula* | *Suffix Name* | *Prefix Name* |
|---|---|---|
| Free radicals | (For naming, *see* Chapter 28) | |
| Cations | -onium | -onio |
| Anions | -ate, -ide | -ato, -ido |
| $-COOH$ | -oic acid | |
| $-COOH$ | -carboxylic acid | Carboxy |
| $-COSH$ | -thioic S-acid | |
| $-COSH$ | -thiocarboxylic S-acid | Mercaptocarbonyl |
| $-CSOH$ | -thioic O-acid | |
| $-CSOH$ | -thiocarboxylic O-acid | Hydroxy(thiocarbonyl) |
| $-CSSH$ | -dithioic acid | |
| $-CSSH$ | -dithiocarboxylic acid | Dithiocarboxy |
| $-C(=NH)OH$ | -imidic acid | |
| $-C(=NH)OH$ | -carboximidic acid | Imidocarboxy |
| $-SO_2OH$ | -sulfonic acid | Sulfo |
| $-SOOH$ | -sulfinic acid | Sulfino |
| $-SOH$ | -sulfenic acid | Sulfeno |
| $-P(O)(OH)_2$ | -phosphonic acid | Dihydroxyphosphinyl |
| $>P(O)OH$ | -phosphinic acid | Hydroxyphosphinylidene |
| $-As(O)(OH)_2$ | -arsonic acid | Dihydroxyarsinyl |
| $>As(O)OH$ | -arsinic acid | Hydroxyarsinylidene |
| $-B(OH)_2$ | -boronic acid | Dihydroxyboryl |
| $-COX$ | -oyl (-yl) halide | |
| $-COX$ | -carbonyl halide | Haloformyl |
| $-CON_3$ | -oyl (-yl) azide | |
| $-CON_3$ | -carbonyl azide | Azidoformyl |
| $-CONH_2$ | -amide | |
| $-CONH_2$ | -carboxamide | Carbamoyl |
| $-CONHCO-$ | -imide | |
| $-CONHCO-$ | -dicarboximide | Iminodicarbonyl |
| $-C(=NH)NH_2$ | -amidine | |
| $-C(=NH)NH_2$ | -carboxamidine | Carbamimidoyl |
| $-C\equiv N$ | -nitrile | |
| $-C\equiv N$ | -carbonitrile | Cyano |
| $-CHO$ | -al | Oxo |
| $-CHO$ | -carbaldehyde | Formyl |
| $-CHS$ | -thial | Thioxo |
| $-CHS$ | -carbothial | Thioformyl |
| $>C=O$ | -one | Oxo |
| $>C=S$ | -thione | Thioxo |
| $-OH$ | -ol | Hydroxy |
| $-SH$ | -thiol | Mercapto |
| $-NH_2$ | -amine | Amino |
| $=NH$ | -imine | Imino |

[b]The functional groups in part B are listed in descending order of precedence for citation as suffixes (*see* Chapter 10). In these formulas an italic C indicates that the carbon atom so designated is not included in the corresponding suffix or prefix name. (For a more comprehensive order of precedence, developed and used by *Chemical Abstracts, see* "Introduction to the Subject Index," *Chem. Abstr.* (1967), **66**, 261-271.

# 9

# Other Nomenclature Systems
# for Functional Compounds

Although substitutive nomenclature as described in Chapters 8 and 10 is recommended as the single best system of naming organic compounds, four other methods, each having limited capability, have found varying degrees of use. These, roughly in order of importance, are radicofunctional nomenclature, conjunctive nomenclature, additive nomenclature, and subtractive nomenclature.

*Recommended Nomenclature Practice*

**Radicofunctional Names.** In forming a radicofunctional name one species of functional group is selected and expressed as the final word of a multiword name. The preceding word or words, usually arranged in alphabetical order, describe the remainder of the molecule.

| | |
|---|---|
| $C_2H_5OH$ | Ethyl alcohol |
| $(ClCH_2CH_2)_2O$ | Bis(2-chloroethyl) ether |
| $CH_3SCH_2CH_3$ | Ethyl methyl sulfide |
| $CH_3COC_6H_5$ | Methyl phenyl ketone |
| $CH_3CH_2COOCH_3$ | Methyl propionate |

An order of precedence of functions is provided in the 1965 IUPAC Rules (Rule C-22) for use in selecting the functional group to be cited as the final word of the name when more than one species of function occurs in the molecule.

| | |
|---|---|
| $HOCH_2CH_2CH_2CN$ | 3-Hydroxypropyl cyanide |
| | *preferred to* |
| | 3-Cyanopropyl alcohol |

**Conjunctive Names.** This type of nomenclature is used when the principal group in the molecule occurs in a side chain attached to a ring system and for some special reason (*see* Discussion) one does not wish to name the ring system as a substituting group.

A conjunctive name is formed simply by connecting without hyphenation the names (as compounds) of the component portions, one cyclic and one acyclic, into which the molecule can be conceptually divided. Such juxtaposition of the two names implies that mutual substitution of hydrogen has occurred in the two component compounds. The name of the cyclic compound is cited first, followed by that of the acyclic compound.

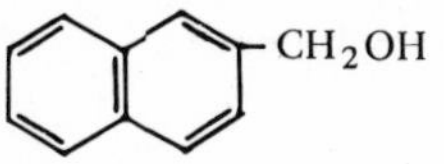

CH$_2$OH                  2-Naphthalenemethanol

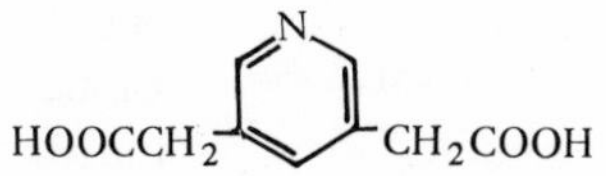

HOOCCH$_2$ — CH$_2$COOH     3,5-Pyridinediacetic acid

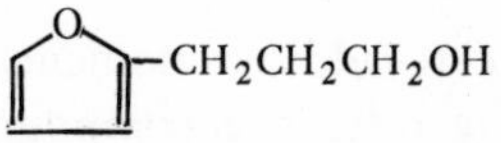

CH$_2$CH$_2$CH$_2$OH       2-Furanpropanol

In conjunctive names the position of the principal functional group arbitrarily defines the "terminal" position of the side chain, and locations of substituents on the side chain are then indicated by Greek letters. The letter α is assigned to the "terminal" carbon atom, and lettering continues along the chain to its point of connection to the ring.

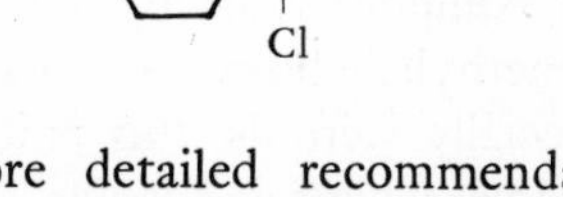

CH—CHCH$_3$
         |     |
         Cl   OH          β-Chloro-α-methyl-2-naphthaleneethanol

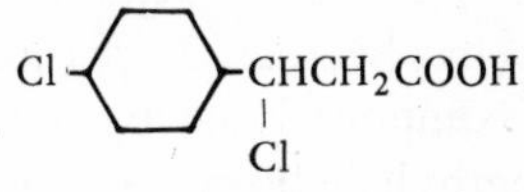

Cl — CHCH$_2$COOH
         |
         Cl              β,4-Dichlorocyclohexanepropionic acid

More detailed recommendations for forming conjunctive names are provided in the 1965 IUPAC Rules (Rules C-51 through C-58).

   Additive Names.   As the term implies, an additive name is formed by citing atoms or groups that have been added conceptually to a reference compound to convert it to the compound being named.

   Hydrogen so added is denoted by the prefix *hydro,* but any other species of atom or group added is expressed as a separate word following the name of the reference compound.

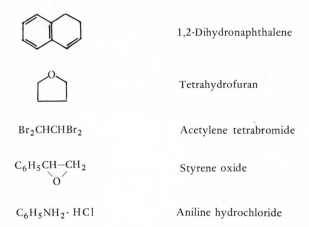

1,2-Dihydronaphthalene

Tetrahydrofuran

$Br_2CHCHBr_2$                    Acetylene tetrabromide

$C_6H_5CH-CH_2$
          \ /
           O                      Styrene oxide

$C_6H_5NH_2 \cdot HCl$            Aniline hydrochloride

**Subtractive Names.** Similarly in principle to additive nomenclature, subtractive names are formed from those of a reference compound by conceptually removing atoms or groups to produce the compound being named.

The prefix *de* followed by the name of an atom or group (other than hydrogen) indicates that the latter has been removed and its place taken by hydrogen—*e.g.*, de-*N*-methylmorphine.

The prefix *didehydro* signifies removal of two hydrogen atoms, with formation of a double bond—*e.g.*, 7,8-didehydrocholesterol (sometimes improperly shortened to "7-dehydrocholesterol").

The prefix *nor* as used in specialized nomenclature of terpene hydrocarbons indicates removal of all methyl groups attached to a ring system—*e.g.*, norpinane. In trivial names of other natural products, especially steroids, *nor* as a prefix has been used to denote shortening of a side chain or contraction of a ring—*e.g.*, 23-nor-5$\beta$-cholanoic acid.

In additive nomenclature the prefix *homo* has been used occasionally to indicate addition of a $CH_2$ group to a compound having a trivial name—*e.g.*, homopiperonylic acid for (3,4-methylenedioxyphenyl)acetic acid, but except for natural products, especially steroids, this practice should be abandoned in favor of systematic names wherever possible.

The prefix *anhydro* is used in specialized nomenclature of carbohydrates to indicate loss of the elements of water from within a molecule—*e.g.*, 3,6-anhydro-$\beta$-D-glucopyranose; *deoxy* denotes replacement of the OH group by H—*e.g.*, 6-deoxy-$\alpha$-D-glucopyranose.

## Discussion

Generally, use of the above methods of nomenclature should be avoided so far as possible except for the use of additive names involving the prefix *hydro* in naming partially or fully saturated ring systems. Systematically formed substitutive names are available and should be used for compounds that in the past have commonly been given names of the radicofunctional, conjunctive, or additive kind—*e.g.,* ethanol (for ethyl alcohol, (methylthio)ethane (for ethyl methyl sulfide), 2-naphthylmethanol (for 2-naphthalenemethanol), 1,1,2,2-tetrabromoethane (for acetylene tetrabromide), phenyloxirane (for styrene oxide), anilinium chloride (for aniline hydrochloride), and many others. In particular, subtractive nomenclature has found very limited use and is seldom needed except with trivial names of complex naturally occurring compounds; however, the subtractive principle is still employed in officially accepted special nomenclature systems devised for terpene hydrocarbons and carbohydrates.

Following a common practice in inorganic chemical nomenclature, replacement of oxygen by sulfur in organic compounds is often indicated by use of the prefix *thio*—*e.g.,* thioacetic acid, thiophenol, etc. The corresponding prefixes *seleno* and *telluro* are used in the same manner. However, such nomenclature has led to many ambiguous names (thioglycolic acid is an example) and should be discontinued as far as possible in naming specific compounds. On the other hand, generic names using *thio* are acceptable—*e.g.,* thioether and thioketone, − as is the use of thio as an adjective (separate word) meaning replacement of oxygen by sulfur—*e.g.,* thio derivative and thio analog. However, for the latter purpose the terms sulfur derivative and sulfur analog would serve equally well. The concept of oxygen replacement is also involved in names such as benzophenone oxime wherein a nitrogen-containing group is regarded as having replaced a carbonyl oxygen atom (*see* Chapter 32). Here again, systematically formed substitutive names such as (diphenylmethylene)hydroxylamine are much to be preferred. In brief, use of the replacement concept in organic nomenclature should be limited, as far as possible, to replacement of carbon as described in Chapters 6 and 7.

Special comment is appropriate concerning conjunctive nomenclature since this method is used extensively by *Chemical Abstracts,* primarily to provide improved arrangement of subject index entries (inverted form) for certain types of compounds. Because of this practice by *Chemical Abstracts* chemists have used conjunctive names widely in general-purpose nomenclature even though in nearly all cases substitutive names are readily available. Except for indexing, conjunctive names offer no significant advantages, and unless one is thoroughly acquainted with *Chemical Abstracts* indexing rules, one cannot always be certain as to when or when not to use the conjunctive type of name. Thus, substitutive names are generally to be preferred.

# 10

# Compounds of Mixed Function

A s stated at the beginning of Chapter 8, compounds of mixed function are those which contain more than one kind of function. In that chapter, which describes the basic principles of substitutive nomenclature, only compounds of simple function are considered. Additional principles, which are required in order to arrive at a single preferred name when the given compound is one of mixed function, are presented and discussed in this chapter.

Later chapters, each dealing with one or more functional classes of compounds, provide specific recommendations for naming compounds of simple function that can be combined with the general recommendations of this chapter in constructing systematic substitutive names for compounds of mixed function.

*Recommended Nomenclature Practice*

In naming a **compound of mixed function** systematically, if all the kinds of **functional groups** present are among those **always named as prefixes** (first part of Table 8.1), then the parent compound is a hydrocarbon, a hetero chain, or a heterocyclic system, and all the functional groups are named as substituents.

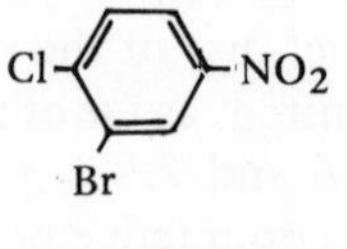

Cl—⟨benzene ring⟩—NO₂
Br

2-Bromo-1-chloro-4-nitrobenzene
(Parent compound: *benzene*)

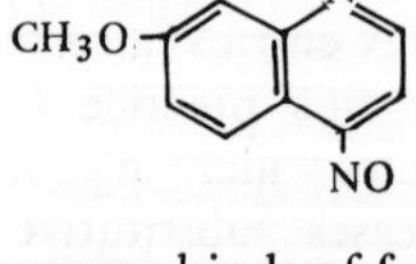

CH₃O—⟨quinoline ring⟩—N
NO

7-Methoxy-4-nitrosoquinoline
(Parent compound: *quinoline*)

If one or more kinds of **functional groups having suffix names** (second part of (Table 8.1) is present, the kind standing highest in the **Order of Precedence** as prescribed by Table 8.1 is selected and used in deriving the name of the parent compound. This highest-ranking kind of functional group is called the **principal function**; the remaining kinds are called **subordinate functions.** The term "principal group" is used with the same meaning in the 1965 IUPAC Rules.

80

The principal function having been selected, the chain or ring system carrying the maximum number of such groups, together with those groups, is designated the parent compound and is then named by the specific principles recommended for that functional class (*see* later chapters).

If two or more different kinds of chains or ring systems each carry the same maximum number of the principal functional group, a choice among them is made using the seniority principles provided in 1965 IUPAC Rules C-13 and C-14 (*see also* Introduction to the *Chemical Abstracts* Subject Index, **66** (1967, p. 5I).

If two or more identical chains or ring systems each carry the same maximum number of the principal functional groups, a special treatment designed for use in naming symmetrical structures may be applicable (*see* Chapter 11).

Once the parent compound has been named, the remaining groups (functional and nonfunctional) are treated as substituents and are named as prefixes, arranged in alphabetical order.

$HOCH_2CH_2CH_2NH_2$ Principal function:      *alcohol*   (−OH)
Subordinate function: *amine*      (−NH$_2$)
Parent compound:      *1-propanol*

Complete name:   **3-Amino-1-propanol**

COOH

SCH$_2$COOH

OCH$_2$CH$_3$

Principal function:      *carboxylic acid* (−COOH)
Subordinate functions: *ether*   (−O−),
*sulfide*   (−S−)
Parent compound:      *benzoic acid*

Complete name:   **3-[(Carboxymethyl)thio]-4-ethoxybenzoic acid**

$H_2NCH_2CH_2CH_2CONHCH_2CH_2OH$

Principal function:      *amide*   (−CONH−)
Subordinate functions: *alcohol*  (−OH),
*amine*   (−NH$_2$)
Parent compound:      *butyramide*

Complete name:   **4-Amino-N-(2-hydroxyethyl)butyramide**

As illustrated in the second example, the larger of the two alternative parent compounds is chosen,−*i.e.*, benzoic acid rather than acetic acid. As shown in the third example, prefix names for subordinate functions are arranged alphabetically; the fact that alcohol stands higher than amine in the Order of Precedence (Table 8.1) plays no further part once the principal function (amide) has been established.

*Discussion*

For indexer and chemist alike, the most challenging aspect of organic nomenclature is the systematic naming of compounds of mixed function. As the structure of the molecule to be named becomes more complex, this task becomes correspondingly more difficult, particularly when a single preferred name is sought. The principles presented in this chapter are those used by *Chemical Abstracts* and recommended by IUPAC.

Fundamentally, the systematic naming of compounds containing more than one kind of functional group requires that one kind of group be singled out and used in naming a portion of the molecule. This selected portion of the original molecule is called the parent compound. The remaining parts of the molecule (other functional groups, chains, and rings) are then treated as substituents and given prefix names. If a single preferred name is required, as in the *Chemical Abstracts* Subject Indexes, some standardized means for selecting the parent compound must be used. This is provided by the Order of Precedence of Functions established by *Chemical Abstracts,* or by the 1965 IUPAC Order of Precedence, which is generally similar (*see* footnote, Table 8.1)

It should be emphasized that in naming compounds of mixed function, adherence to an official order of precedence of functions is not always necessary. In certain instances an author may find it advantageous to name a compound or series of compounds differently from exact IUPAC or *Chemical Abstracts* systematic procedure in order to present his information more effectively. As long as the parent compound and appended substituent groups are named correctly, any one of the structural components of a complex molecule may be selected for emphasis— *e.g.,* *p*-hydroxyaniline is acceptable, although this compound appears in the *Chemical Abstracts* Subject Indexes as *p*-aminophenol. Nevertheless, authors are encouraged, whenever such deviation is not of major importance, to use preferred IUPAC or *Chemical Abstracts* names, particularly in articles submitted to American Chemical Society journals.

In the same vein, when dealing with especially complex structures— *i.e.,* dyes, steroids, carbohydrates, and other large molecules—an author may be justified in using short trivial names rather than very long, fully systematic names (*see also* Chapter 30). However, such trivial names should, if possible, be those which have appeared in *Chemical Abstracts* indexes or received prior usage by other workers in the field. Premature and unnecessary coining of new trivial names is a disservice both to indexers and to readers of the chemical literature. Proper use of established trivial names is a boon to all, but it is recommended that when trivial names are employed in a publication the corresponding systematic names and/or structural formulas also be included early in the text.

To return to systematic names for compounds of mixed function, there are a few fundamental principles which should be discussed briefly.

**Nonfunctional Treatment of Unsaturation and Hetero Atoms.** As mentioned at the start of Chapter 8, carbon-to-carbon unsaturation (double, triple, or benzenoid) and hetero atoms are considered nonfunctional for nomenclature purposes. Thus, 3-pentenoic acid and 4-pyridine-carboxylic acid are regarded as compounds of simple function. Also, when a hetero atom occurs as an integral part of a functional group built into a ring, that function loses its identity for naming purposes; for example, 2-pyrrolidinone, although a cyclic amide, is named as a ketone. However, a few such functional groups, in particular *anhydride,* have by custom retained their functional names even when contained in a ring.

**Treatment of "Like Things Alike."** Insofar as possible, like treatment should be given to all groups of a given kind occurring in a single molecule. Thus, if the principal function is carboxylic acid, the parent compound should be chosen so as to include as many COOH groups as possible:

$$\text{HOOCCH}_2\text{CH} \Big\langle \begin{array}{l} \text{CH}_2\text{CH}_2\text{COOH} \\[2mm] \text{CH}_2\text{CH}_2\text{CH}_2\text{CH}_2\text{NH}_2 \end{array}$$

3-(4-Aminobutyl)hexanedioic acid, *preferred to*
8-Amino-4-(carboxymethyl)-octanoic acid

The principle of "like things alike" is also the basis for a special approach to the naming of functionally symmetrical compounds as described in Chapter 11.

**Only the Principal Function Named as a Suffix.** Multiple suffixes should be avoided. Although by usage over the years, a few exceptions to this rule have become familiar to chemists, names such as 1-phenol-4-sulfonic acid and anthraquinone-2-carboxylic acid are not recommended and should be abandoned. The corresponding systematic names—4-hydroxybenzenesulfonic acid and 9,10-dihydro-9,10-dioxo-2-anthracene-carboxylic acid—are preferred; *cf.* Chapter 15.

**Choice of Chain or Ring System in Parent Compound.** After the principal functional group has been established, it may turn out that two or more chains or rings each carry the same maximum number of these groups. If so, a choice of chain or ring system must be made. Here, a good general rule is to select the largest system (chain or ring) giving priority to heterocyclic systems no matter what their size. As mentioned above under Recommended Nomenclature Practice, detailed seniority rules for selecting the parent chain or ring system are provided by both IUPAC and *Chemical Abstracts.*

# 11

# Functional Compounds
# of Symmetrical Structure

**M**any **polyfunctional compounds**, particularly those in which both cyclic and acyclic components are present, possess sufficient symmetry in their structures to warrant treatment by a specialized extension of substitutive nomenclature. This chapter describes the naming of compounds in which the structural unit (chain or ring) carrying the principal function occurs more than once in the molecule, each such replicated unit being linked to its counterparts through a single central structure comprising one or more chains and/or rings. The special "inside out" treatment recommended below for such functionally symmetrical compounds makes possible both retention of the suffix name for the principal function and application of the general principle of organic nomenclature whereby, whenever possible, like groups are treated alike.

Naming of **nonfunctional symmetrical structures**— *i.e.*, cyclic-acyclic hydrocarbons and cyclic-acyclic heterocyclic systems whose functional character arises only from hetero atoms which are part of a ring — is described in Chapters 3 and 6, respectively.

*Recommended Nomenclature Practice*

If the symmetrically replicated chain or ring system carries only those kinds of **functional groups that are always named as prefixes** (first part of Table 8.1), the most centrally located chain or ring system is chosen as the parent compound, and no specialized treatment is required.

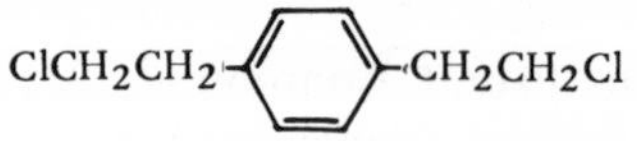

$ClCH_2CH_2$—⟨ ⟩—$CH_2CH_2Cl$     1,4 - Bis(2-chloroethyl)benzene

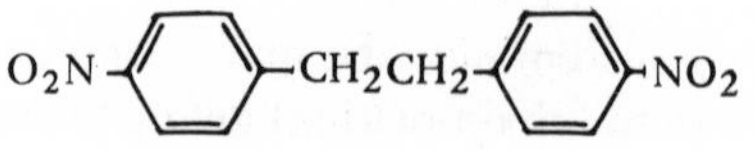

$O_2N$—⟨ ⟩—$CH_2CH_2$—⟨ ⟩—$NO_2$     1,2 - Bis(4-nitrophenyl)ethane

If the symmetrically replicated chain or ring system carries one or more kinds of **functional group for which suffix names are provided** (second part of Table 8.1), one of these is selected as the principal function and used in deriving the name of the parent compound. Naming of the parent compound is completed by attaching the usual prefix names for substituents, except that the substituting central portion of the molecule connecting the several replicates of the parent compound is named by use of special prefix names denoting more than one point of attachment per substituting group. The position in each parent compound unit at which it is attached to the central connecting structure is indicated by a number locant, using prime markings to distinguish among the several parent compound units present.

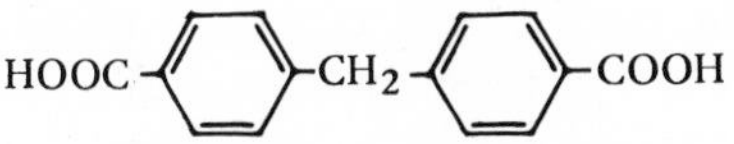

HOOC—⟨⟩—$CH_2$—⟨⟩—COOH     4,4′-Methylenedibenzoic acid

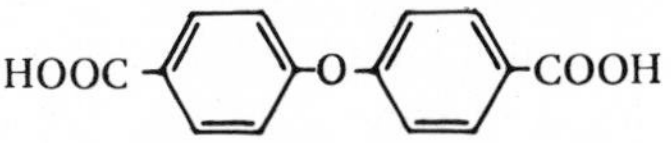

HOOC—⟨⟩—O—⟨⟩—COOH     4,4′-Oxydibenzoic acid

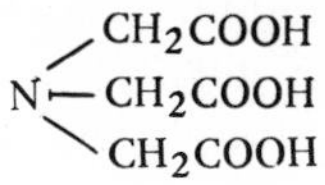

$CH_2COOH$
N—$CH_2COOH$     Nitrilotriacetic acid
$CH_2COOH$

$HOCH_2CH_2$—⟨⟩—$CH_2CH_2OH$     2,2′-(1,4-Phenylene)diethanol

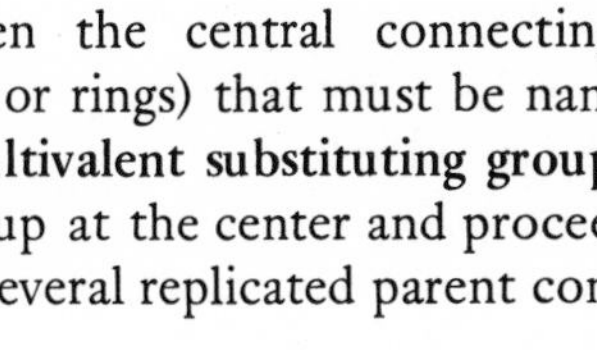

$H_2NCH_2CH_2$   $CH_2CH_2NH_2$

    2,2′,2″,2‴-(1,4,5,8-Naphtha-lenetetrayl)tetrakis(ethylamine)

$H_2NCH_2CH_2$   $CH_2CH_2NH_2$

When the central connecting structure comprises several groups (chains or rings) that must be named individually, each of these is named as a **multivalent substituting group** (*see* Table 11.1). Citation begins with the group at the center and proceeds outward to the points of attachment to the several replicated parent compound units.

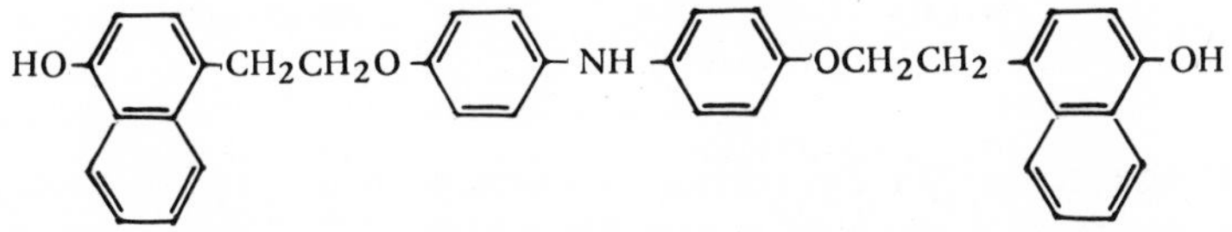

HO—⟨⟩—$CH_2CH_2O$—⟨⟩—NH—⟨⟩—$OCH_2CH_2$—⟨⟩—OH

4,4′-[Iminobis(1,4-phenyleneoxyethylene)]di-1-naphthol

The respective points of attachment of replicated parent compound units need not be the same for each unit.

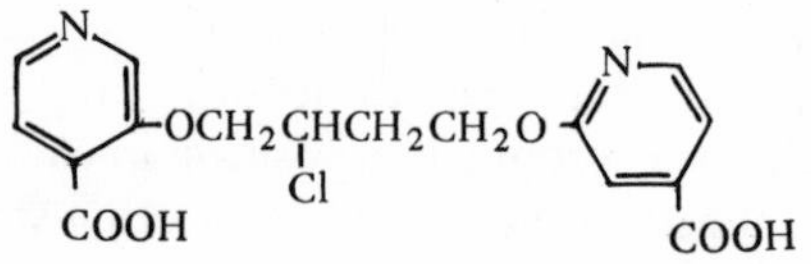

2,3'-[(3-Chlorotetramethylene)dioxy] bis(4-pyridinecarboxylic acid)

## Discussion

The "inside out" method recommended above for naming polyfunctional compounds of symmetrical structure should be used only in cases where the outermost replicated structural unit contains a functional group that can be designated the principal group and named by use of a suffix. However, *Chemical Abstracts* has taken advantage of the technique in forming index names for symmetrical heterocyclic compounds even when no such principal functional group is present—*e.g.*, 2,2'-methylenedipyridine (for di-2-pyridylmethane). Except for indexing purposes such as this, where it is considered important to use the name of the heterocyclic system as the main subject entry, extension of the "inside out" method of naming to nonfunctional compounds is neither desirable nor warranted.

A further requirement for proper use of the "inside out" method is that in the central connecting structure of the symmetrical compound being named, each "spoke of the wheel" must be identical. However, the centermost chain or ring system—*i.e.*, the "hub"—need not be symmetrical within itself. These principles are illustrated in the examples shown above.

### Table 11.1. Typical Multivalent Groups Used in Naming Compounds of Symmetrical Structure

| Group | Prefix Name |
|---|---|
| $-N=N-$ | Azo |
| 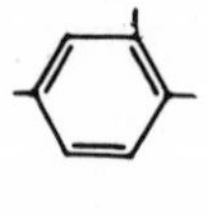 | 1,2,4-Benzenetriyl |
| $C_6H_5CH-$ | Benzylidene |
| $C_6H_5C\!\!<$ | Benzylidyne |
| $-CH_2CH=CHCH_2-$ | 2-Butenylene |
| $-COCH_2CH_2CO-$ | Butanedioyl |
| $-C=O$ | Carbonyl |

## Table 11.1 Continued

| Group | Prefix Name |
|---|---|
| $-NHCONH-$ | Carbonyldiimino |
| $-OC(O)O-$ | Carbonyldioxy |
| 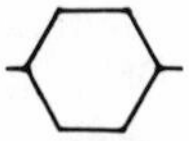 | 1,4-Cyclohexylene |
| 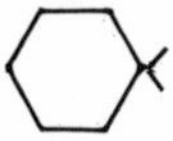 | Cyclohexylidene |
| $-NHNH-$ | 1,2-Diazanediyl |
| $-OO-$ | Dioxy |
| $-SS-$ | Dithio |
| $-COCO-$ | Ethanedioyl |
| $-CH{=}CH-$ | Ethenylene |
| $-CH_2CH_2-$ | Ethylene |
| $-OCH_2CH_2O-$ | Ethylenedioxy |
| $CH_3\overset{\mid}{C}H-$ | Ethylidene |
| $CH_3\overset{\mid}{\underset{\mid}{C}}-$ | Ethylidyne |
| $-C{\equiv}C-$ | Ethynylene |
| $-NH-$ | Imino |
| $(CH_3)_2\overset{\mid}{C}-$ | Isopropylidene |
| $-CH_2-$ | Methylene |
| $-OCH_2O-$ | Methylenedioxy |
| $H\overset{\mid}{\underset{\mid}{C}}-$ | Methylidyne |
| $-\overset{\mid}{\underset{\mid}{N}}-$ | Nitrilo |
| $-O-$ | Oxy |
| $-CH_2OCH_2-$ | Oxydimethylene |
| $-CH_2CH_2OCH_2CH_2-$ | Oxydiethylene |
| $H\overset{\mid}{P}-$ | Phosphinidene |
| $(O)\overset{\mid}{\underset{\mid}{P}}-$ | Phosphinylidyne |
|  | 1,4-Phenylene |
| $C_6H_5\overset{\mid}{N}-$ | Phenylimino |
| $-CH_2\overset{\mid}{C}HCH_2-$ | 1,2,3-Propanetriyl |
| $-CH_2CH_2\overset{\mid}{C}H-$ | 1,1,3-Propanetriyl |

## Table 11.1  Continued

| Group | Prefix Name |
|---|---|
| $-CH_2CH=CH-$ | Propenylene |
| $-Se-$ | Seleno |
| $-S(O)-$ | Sulfinyl |
| $-SO_2-$ | Sulfonyl |
| $-S-$ | Thio |

# Positional and Multiplying Affixes

**F**or any but the most simple molecular structures, it is necessary in constructing a systematic name to specify the location of substituting groups, unsaturation, hetero atoms, and any other structural modifications whose positions are not already implied in the name of the parent chain or ring system. In addition, when more than one (or more than one kind of) such group or modification is present, an affix denoting the number of each kind is required. Symbols specifying location are called **locants** and those describing plurality are termed **multiplying affixes**. Depending on its meaning and on the need for other descriptive affixes in the completed name, an affix may take the form of a **prefix**, an **infix**, or a **suffix**.

*Recommended Nomenclature Practice*

   **Positions** of substituting groups and other structural modifications in organic chain and ring systems are designated preferably by the use of Arabic numerals. Detailed rules for assigning such numbers are given in Chapters 2-7. Except for fused-ring systems, where fixed numbering (*see* Discussion section) applies and takes precedence, priority in assigning lowest locants is given successively to the following structural modifications:

   1. Hetero atoms in systems named by replacement nomenclature (*see* Chapters 6 and 7).

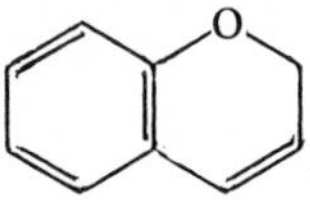

2*H*-1-Oxanaphthalene

$CH_2{=}CHOCH_2CH_2OCH_2CH_2OCH_3$     2,5,8-Trioxadec-9-ene

   2. Indicated hydrogen (*see* Chapter 4).

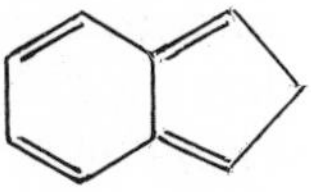

2*H*-Indene

3. The principal function (in parent compounds) or the point of attachment of a substituting group.

$$HOCH_2CH=CHCOOH \qquad \text{4-Hydroxy-2-butenoic acid}$$

4. Double bonds.   $CH_2=CHCH_2C≡CH$   1-Penten-4-yne

5. Triple bonds.   $HC≡CCH_2CH_2Cl$   4-Chloro-1-butyne

6. Substituents designated by prefix names that are earlier alphabetically.

$$ClCH_2CH_2CH_2NO_2 \qquad \text{1-Chloro-3-nitropropane}$$

The expression "lowest locant" refers to that series of locants which contains the lowest individual number at the first point of difference when compared term by term with an alternative series. For example, 1,-5,5 is lower than 2,4,4; 1,1,3 is lower than 1,2,2; and 1,1,3,4 is lower than 1,2,2,3.

When positions in a substituting group must be designated, the appropriate locants are placed within parentheses or other enclosing marks along with the name of the substituted group, as in (2-chloroethyl)-benzene.

**Element symbols**, italicized, are used to denote attachment to an atom that is not numbered as part of a chain or ring, as in *N*-ethylaniline and *O,O,S*-triethyl phosphorodithioate.

**Numerals** representing location of **unsaturation** or **functional groups** are placed before the name stem, except that such locants are placed between the name stem and the ending in naming bridged and spiro hydrocarbon ring systems and in replacement nomenclature.

| | |
|---|---|
| 2-Hexene | *not* Hex-2-ene |
| | *not* Hexene-2 |
| 1-Chloro-2-hexene | *not* 1-Chlorohex-2-ene |
| | *not* 1-Chlorohexene-2 |
| 1-Hexen-3-yne | *not* Hex-1-en-3-yne |
| | *not* 1-Hexenyne-3 |
| 1,6-Hexanediol | *not* Hexane-1,6-diol |
| | *not* Hexanediol-1,6 |
| Bicyclo[2.2.2]oct-2-ene | *not* Bicyclo[2.2.2]-2-octene |
| | *not* Bicyclo[2.2.2]octene-2 |

When locants for (1) unsaturation or hetero atoms and (2) the principal function are required in the same name, the locant(s) for unsaturation or hetero atoms immediately precedes the name of the parent compound, and the locant(s) for the functional group immediately precedes the suffix—*e.g.,* 2-hexen-1-ol, not 2-hexenol-1; 4-imidazoline-2-thione. In naming substituting groups, the locant indicating the position of attachment is placed just before the name of the stem unless the name of the corresponding parent compound contains one or more locants. In that case the locant for the position of attachment immediately precedes the suffix:—*e.g.,* 2-naphthyl; azacyclohept-2-yl; 1,3-oxathiolan-2-yl.

Names sometimes contain several different types of locants. In such names, the preferred order of listing is: Roman letters, Greek letters, Arabic numerals. As noted below, however, use of Greek letters as locants is not recommended.

**Multiplying affixes** are used to denote how many of a given kind of substituent groups or structural modifications are being cited. The prefixes *di, tri,* etc. are used with names of simple groups and *bis, tris,* etc. with names of complex (substituted) groups; with the latter prefixes, enclosing marks are added to assure clarity, as in bis(2-chloroethyl)amine. For the sake of euphony, *bis, tris,* etc. are also used when the name of the group itself begins with a multiplying prefix, as in bis(4-biphenylyl)methanol. The prefixes *bi, ter, quater,* etc. are used exclusively to describe ring assemblies (*see also* Chapter 5)—*e.g.,* biphenyl, quaterpyridine.

### Table 12.1  Multiplying Prefixes[a]

| Numerical Designation | For Simple Groups | For Complex Groups | For Ring Assemblies |
|---|---|---|---|
| 2 | di | bis | bi |
| 3 | tri | tris | ter |
| 4 | tetra | tetrakis | quater |
| 5 | penta | pentakis | quinque |
| 6 | hexa | hexakis | sexi |
| 7 | hepta | heptakis | septi |
| 8 | octa | octakis | octi |
| 9 | nona | nonakis | novi |
| 10 | deca | decakis | deci |

[a]Prefixes denoting multiplicity in excess of 10 are shown in Table 2.1.

**Isotopically labeled compounds** are named by affixing one or more expressions to the regular name of the compound to show the nature and location of the labelling, as specified in the modified Boughton system (*1*). Each expression consists of (1) any necessary locant numeral or numerals, italicized, (2) the atomic symbol, italicized, with a preceding superscript designating the mass number (except that *d* and *t* are preferred over $^2H$ and $^3H$, respectively), and (3) a subscript (other than unity) showing the number of such atoms. Each expression is added as a suffix to the regular name, or, if necessary for specificity, as a suffix to that part of the name designating the portion of the molecule containing the special atom.

$CH_3CH_2OD$                                    Ethanol-$O$-$d$

$DCH_2CHDOH$                                  Ethanol-$1,2$-$d_2$

$(C_2H_5)_2{}^{15}NCO^{35}S^{35}SCO^{15}N(C_2H_5)_2$     Bis(diethylcarbamoyl-$^{15}N$)
                                                disulfide-$^{35}S_2$

## Discussion

Greek letters have long been used to designate positions in a side chain attached to a ring, the side chain being treated as part of the parent compound, as in α-chloroethylbenzene. However, naming the side chain as a substituted substituent, as in (1-chloroethyl)benzene, is much to be preferred. The use of Greek letters to denote positions relative to a functional group, as in β-chloropropionic acid, is obsolete, except in conjunctive names (*see* Chapter 9). Use of Greek letters in class names, such as β-diketones and α-amino acids, is still acceptable.

Use of other prefixes that show relative positions of two or more atoms or groups has decreased greatly in recent years in favor of Arabic numerals. The most common prefixes that have been so used, along with some others describing structural arrangement, are listed in Table 12.2. Not included in this list are prefixes designating configurational relationships, unless they also have other meanings (an example is *endo*), in which case the definition given in Table 12.2 is one concerned with the location of a group and not with configuration. For a description of terms relating to configuration, *see* Chapter 14.

### Table 12.2. Nonnumerical Positional Prefixes

*ac-*   (for alicyclic) Designates substitution at unspecified position(s) in the alicyclic portion of the molecule, as in *ac*-chloro-1,2,3,4-tetrahydronaphthalene.

*α-*   (for alpha) Relates to or designates an end group or position in the absence of a functional group. Use with a functional group is not recommended (*see* text).

*ar-*   (for aromatic) Indicates substitution at unspecified position(s) in the aromatic portion of the molecule, as in *ar*-chlorotoluene.

*as-*   (for asymmetric) Sometimes used in the same sense as *unsym-* (unsymmetrical), as in *as*-trichlorobenzene.

*endo-*   (Greek, within) Indicates attachment as a bridge within a ring, as in 1,4-*endo*-methyleneanthracene.

*exo-*   (Greek, outside) Denotes attachment in a side chain, as in *exo*-chloro-*p*-menthane.

*gem-* (for geminate) Indicates that two groups are attached to the same atom, as in the *gem*-dimethyl grouping in camphor.

*m-* (for meta) Represents the relation of the 1- and 3- positions in benzene or substituted benzenes, as in *m*-dibromobenzene.

*ω-* (for omega) Relates to or designates an end group or position in the presence of a functional group or when used with $\alpha$ in the sense described above, as in *ω*-chloroalkanols; $\alpha, \omega$ -dichloroalkanes.

*o-* (for ortho) Represents the relation of the 1- and 2- positions in benzene or substituted benzenes, as in *o*-dibromobenzene.

*p-* (for para) Represents the relation of the 1- and 4- positions in benzene or substituted benzenes, as in *p*-dibromobenzene.

*sym-*, (for symmetrical) Designates symmetry in structural formula, especially
*s-* derivatives in which groups are substituted symmetrically in the molecule, as in *sym*-dichloroethylene, *s*-triazine.

*unsym-* (for unsymmetrical) Designates a form not symmetrical or lacking symmetry, as in *unsym*-dichloroethane, $CH_3CHCl_2$.

*vic-*, *v-* (for vicinal) Designates adjoining positions in a chain or ring, as in *vic*-triazine.

*?-*, *x-* Indicates that the position of a substituent is unknown, as in *?*-chlorophenanthrene, *x*-chlorophenanthrene.

Predominant practice in the United States is to use primed numbers only when two or more identical parts of a parent compound or substituting group carry the same numbers—*e.g.*, certain ring assemblies (*see* Chapter 5)—or when the same parent structure occurs more than once in the same molecule (*see* Chapter 11).

The term **fixed numbering** is applied to those polycyclic hydrocarbons or heterocyclic compounds that may be numbered in only one manner, regardless of structural modifications or substituents.

Aceanthrylene (numbering fixed)

Other polycyclic systems that may be numbered starting at any one of two or more carbon atoms—*e.g.*, naphthalene—are said to have **partially fixed numbering**.

A locant is omitted from a name when no ambiguity results, as in chlorobenzene. However, when two or more groups are named as substituents, the positions of all such groups should be designated. For example, 1-chloro-4-nitrobenzene is preferred over "4-nitrochloro-benzene."

The placement (in a name) of locants for unsaturation and for the principal function varies somewhat from country to country. In the United States, the standard practice is that recommended above.

### Table 12.3. Examples of Acceptable Usage

1.  $CH_3CHClCH_2CHO$

3-Chlorobutanal
3-Chlorobutyraldehyde

2.  $HOOCC{\equiv}CCH{=}CHCOOH$

2-Hexen-4-ynedioic acid

3.  $HOOCC{\equiv}CCH_2CHBrCOOH$

5-Bromo-2-hexynedioic acid

4.  $BrCH_2CH_2CH_2Cl$

1-Bromo-3-chloropropane

5.  $CH_3CH_2CH_2\underset{\underset{CH_2CH_2Cl}{|}}{CH}COOH$

2-(2-Chloroethyl)pentanoic acid

6.  $CH_3CH_2CH_2\underset{\underset{CH_2CH_2Cl}{|}}{CCl}COOH$

2-Chloro-2-(2-chloroethyl)pentanoic acid

7.

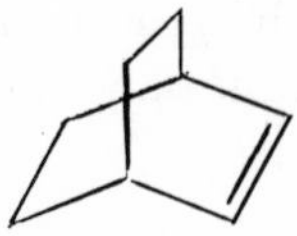

Bicyclo[2.2.2]oct-2-ene

8.

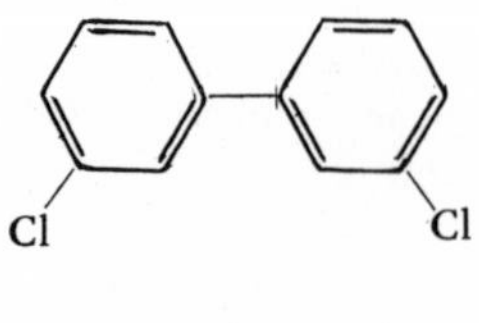

3,3'-Dichlorobiphenyl

9.

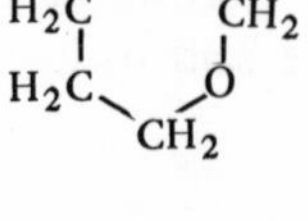

1,3-Dioxacyclohexane
1,3-Dioxane
m-Dioxane

10.

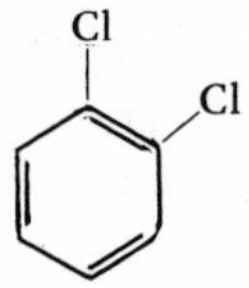

1,2-Dichlorobenzene
*o*-Dichlorobenzene

11.

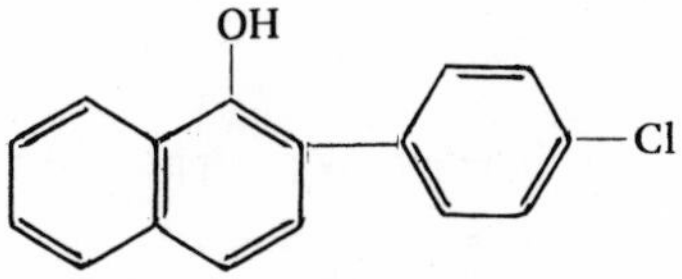

2-(4-Chlorophenyl)-1-naphthalenol
2-(*p*-Chlorophenyl)-1-naphthol

12.

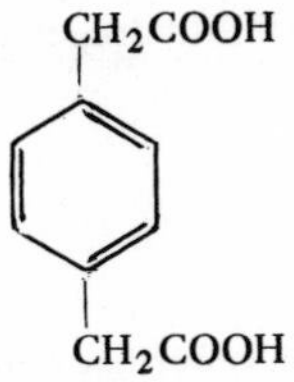

1,4-Phenylenediethanoic acid
*p*-Benzenediacetic acid

13.

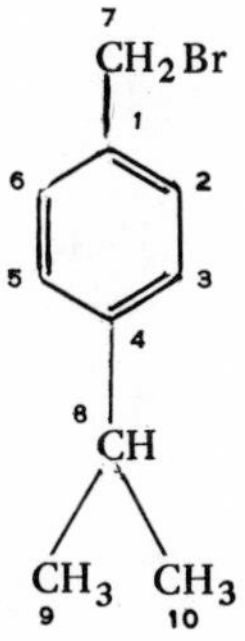

1-(Bromomethyl)-4-isopropylbenzene
7-Bromo-*p*-cymene (numbering shown)

14.

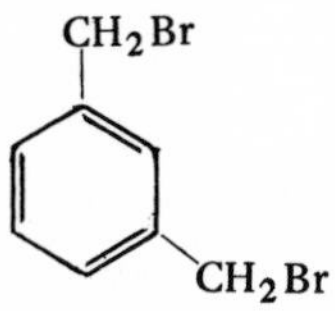

1,3-Bis(bromomethyl)benzene
α,α''-Dibromo-*m*-xylene

15.

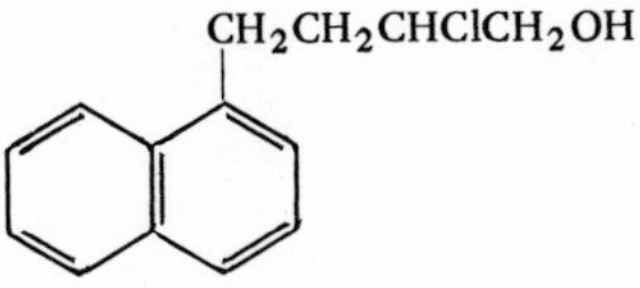

2-Chloro-4-(1-naphthyl)-1-butanol
β-Chloro-1-naphthalenebutanol

16.        $C_6H_5CH\!\!=\!\!\overset{\displaystyle |}{\underset{\displaystyle Cl}{C}}C_6H_5$        1-Chloro-1,2-diphenylethene
                                                            α-Chlorostilbene

17.        $C_6H_5N(C_2H_5)_2$        *N,N*-Diethylbenzenamine
                                           *N,N*-Diethylaniline

18.        $(C_2H_5NH)_2P(O)C_6H_5$        *N,N'*-Diethyl-*P*-phenylphosphonic diamide

19.        $(C_2H_5O)_2P(O)SH$        *O,O* -Diethyl *S*-hydrogen phosphorothioate

20.

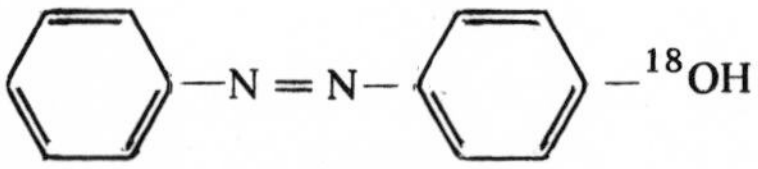

4-(Phenyldiazenyl)benzenol-$^{18}O$
*p*-(Phenylazo)phenol-$^{18}O$

21.

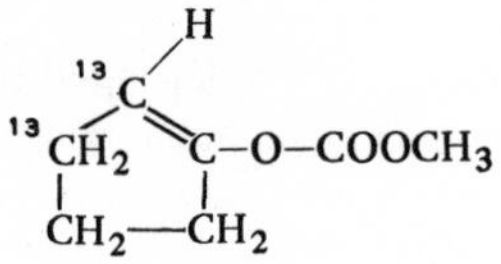

1-Cyclopenten-1-yl-*2,3*-$^{13}C_2$        methyl
  carbonate

## Literature Cited

1. Crane, E. J., *Ind. Eng. Chem., News Ed.* (1935) **13**, 200-1.

# Construction of Systematic Names

S ystematic names descriptive of molecular structure more often than not contain one or more kinds of punctuation and enclosing marks. Correct incorporation of such marks constitutes an important step in constructing a chemical name. Omission or incorrect usage of a comma or parentheses may prove to be as serious an error as the failure to select the right word or prefix name to designate a given structural unit or group. Of equal importance are the established conventions for word separation, elision of vowels, and use of italics.

*Recommended Nomenclature Practice*

**Word Separation.** Systematic names formed by applying the principles of **substitutive nomenclature** are **single words**, except for compounds named as *acids*.

| | |
|---|---|
| $ClCH_2CH_2Cl$ | 1,2-Dichloroethane |
| $(C_6H_5)_2CHCH_2CH_2OH$ | 3,3-Diphenyl-1-propanol |
| $CH_3CH_2CH_2COOH$ | Butanoic acid |
| $C_6H_5SO_2OH$ | Benzenesulfonic acid |

The same is true for the most part in the other fundamental methods of naming functional compounds (*see* Chapter 9). However, a major exception is **radicofunctional nomenclature**, wherein properly formed names consist of **more than one word.**

| | |
|---|---|
| $CH_3CH_2OH$ | Ethyl alcohol |
| $CH_3COOCH_2CH_3$ | Ethyl acetate |
| $CH_3OCH_2CH_2CH_3$ | Methyl propyl ether |
| $CH_3CH_2COCl$ | Propionyl chloride |
| $CH_3COC_6H_5$ | Methyl phenyl ketone |
| $CH_3CH_2SCH_2CH_3$ | Diethyl sulfide |

In addition to those in the above examples, functional class names that stand as separate words when used to form names of specific compounds include *anhydride, bromide, cyanide, fluoride, hydroperoxide, hydroxide, iodide, oxide, peroxide, sulfone, sulfoxide, and sulfoximine.* However, the class name *amine* is not used radicofunctionally and therefore never stands alone in a chemical name (*see* Discussion section).

A few other kinds of multiword names, not readily categorized here, will be encountered elsewhere in this book.

**Elision of Vowels.** Generally, to avoid ambiguity, vowels, whether pronounced or silent, are retained in the construction of systematic names of organic compounds. This practice often produces double vowels, since names for component groups and modifying prefixes often begin or end with a vowel.

| | |
|---|---|
| Cyclooctane | Trioxaazaoctadecane |
| Chloroacetic acid | Undecatetraene |
| Indenoisoindole | Thiopheneacetic acid |
| Anthraimidazole | Naphthalenetetraacetic acid |

However, it has become accepted practice to elide the vowels *a, e,* and *o* in the following circumstances:

(1)   Preceding the suffix name of a functional group.

| | |
|---|---|
| Ethanol | *not* Ethaneol |
| Hexanamine | *not* Hexaneamine |
| Pyridinone | *not* Pyridineone |
| Anthracenetetrol | *not* Anthracenetetraol |

(2)   When both ethylenic and acetylenic unsaturation occur in the same chain and ring system and hence are named sequentially as suffixes.

| | |
|---|---|
| Hexenyne | *not* Hexeneyne |
| Cyclodecadienyne | *not* Cyclodecadieneyne |

(3)   In fusion names for polycyclic hydrocarbons and heterocycles.

| | |
|---|---|
| Benzanthracene | *not* Benzoanthracene |
| Naphthacridine | *not* Naphthoacridine |

(4)   In Hantzsch-Widman names for heterocyclic systems.

| | |
|---|---|
| Oxazoline | *not* Oxaazoline |
| Dioxarsenin | *not* Dioxaarsenin |

Elision of vowels is not affected by inserting locants or other modifiers into the name.

| | |
|---|---|
| 4-Hexen-2-one | *not* 4-Hexene-2-one |
| Naphth[1,2-*d*]oxazole | *not* Naphtho[1,2-*d*]oxazole |

**Punctuation Marks.** In systematic names the most commonly used punctuation marks by far are hyphens and commas; colons and periods are encountered much less often, and other punctuation marks such as the semicolon, exclamation point, apostrophe, quotation mark, and long dash are not used. **Hyphens** are used to connect numbers and letters serving as locants or descriptive symbols to the syllabic portion of the name.

|  |  |
|---|---|
| 2-Chloroethanol | 1-Bromo-4-nitrobenzene |
| *N*-Methylaniline | Propanethioic *S*-acid |
| 4*H*-1,3-Dithianaphthalene | *s*-Triazine |

Configurational prefixes, whether syllabic or in the form of single letters, are set off with hyphens, as are the structure-defining prefixes *sec* and *tert*.

|  |  |
|---|---|
| α-D-Glucopyranose | *sec*-Butyl alcohol |
| *trans*-2-Butenoic acid | *tert*-Butyl iodide |
| *cis-exo*-1,3-Dichloronorbornane |  |

However, the structure-defining prefixes cyclo, spiro, iso, and neo are considered to be an integral part of the name in which they are used and are used and therefore are not separated nor written in italics.

|  |  |
|---|---|
| Cyclopropane | *not cyclo*-Propane |
| Isobutane | *not iso*-Butane |
| Neopentane | *not neo*-Pentane |

Commas are used to separate the individual members of a series of locants.

|  |  |
|---|---|
| 1,1,2,2-Tetrachloroethane | 1,3-Butadiene |
| 1,2,3-Benzenetriol | 2,5-Dithiaheptane |
| β,4-Dichlorocyclohexanepropionic acid |  |
| *N,N*-Dimethyl-3,5-dinitroaniline |  |
| Dicyclopenta[*a,f*]naphthalene |  |

No comma is used, however, within a locant expression that consists of more than one kind of symbol.

|  |  |
|---|---|
| 2*H*-Pyrrole | *not* 2,*H*-Pyrrole |
| 5β,17α-Pregnane | *not* 5,β,17,α-Pregnane |

**Periods** are used within brackets to form names for bridged and spiro alicyclic hydrocarbons (*see* Chapter 2).

|  |  |
|---|---|
| Bicyclo[3.2.0]heptane | Spiro[4.5]decane |

**Colons** are used to separate two or more series of locants when each such series refers to the same species of substituting group.

1,2,4,5-Benzenetetracarboxylic acid 1,2:4,5-dianhydride
1,4:5,8-Dimethanonaphthalene

**Enclosing Marks.** Parentheses ( ) and square brackets [ ] are used as demarcation symbols in conjunction with locants when the latter relate to names of complex groups.

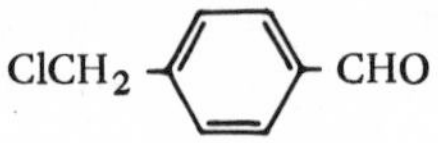

$$ClCH_2 - \langle \bigcirc \rangle - CHO$$

4-(Chloromethyl)benzaldehyde

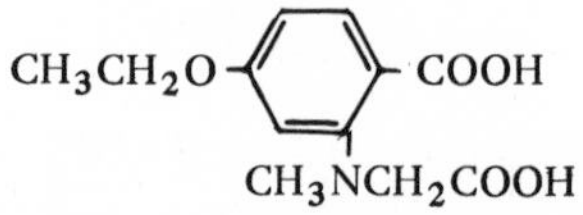

$$CH_3CH_2O - \langle \bigcirc \rangle - COOH$$
$$CH_3NCH_2COOH$$

2-[(Carboxymethyl)methylamino]-4-ethoxybenzoic acid

$$CH_3SCH_2CH_2NCH_2CH_2OH$$
$$CH_3$$

2-[Methyl[2-(methylthio)ethyl]amino]ethanol

As indicated by the above examples, **parentheses** are used first, then **brackets**, and brackets again, if needed. Enclosing marks do not replace hyphens, nor do they otherwise affect the use of punctuation marks. Brackets also have other, more specialized uses—*i.e.*, in naming bridged and spiro alicyclic hydrocarbons (Chapter 2) and in fusion names for heterocyclic systems (Chapter 6).

**Italics.** English letters used as locants, whether capital or lower-case, are printed in italic (*see* Chapter 12) except when they merely modify a number locant.

| | |
|---|---|
| *N*-Methylaniline | *O*-Ethyl phosphorothioate |
| *p*-Dichlorobenzene | *s*-Triazine |
| Dibenz[*a,b*]anthracene | 4a,8a-Dihydronaphthalene |

Configurational and other structure-defining prefixes (except for D and L; *see* Chapter 14), when set off from the rest of the name with a hyphen, are italicized.

| | |
|---|---|
| *meso*-Tartaric acid | *cis*-9-Octadecenoic acid |
| *sec*-Butyl alcohol | *trans*-1-Bromo-4-methylcyclohexane |
| *tert*-Butyl chloride | *ar*-Chlorotoluene |

The capital letter H used as a prefix to denote indicated hydrogen is italicized.

| | |
|---|---|
| 3*H*-Fluorene | 1*H*-1,3-Diazepine |

Suffixes and accompanying locants used to denote the presence of isotopic atoms are italicized.

| | |
|---|---|
| Benzene-*d* | Ethylamine-*1*-$^{14}C$ |
| Methane-*t* | Pyrimidine-*1*-$^{15}N$ |

## Discussion

In the semitechnical literature, in patents, and in advertising copy, errors are often made in constructing systematic names of organic compounds. One of the most common mistakes is that of separating substitutive names into two or more words, either standing alone or connected by hyphens—*e.g.,* "chloro benzene" or "chloro-benzene" for chlorobenzene. In particular, *amine* is frequently written as a separate word, probably by supposed analogy to *alcohol, ketone,* and other radicofunctional class names. Correct use of *amine* is substitutive only, either as a functional suffix—*e.g.,* methanamine—or as a parent compound denoting $NH_3$—*e.g.,* methylamine. The whole problem of word separation, which is discussed further in Chapter 15, can be held to a minimum by using substitutive single-word names whenever possible.

Conventions for eliding vowels in organic chemical names have been established to simplify spelling and improve euphony. These conventions are, in effect, rules for spelling and hence must be committed to memory or checked by reference to an official source such as the Subject Index of *Chemical Abstracts.* Elision of syllables—*e.g.,* "2-pyridone" for 2-pyridin-one—has been practiced to a limited extent in the past but is no longer recommended.

Authors should take care not to overpunctuate in constructing systematic chemical names. There is often a tendency to use hyphens where they are not needed, particularly immediately preceding or following enclosing marks. Examples: ethylbis(fluoromethyl)amine, where no hyphens are needed; and 2-(2-chloroethyl)benzoic acid, not "2-(2-chloroethyl)-benzoic acid." Interposition of enclosing marks does not create a need for hyphenation if the latter would not otherwise be required.

**Prime marks** are used to distinguish among two or more corresponding series of locants denoting positions within a single molecule—*e.g.,* 4,4',-4"-nitrilotribenzoic acid (*see also* Chapter 12).

A **question mark** may be used to denote uncertainty in the position of a substituting group — *e.g.,* ?-chlorotoluene. However, this practice, unless clearly defined by the user, does not distinguish the case where all chlorine substitution is at the same (though unknown) position from the case where a mixture of isomers is involved. The letter locant *x* (italicized) has been used in the same manner (*see* Chapter 12).

Currently, parentheses and square brackets are the only enclosing marks officially used in systematic organic chemical names. In years past, braces { } were used as "third-stage" enclosing marks, following parentheses and brackets in that order. The decision to abandon braces was made for simplification since a relatively small proportion of names require three stages of enclosing marks.

Italics are used for letter locants and most configurational prefixes, chiefly to facilitate arrangement of chemical names in indexes, but also to emphasize the separation of such letters and word prefixes from the main portion of the name. However, the structure-defining prefixes iso, neo, spiro, cyclo, bicyclo, etc. are not separated from the rest of the name and are not italicized in current practice (*see* 1965 IUPAC Rules A-2.1 and A-2.25). Until after World War II, accepted British nomenclature called for "*iso*-butyl" which is consistent with *sec*-butyl and *tert*-butyl, as well as "*cyclo*-hexane", "*spiro*-octane", and the like, but this usage has now been officially abandoned for sake of consistency with established *Chemical Abstracts* practice. It may be argued that the prefixes cyclo and spiro alter the molecular formula of the reference hydrocarbon—*e.g.,* cyclohexane is $C_6H_{12}$ but hexane is $C_6H_{14}$—and that therefore cyclo and spiro are not simply prefixes used to distinguish isomers; however, it must be admitted that logic supports the earlier British position in favor of italicizing iso and neo.

# Stereoisomers

This chapter deals with the names by which individual **configurational isomers** are designated and distinguished from each other. According to recently published IUPAC Tentative Rules (*1*), "The configuration of a molecule of defined constitution is the arrangement of its atoms in space without regard to arrangements that differ only as after rotation about one or more single bonds." The nomenclature of **conformational isomers** is not considered here. Likewise, with few exceptions, those terms in stereochemistry that are not parts of the names of individual compounds— *e.g.,* diastereoisomers, enantiomers, and quasi-racemates — are not treated.

The chapter is divided into two parts, the first dealing with stereoisomers differing from each other only as an object differs from its mirror image, and the second dealing with stereoisomers whose structures differ only with respect to the arrangement of certain "rigidly" positioned atoms or groups relative to a specified plane of reference. The former have commonly been referred to as "optical isomers" and the latter as "geometric isomers," but the newer terms **chiral isomers** and **cis-trans isomers**, respectively, are now preferred. All chiral molecules exhibit the phenomenon of optical activity, and most of them contain at least one **asymmetric** carbon atom—*i.e.,* a carbon atom attached to four different atoms or groups.

## Chiral Isomers

### Recommended Nomenclature Practice

**Absolute configuration** relative to an asymmetric carbon atom is best expressed by the recently developed (*R*)-(*S*) system (Latin: *rectus*, right hand; *sinister*, left hand). In this method (*1,3*) the four different groups comprising the tetrahedral chiral structure are first arranged in descending order of precedence, *a, b, c, d,* by applying the (*R*)-(*S*) sequence rule (abridged version given below). Next, the tetrahedral structure is visualized as oriented so that the lowest-ranking group, *d*, is at the bottom apex. The tetrahedron is then viewed from the top. If reading from *a* to *b* to *c* on the formula requires clockwise motion around the triangular top, the configuration is designated by the prefix (*R*), if counterclockwise, (*S*).

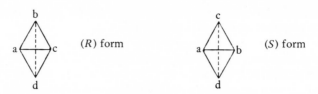

The $(R)$-$(S)$ sequence rule as given here is highly abridged; for more general and detailed information the literature should be consulted $(1,2,3)$.

(1) Groups are arranged in order of decreasing atomic number of the atom directly attached to the atom at the chiral center.

(2) If two atoms attached to the central atom are the same, the one substituted with atoms of higher atomic number takes precedence. If two atoms are equivalent in that respect, the one with more substituents of high atomic number comes first. If the second atom out permits no choice, the third is examined, etc.

(3) If an atom is attached to another atom by a multiple bond, both atoms are regarded as being replicated (italicized symbols in formulas below):


—CH=CH—                regarded as        —CH—CH—
                                           |    |
                                           $C$    $C$

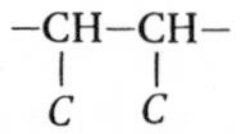

$>$C=O                  regarded as 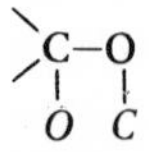        $>$C—O
                                           |   |
                                           $O$   $C$


—C≡N                   regarded as        —C————N
                                          $/\$     $/\$
                                          $N$ $N$   $C$ $C$

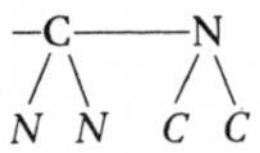

A benzene ring is regarded as having the Kekulé structure. Multiple singly bound atoms take precedence over their doubly or triply bound counterparts,—$e.g.$ C(-OR)$_2$ outranks C=O.

(4) A missing substituent on a tetracovalent atom is regarded as being of atomic number zero.

(5) An isotope of higher mass number takes precedence.

(6) In choosing among the groups $a,a',b,c$ where the groups $a$ and $a'$ are

stereoisomeric, an (*R*) group precedes an (*S*) group, and a *Z* (*cis*) group precedes an *E* (*trans*) group.

Thus in hydroxyphenylacetic acid, $C_6H_5CHOHCOOH$, the *a* group is identified as OH, *b* as COOH, *c* as $C_6H_5$, and *d* as H. The formula is then visualized as corresponding to one of the two following views. In one of these the reading from *a* to *b* to *c* goes clockwise and in the other, counterclockwise.

(*R*)-Hydroxyphenylacetic acid          (*S*)-Hydroxyphenylacetic acid

The (*R*)-(*S*) system is applied to compounds containing two or more chiral centers in the same way, configuration being designated for each center.

$$
\begin{array}{c}
CH_3 \\
| \\
Cl-C-COOH \\
| \\
H-C-COOH \\
| \\
CH_3
\end{array}
$$

(2*R*,3*R*)-2-Chloro-2,3-dimethylsuccinic acid

Although the (*R*)-(*S*) system is to be preferred, absolute configuration may also be described in some molecules containing a single chiral center by the older D-L system, in which a Fischer projection formula (*see* Discussion) is oriented so that the No. 1 carbon atom of the main chain is at the top. The prefix D is used to represent the absolute configuration of the isomer in which the functional (determining) group is on the right side of the carbon atom at the chiral center (the asymmetric

$$
\begin{array}{c}
CH_2OH \\
| \\
H-C-OH \\
| \\
CH_2 \\
| \\
CH_2OH
\end{array}
\qquad
\begin{array}{c}
CH_2OH \\
| \\
HO-C-H \\
| \\
CH_2 \\
| \\
C_6H_5
\end{array}
\qquad
\begin{array}{c}
CH_3 \\
| \\
Cl-C-Br \\
| \\
CH_2 \\
| \\
CH_3
\end{array}
$$

D-1,2,4-Butanetriol          L-3-Phenyl-1,2-propanediol          (?)-2-Bromo-2-chlorobutane carbon atom), and L, that of the isomer in which it is on the left. This treatment can be applied when there *is* a main chain and one of the groups attached to the asymmetric carbon atom is hydrogen, but it is useless when both the attached groups are functional since there is then no basis for choice of the determining group.

In specialized carbohydrate nomenclature (4), the absolute configuration of the highest-numbered asymmetric carbon atom of a sugar (the *only* such carbon in the simplest sugar, glyceraldehyde) is designated by the D or L prefix. This prefix is attached to a trivial name which implicitly specifies configurations at the other chiral centers, as in D-arabinose.

CHO

HO—C—H

CH$_2$OH

L-Glyceraldehyde

β-D-Arabinose

As used in amino acid nomenclature (5), the prefixes D and L designate absolute configuration, not of the highest-numbered asymmetric carbon atom, but of the α-carbon atom. Ambiguity may thus arise unless the prefixes are amplified to D$_s$ and L$_s$ (the subscript *s* referring to *serine* as the standard):

COOH

H—C—NH$_2$

CH$_2$OH

D-Serine

COOH

H—C—NH$_2$

HO—C—H

CH$_3$

D$_s$-Threonine

Amino acids may also be named like carbohydrates, with D and L prefixes designating the configuration of the highest-numbered asymmetric carbon atom, and using glyceraldehyde as the standard, if the prefixes are amplified to D$_g$ and L$_g$. Then D$_s$-threonine and L$_g$-threonine are alternative names for the same structure.

Absolute configuration of cyclitols (6) may also be designated by D and L prefixes, used with either a trivial name (*e.g.*, inositol) or numerals designating relative configuration (*see* page 108). The configuration of the *lowest*-numbered asymmetric carbon atom (usually at Position 1) is designated as D or L, or if necessary D$_c$ or L$_c$. To show that the No. 1 carbon atom is referred to, it is advisable to add the numeral 1 as a prefix.

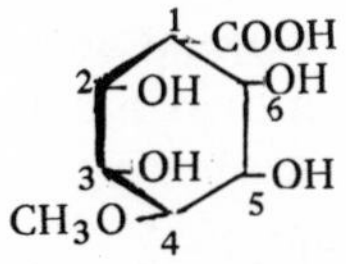

1D-(1,2,3,5,6/4)-2,3,5,6-Tetrahydroxy-4-
methoxycyclohexanecarboxylic acid

An alternative system of expressing absolute configuration is acceptable for steroids (7) and some other cyclic compounds. The ring system,
conceived as flat, is placed so that the cyclopentane ring or the principal
functional group is at the upper right. The location of each atom or group
under consideration is described as α when it lies behind this plane of the
ring system and β when it lies in front of it; an unknown configuration is
described by the letter ξ. The appropriate Greek letter is suffixed to the
number locant for the atom or group. Pictorial conventions for representing these relative positions in structural formulas are described in the
Discussion section.

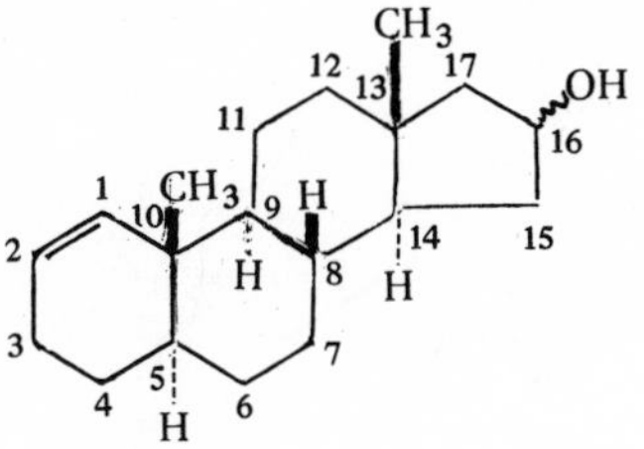

5α-Androst-1-en-16ξ-ol

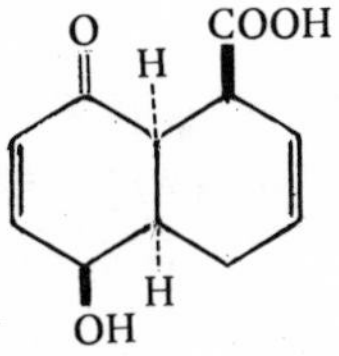

1,4,4aα,5,8,8aα-Hexahydro-5β-hydroxy-
-8-oxo-1β-naphthalenecarboxylic acid

It should be remembered that many of the configurations present in
steroids and terpenes are implicit in their trivial names.

In the perspective ring formulas for sugars of the D series (4), if the
ring is oriented so that the ring oxygen is at the upper right and the No. 1
carbon atom is at the middle right side, the form that has the 1-hydroxy
group beneath the plane of the ring is called the α-isomer and the one with
the hydroxy above is called the β-isomer. Conversely, in the L sugars, the

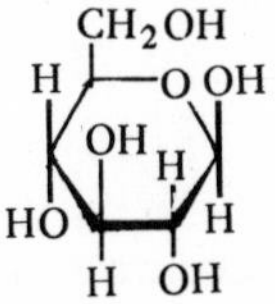

β-D-Glucopyranose

form that has the 1-hydroxy group above the plane of the ring is the
α-isomer, etc.

**Relative configuration** of multiple chiral centers in a molecule is best
expressed by the prefixes $R^*$ and $S^*$, the one first cited being arbitrarily
considered to be $R$ and the others being labeled accordingly (*1*). In
complex cases the stars may be omitted and the prefix *rel* added to the
whole name. If only relative configuration of whole molecules is known,
these are distinguished by using the prefix (+) or (−) describing the
observed direction of optical rotation.

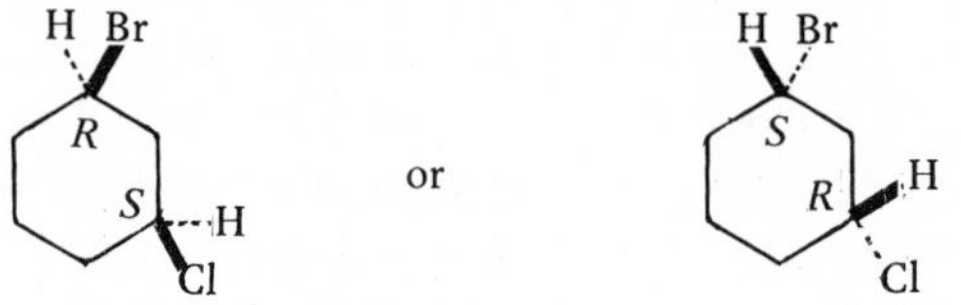

(1$R^*$,3$S^*$)-1-Bromo-3-chlorocyclohexane

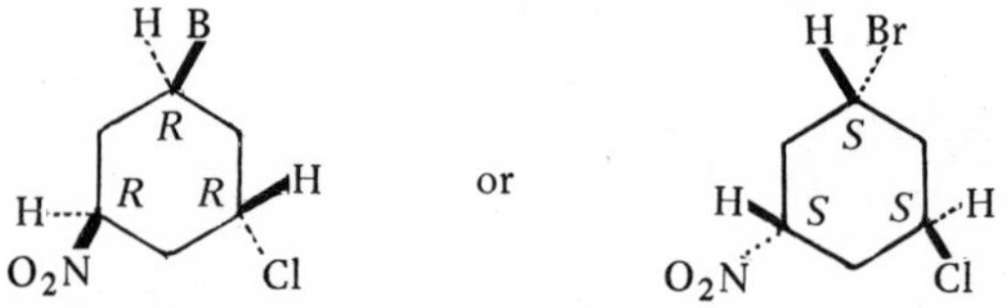

*rel*-(1$R$,3$R$,5$R$)-1-Bromo-3-chloro-5-nitrocyclohexane

For carbohydrates, the relative configuration of a group of consecutive but not necessarily contiguous chiral carbon atoms (up to four) is
designated by special prefixes derived from the corresponding aldose
names (*4*).

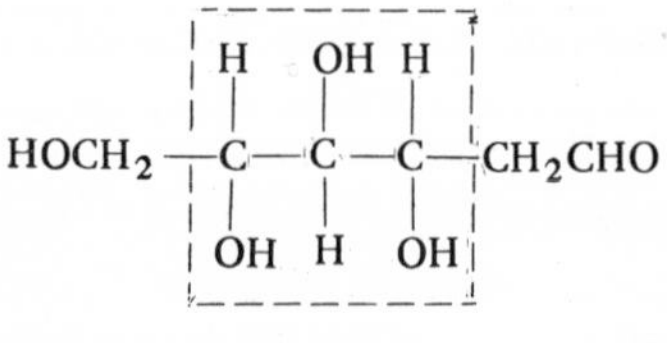

2-Deoxy-D-*xylo*-hexose
(D-*xylo* configuration enclosed)

Relative configuration of cyclitols may be expressed by listing the
positions of groups on one side of the plane of the ring (hypothetically
flattened if necessary), then those of groups on the other side (*6, 8*).

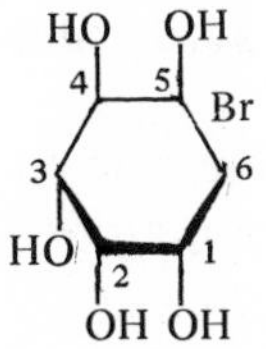

(1,2,3/4,5,6)-6-Bromo-1,2,3,4,5-cyclohexanepentol


## Discussion

The IUPAC Tentative Rules on Stereochemistry (*1*) should be consulted for definitions of terms not part of the names of individual compounds (*e.g.*, diastereoisomers, racemic) and for nomenclature of conformational isomers.

Isomers differing because of the presence of chiral structures have long been known generically as "optical isomers" because they usually differ from each other in the direction and extent of rotation of the plane of plane-polarized light, but this term does not describe structure and is otherwise objectionable (*2*). Similarly, such isomers were originally distinguished by the prefixes *d* (dextro), *l* (levo) and *dl* (racemic), all referring to observed optical rotation. However, these letter prefixes were later applied for both operational and configurational descriptions, and this has caused so much confusion that their use should now be abandoned. They persist only in a few trivial names such as *d*-camphor. The prefix *meso* has long been used to describe stereoisomers in which the presence of identical chiral centers of opposite configuration produces a total structure having a plane, center, or alternating axis of symmetry (and therefore optically inactive), but the term now appears superfluous except in a generic sense.

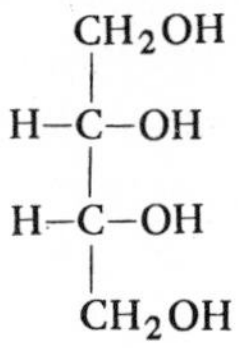

(2*R*,3*S*)-1,2,3,4-Butanetetrol
formerly *meso*-1,2,3,4-Butanetetrol


The representation of three-dimensional structures on paper is indispensable in stereochemistry but poses some problems. Conventions must be established to depict three dimensions in two. In one system of perspective formulas representing absolute configuration, heavier lines represent bonds extending toward the reader, and broken ones, bonds away from the reader.

Of the several types of projection formulas, the Fischer type is used here. In it the asymmetric carbon atom is placed at the center, horizontal lines represent bonds to the two groups lying above the plane of the paper, and vertical lines represent bonds to the two groups below the plane of the paper.

The only truly general method of describing configuration is the $(R)$-$(S)$ system. While the D-L system, which originated in carbohydrate chemistry, is still used, it is not general, at least without special rules for extension to multiple chiral centers ($9$). Semisystematic nomenclature based on the use of italic capital letters $D$—$L$ ($10$) is even more specialized.

In designating the stereochemistry of lipids, the top carbon atom in the Fischer projection formula for glycerol that shows the secondary hydroxy group on the left side of the vertical carbon chain may be designated as No. 1 and the prefix *sn* (for stereospecifically numbered) placed just before the term signifying glycerol ($10$).

$$CH_2OPO_3H_2$$
$$|$$
$$HO-C-H$$
$$|$$
$$CH_2OH$$

*sn*-Glycerol 1-(dihydrogen phosphate)

When a compound has the structure R-Cab-Cac-R' — *i.e.*, when it contains two adjoining chiral carbon atoms each a part of the main chain of the molecule (IUPAC system) and carrying at least one atom or group the same as on the other—the isomer in which the Fischer projection formula shows the like groups on the same side has long been designated by the prefix *erythro,* and the one in which they are on opposite sides, by the prefix *threo.* However, like the D-L system, the *erythro-threo* designation of configuration often cannot be applied, or at least not without ambiguity—*e.g.,* when there is no identifiable main chain or when no two groups on the asymmetric carbon atoms are alike. Moreover, some authorities do not base *erythro-threo* names on the main chain. Thus the $(R)$-$(S)$ or $(R^*)$-$(S^*)$ system is to be preferred.

$$COOH$$
$$|$$
$$H-C-Cl$$
$$|$$
$$Cl-C-H$$
$$|$$
$$CH_3$$

L-*threo*-2,3-Dichlorobutyric acid

There has been some extension of the $\alpha,\beta$ system of designating steroid configuration to simpler compounds, especially terpenes. By this method the hydrocarbon represented in the following example would be called l$\alpha$-ethyl-3$\beta$-methyl-5$\alpha$-propylcyclohexane. However, since no rule is provided specifying orientation of the structure, the name could just as well be 1$\beta$-ethyl-3$\alpha$-methyl-5$\beta$-propylcyclohexane.

$$CH_3CH_2CH_2 \quad CH_2CH_3$$

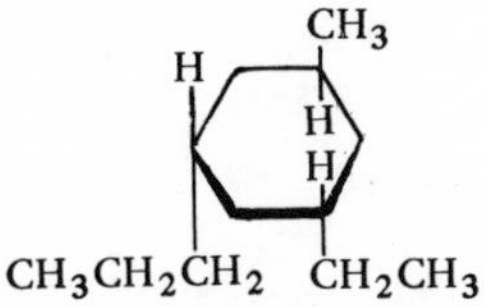

Most of the stereochemical names and symbols discussed here, and a number of others, are explained and used in Beilstein's *Handbuch (11)*.

### Cis-trans Isomers

Isomerism of the *cis-trans* type is encountered in acyclic olefins in which the doubly bound carbon atoms bear nonidentical groups, in monocyclic structures whose substituents lie outside the plane of the ring, and in polycyclic systems having saturated bridgeheads.

### Recommended Nomenclature Practice

In specifying configuration about a **double bond**, it is necessary to establish, for each of the doubly linked atoms, which attached group has precedence according to the sequence rule devised for the $(R)$-$(S)$ system *(1,3)*. When this is done, that configuration in which the two vicinal groups of higher priority are closer together is designated as Z (German, *zusammen*, together) and that in which they are farther apart as E (German, *entgegen*, opposite) *(12, 13)*.

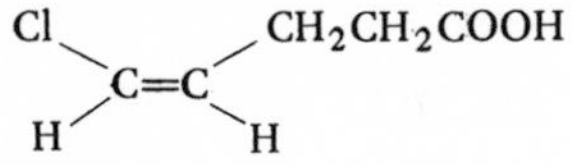

(Z)-5-Chloro-4-pentenoic acid

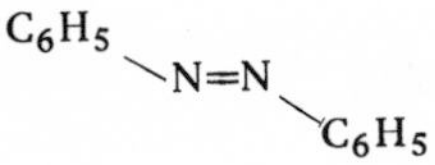

(E)-Diphenyldiazene
(E)-Azobenzene (old name)

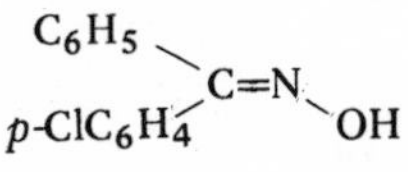

(Z)-[(4-Chlorophenyl)phenylmethylene]-hydroxylamine
(Z)-4-Chlorophenyl phenyl ketone oxime (old name)

The (R)-(S) sequence rule may be used repeatedly to rank groups about multiple double bonds so that a unique set of Z and E descriptors is generated for each compound (1,13). In such cases, in numbering the chain, Z groups take precedence over E groups, and locants are used with the Z and E prefixes as well as in the rest of the name.

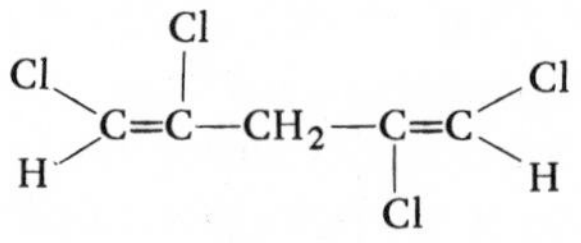

(1Z,4E)-1,2,4,5-Tetrachloro-1,4-pentadiene

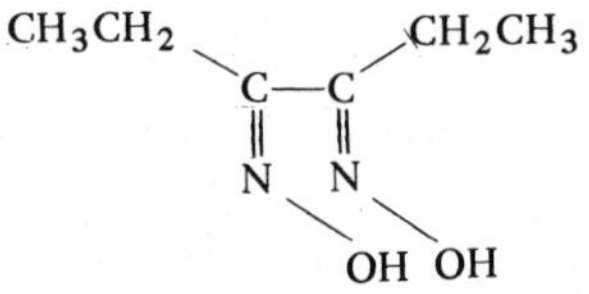

(Z,E)-N,N'-Hexane-3,3,4,4-tetrayldihydroxylamine

(Z,E)-3,4-Hexanedione dioxime (old name)

In specifying cis-trans configuration in a **monocyclic compound**, the plane used is that of the ring, hypothetically flattened if necessary. In compounds having only two substituents, each at a different ring position, relative configuration is denoted by the prefix cis or trans according to whether the substituents are on the same or opposite sides, respectively, of the ring plane (1).

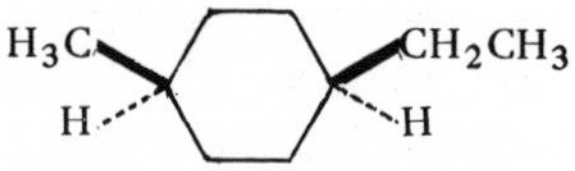

cis-1-Ethyl-4-methylcyclohexane

In compounds having one substituent and one hydrogen atom at each of more than two positions of the ring, relative configuration is denoted by adding the prefix r (for reference) to the lowest number locant and the

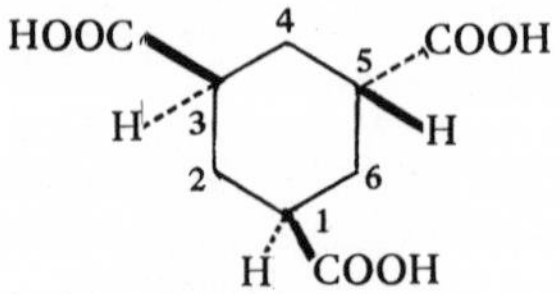

r-1,c-3,t-5-Cyclohexanetricarboxylic acid

prefix *c* (for *cis*) or *t* (for *trans*) as appropriate to the other number locants (*1*). When alternative numberings of the ring are possible, that one is chosen that gives a *cis* designation at the first point of difference.

If two different substituents are attached to the same position of the ring, the reference group is chosen to be, in order of preference, (1) the principal function (*see* Chapter 10) or (2) the group having precedence according to the (*R*)-(*S*) sequence rule; the relative configuration at each other position is then specified by citing the substituent in each case that has precedence according to the sequence rule (*1*).

*t*-3-Bromo-3-chloro-1-methyl-*r*-1-cyclopentanecarboxylic acid

*r*-1-Bromo-1-chloro-*c*-3-ethyl-3-methylcyclopentane

In **fused-ring systems** the ring fusion may be *cis* or *trans*, as in the decahydronaphthalenes.

*trans*-Decahydronaphthalene                          *cis*-Decahydronaphthalene

When more than one pair of bridgehead atoms is present, the relation between the nearest atoms of bridgehead pairs is expressed by the prefixes *cisoid* and *transoid* (*1*). If there is a choice among nearest atoms, the pair containing the lower-numbered atom is selected.

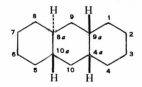

*trans-cisoid-trans*-Perhydrophenanthrene

*cis-cisoid*-4a,10a-*trans*-Perhydroanthracene

Configuration in **bridged-ring systems** wherein each bridge contains at least one atom is described by special prefixes that designate the relative positions of the main (senior) bridge and a given substituent atom or group with respect to the planar portion of the ring to which they are both attached. The main bridge is chosen by the following criteria, applied successively as necessary: (1) bridges containing hetero atoms, (2) bridges containing fewer atoms, (3) saturated bridges, (4) bridges carrying a smaller number of substituents, and (5) bridges carrying substituents lower according to the (*R*)-(*S*) sequence rule (*3*). In the formulas that follow, the main bridge is shown in heavy lines, which here do not have any three-dimensional significance.

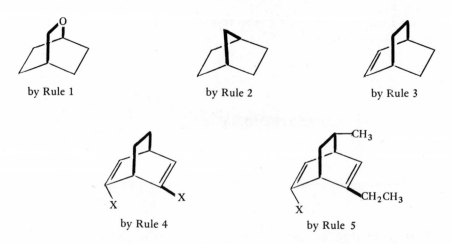

by Rule 1        by Rule 2        by Rule 3

by Rule 4        by Rule 5

A substituent atom or group not on the main bridge is designated by the prefix *exo* (preceding the locant) when it is on the same side of the reference plane as the bridge, and by *endo* when it is on the opposite side.

*endo*-2-Chlorobicyclo[2.2.1]heptane
*endo*-2-Chloronorbornane

*exo*-Bicyclo[2.2.2]oct-5-en-2-ol

However, when the atom or group to be described *is* on the main bridge, obviously that bridge cannot serve as a reference point, and another reference point must be selected. This secondary bridge is chosen by the following criteria, applied successively as before: (1) bridges containing hetero atoms, (2) bridges carrying a larger number of substituents, (3) bridges carrying substituents higher according to the sequence rules, (4) unsaturated bridges, and (5) bridges containing more atoms. The atom or group on the main bridge is then described as *syn* when it is in the position nearer the secondary bridge, and *anti* when it is in the one farther away.

*syn*-7-Methylbicyclo[2.2.1]hept-2-ene

## Discussion

Until the advent of the *Z-E* system, there was no *general* way to distinguish *cis-trans* ("geometric") isomers by nomenclature because there was no way to define the necessary point or points of reference except in simple cases. Nevertheless, the *cis-trans* names, which are recommended for use with rings, have long been used to designate configuration about carbon-carbon double bonds and are still acceptable for such use if they are unambiguous (*cf. 14*), as in *cis*-1,2-dichloroethene. Incidentally, configuration about rings may often be described by either the (*R*)-(*S*) system or the systems for naming *cis-trans* isomers, as convenient, and as illustrated in some of the examples in Table 14.1.

Configuration about a carbon-nitrogen double bond has long been denoted by the prefixes *syn* (corresponding to *cis*) and *anti* (corresponding to *trans*). This is still acceptable only when ketone derivatives ($R^1 R^2 C=N-$) are excluded from such naming since the choice between $R^1$ and $R^2$ as the point of reference is then uncertain. The old names for dioxadimes—*i.e., syn* (for *Z,Z*), *anti* (for *E,E*), and *amphi* (for *Z,E* and *E,Z*)—are not recommended. The isomeric diazoates are still often referred to as "normal diazoates" (or less desirably, "diazotates") and "isodia-zoates" without commitment as to structure, but the evidence favors designating the "normal" benzenediazoate ion, for example, as *Z*-benzene-diazoate ion, and the "isodiazoate" as the *E* form.

The prefixes *syn* and *anti* should no longer be used for *cisoid* and *transoid*.

The prefix *endo* was proposed to describe the "inner" location of a group when it is opposite the reference bridge and therefore to some extent "within" the obtuse dihedral angle of the main ring; a group in the *exo* position is "outside" this angle. These terms have been used by most workers in spite of the preference expressed (*15*) for extension of the α,β system (*see* page 107) to bridged rings with larger bridges than in steroids. According to this preference the *endo* position would have been called α, and the *exo* position β.

The rules suggested here for selecting senior bridges and secondary bridges in bridged-ring compounds are new but are believed to state and extend current practice (*16*) for the most part. Some complexity is unavoidable, and modifications by IUPAC are probable.

The notation for *cis-trans* isomers in Beilstein's *Handbuch* (*11*) is peculiar to that publication.

## Table 14.1. Examples Of Acceptable Usage

1.

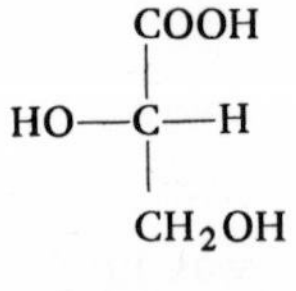

COOH
|
HO—C—H
|
CH₂OH

(*S*)-2,3-Dihydroxypropanoic acid
L-2,3-Dihydroxypropanoic acid

2.

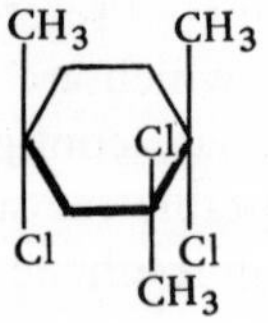

(1*S*,2*S*,4*R*)-Trichloro-1,2,4-trimethyl-
cyclohexane
*r*-1,*t*-2,*c*-4-Trichloro-1,2,4-trimethylcyclo-
hexane

3.

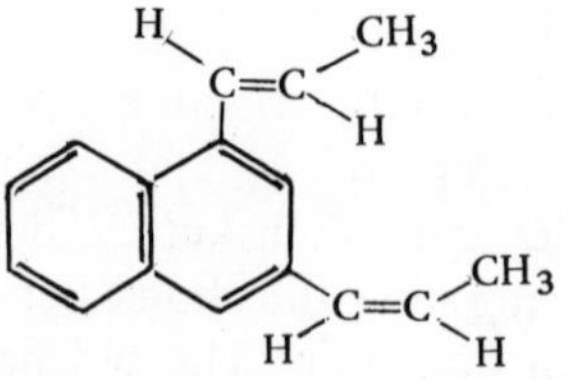

(*E,Z*)-1,3-Di-1-propenylnaphthalene
1-(*trans*-1-Propenyl)-3-(*cis*-1-propenyl)-
naphthalene

4.

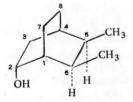

(2R)-(E)-1-Ethylidene-2-methyl-
cyclohexane
(2R)-*trans*-1-Ethylidene-2-methylcyclo-
hexane

5.

*exo*-2-Chlorobicyclo[2.2.1]heptane
(1S,2S,4R)-2-Chloronorbornane

6.

*cis-exo*-2,3-Dichlorobicyclo[2.2.1]heptane
(1S,2R,3S,4R)-*exo*-2,3-Dichloronorbornane

7.

*exo*-2, *endo*-3-Dichlorobicyclo[2.2.1]heptane
(1S,2R,3R,4R)-2,3-Dichloronorbornane

8.

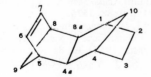

*exo*-5,6-Dimethyl-*endo*-bicyclo[2.2.2]-
octan-2-ol

9.

1,2,3,4,4a,5,8,8a-Octahydro-*exo*-1,4:*exo*-
5,8-dimethanonaphthalene

10.

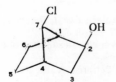

*syn*-7-Chloro-*exo*-bicyclo[2.2.1]heptan-2-ol
*syn*-7-Chloro-*exo*-2-norbornanol
(1R,2S,4S,7R)-7-Chloro-2-norbornanol

11.

*syn*-Bicyclo[2.2.1] hept-2-ene-7-carboxylic acid
*syn*-2-Norbornene-7-carboxylic acid

12.

*anti*-Bicyclo[3.2.1] octan-8-amine

13.

(Z)-1-[(R)-3-Methylcyclohexylidene] - 2-phenylhydrazine
(R)-3-Methylcyclohexanone (Z)-phenylhydrazone   (old name)

## Literature Cited

1. "IUPAC Tentative Rules for the Nomenclature of Organic Chemistry, Section E.; Fundamental Stereochemistry," *J. Org. Chem.* (1970) **35**, 2849.
2. Eliel, E. L., *J. Chem. Ed.* (1971) **48**, 163 and references therein.
3. (a) Cahn, R. S., *J. Chem. Ed.* (1964) **41**, 116, 503 and earlier articles; (b) Cahn, R. S., Ingold, C. K., Prelog, V., *Angew. Chem., Intern.Ed. Engl.* (1966), **5**, 385, 511.
4. "Rules of Carbohydrate Nomenclature" (an approved report of committees), *J Org. Chem.* (1963) **28**, 281.
5. "Addendum to Definitive Rules for the Nomenclature of Natural Amino Acids and Related Substances" (ACS), *J. Org. Chem.* (1963) **28**, 291.
6. "The Nomenclature of Cyclitols" (CON-CBN Tentative Rules), *Eur. J. Biochem.* (1968) **5**, 1; *Biochem. Biophys. Acta* (1968) **165**, 1; *Arch. Biochem. Biophys.* (1968) **128**, 269.
7. "Definitive Rules for the Nomenclature of Amino Acids, Steroids, Vitamins, and Carotenoids" (IUPAC), *J. Amer. Chem. Soc.* (1960) **82**, 5575.
8. Lespieau, R., *Bull. Soc. Chim. Fr.* (1895) (3) **13**, 105; Maquenne, L., "Les Sucres et leur Principaux Dérivés," p. 190, Gauthier-Villars, Carre et Naud, Paris, 1900. Cahn, R. S., Ingold, C. K., Prelog, V., *Experientia* (1956) **12**, 81; McCasland, G. E., Furuta, S., Johnson, L. F., Shoolery, J. N., *J. Amer. Chem.Soc.* (1961) **83**, 2335.
9. McCasland, G. E., "A New General System for the Naming of Stereoisomers," Chemical Abstracts, Columbus, Ohio, 1950; Klyne, W., *Chem. Ind. (London)* (1951) 1022.
10. "The Nomenclature of Lipids" (CBN Tentative Rules), *Eur. J. Biochem.* (1967) **242**, 4845.
11. "Beilstein's Handbuch der organischen Chemie," 4th ed., 3d suppl., Vol 6, part 8, p. IX, 1967.
12. Blackwood, J. E., Gladys, C. L., Loening, K. L., Petrarca, A. E., Rush, J. E., *J. Amer. Chem. Soc.* (1968) **90**, 509.

13. Blackwood, J. E., Gladys, C. L., Petrarca, A. E., Powell, W. H., Rush, J. E., *J. Chem. Doc.* (1968) **8**, 30.
14. IUPAC, Comptes Rendus of the 16th Conference (1951) and the 17th Conference (1953); Epstein, M. B., Rossini, F. D., *Chem. Eng. News* (1948) **26**, 2959; Crane, E. J., *Chem. Eng. News* (1949), **27**, 1303
15. "System of Nomenclature for Terpene Hydrocarbons" (ACS), *Advan. Chem. Ser.* (1955) **14**, 4.
16. Alder, K., Wirtz, H., Koppelberg, H., *Ann. Chem.* (1956) **601**, 138.

# 15

# Common Errors and Poor Practices

A chapter on errors in nomenclature may seem out of place in a guidebook that is devoted to correct principles and procedures. It is axiomatic that one needs only to follow these to avoid errors. This is not always easy, however, since nomenclature is a growing subject. With choices available among the several systems of nomenclature that have been developed, judgment is required in selecting a good name.

Rules are usually concerned with procedures to follow rather than pitfalls to avoid. Also, rules may pertain to a particular system under discussion and say nothing about other systems which also may be acceptable. Hence, this chapter treats some of the common deviations from rules with the hope that it may prove helpful to a better appreciation of good usage.

Fortunately, most of the methods that have enjoyed widespread acceptance have common threads running through them, thus making many nomenclature systems use substitutive prefixes attached to a term that specifies a parent compound. To illustrate: $(CH_3)_4C$ is named tetramethylmethane by the "simple-nucleus" system and dimethyl-propane by the Geneva or IUC or IUPAC system. Another example is $(C_2H_5)_2CHCOOH$, which is named diethylacetic acid by the first plan, 2-ethylbutanoic acid by IUC rules, and 2-ethylbutanoic acid or 2-ethyl-butyric acid by the IUPAC system. One may ask if it is an error to use the simple-nucleus names in view of the later IUPAC rules. Some will maintain that it is, but most chemists would agree that either name is intelligible. It is more a matter of style than of error, and styles change with time. Some names, formerly acceptable, are now so out of style that they are practically abandoned; one is α-$n$-amylene, which now is almost universally replaced by 1-pentene.

Another related example is the caproic-hexanoic pair of names. IUPAC has retained in its system the common names for acids from formic to valeric but has ruled against caproic, caprylic, and capric, primarily because of its decision to change the ending designating many acyl groups from $yl$ to $oyl$. This makes "caproyl" impossible in the system since this name would now represent the acyl groups from both caproic and capric acids. Although use of "caproic acid" cannot be classed as an error, it certainly does not represent good current usage. The fact that IUPAC excluded caproic in favor of hexanoic is increasing the rate at which the former is going out of style.

## Selection of a Principal Function

Organic nomenclature systems have been based generally on function-ality; hence, naming all aliphatics as alkanes or all monocyclic aromatics as derivatives of benzene would be major errors. Chemists are so accustomed to the terms phenol and 2-butanol that the terms "hydroxybenzene" and "2-hydroxybutane" are almost never encountered, but unfortunately other deviations of this kind in naming relatively simple structures are not unusual. One may select almost any issue of a chemical journal and find such lapses. Structures 1 and 2, taken from recent articles, were

$H_2NCH_2CHOHCH_2CH_2CH_2NH_2$

1

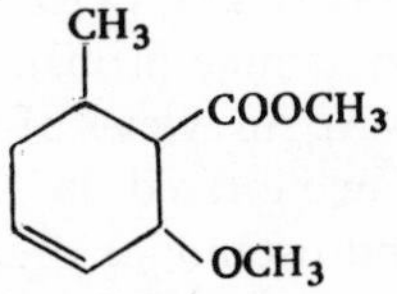

2

assigned names ending in *pentane* and *cyclohexene,* respectively, and structures 3 and 4 were given names ending in *ethylene* and *hexane,*

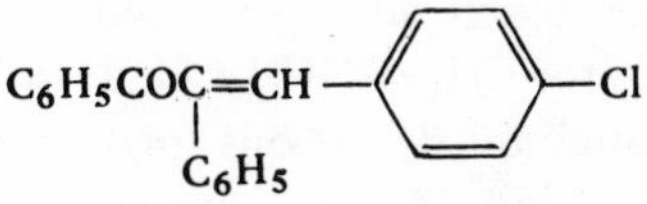

3

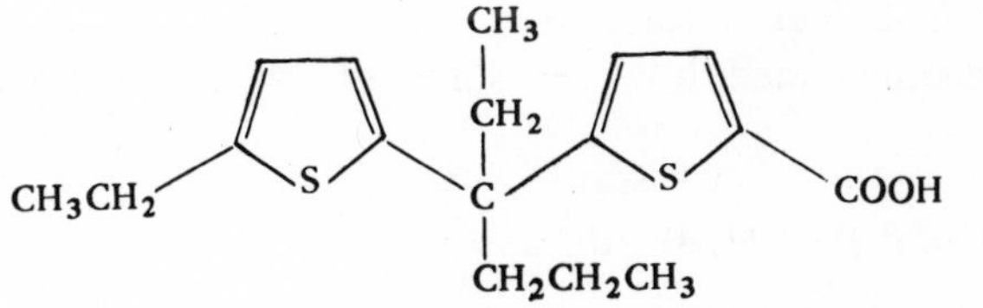

4

respectively. The preferred name for 1 should end in *pentanol:* 1,5-dia-mino-2-pentanol. Since 2 is an *ester,* its name should follow ester terminology: methyl 2-methoxy-6-methyl-3-cyclohexene-1-carboxylate. The ketone function in structure 3 calls for a name ending in *one:* 3-(4-chlorophenyl)-1,2-diphenyl-2-propen-1-one. In conformity with published rules, compound 4 should be named as an *acid,* not as an alkane: 5-[1-ethyl-1-(5-ethyl-2-thienyl)butyl]-2-thiophenecarboxylic acid.

It is industry's privilege to coin any trade name it pleases as long as the name does not represent chemical terminology. Sevin, Orlon, and Acrilan are good trade names, but it is regrettable when erroneous pseudochemical terms are so coined. An example of confusing usage is Telone fumigant, in which the -one ending implies the presence of a ketone, whereas the substance is actually a mixture of chlorinated propenes.

### Selection Among Principal Functions

Although it deviates from established practice to omit designation of a principal function at the end of a name, it is equally bad to include more than one such ending. Thus, "5-hexanonoic acid" is not acceptable for $CH_3COCH_2CH_2CH_2COOH$. The correct name is 5-oxohexanoic acid. The term ethanolamine probably is here to stay, but it is not good systematic nomenclature in view of the avoidable second suffix. The successive suffixes encountered in some names, such as "androstanolone" (for hydroxyandrostanone), are in this same objectionable category.

An order of precedence of functionality was set forth many years ago, somewhat arbitrarily to be sure, to permit a decision on the preferred principal function among two or more possibilities (see Chapter 10). Near the top of the list is carboxylic acids. Therefore, compounds containing this function are generally named as acids, even when other functions are present. For example, the names "3-carboxy-1-propene" and "3-carboxy-1-propanol" would be much less preferred than 3-butenoic acid and 4-hydroxybutanoic acid for $CH_2=CHCH_2COOH$ and $HOCH_2CH_2CH_2COOH$, respectively. Less unusual are deviations involving two functions that are reasonably close in the order of precedence. Thus, in keeping with the fact that alcohol is listed a little ahead of amine, 1,5-diamino-2-pentanol was the name selected above for 1, but the synonym 2-hydroxy-1,5-pentane-diamine might be better for some purposes. Following the listed order is generally recommended, however, since it helps to promote uniformity in naming.

### Single-Word and Multiple-Word Names

One of the most common errors encountered is use of spaces in names that should have none, or no spaces in names requiring them. An example is "alkyl benzene" and "alkyl benzene sulfonic acid" for alkylbenzene and alkylbenzenesulfonic acid, respectively.

If a hydrogen atom of a compound is replaced by some other atom or group, the derived compound is named by prefixing directly the name of the substituent to the name of the unsubstituted compound. For example, 3-ethylpyridine is correct since pyridine is the name of the compound into which ethyl is substituted. On the other hand, butyl alcohol is a two-word name since "alcohol" is not a compound (see Chapter 13). Radicofunc-tional names of ketones, ethers, sulfides, sulfoxides, sulfones, acetals, acid anhydrides and glycosides also contain spaces: dimethyl sulfide, methyl

ethyl ketone, benzaldehyde diethyl acetal, methyl $\alpha$-D-galactoside, etc. Oxide is a separate word in names such as pyridine $N$-oxide.

The term *amine* is not used in radicofunctional names since it is simply a short form for the name *ammonia*—itself a compound. Thus, the name of a substituent is prefixed to the parent name: methylamine, not methyl amine.

A deviation from acceptable usage that is frequently encountered is contraction of a two-word name to a one-word name. An obvious error is methylethyl ketone for methyl ethyl ketone. Such mistakes are most common in class names: "orthoesters" for ortho esters, "thioacids" for thio acids, "chloroacids" for chloro acids, "aminoacids" for amino acids, and "peracids" for peroxy (or per) acids. Additional guidance on whether to space or not is presented by Hurd (*1*).

## Hyphens

Hyphens are used in a chemical name to maintain its status as a single-word name: 2-propanol, not "2 propanol." Frequently, in writing, a dash is used to separate the names of two chemicals in a mixture—*e.g.,* naphthalene—biphenyl. If one of the two chemicals has a two-word name (as acetic acid), this would give rise to three words with only one dash—*e.g.,* naphthalene—acetic acid. Then, mistaking the dash for a hyphen, a reader might interpret "naphthalene—acetic" as a one-word name, equivalent to naphthaleneacetic. This misunderstanding can be avoided by use of a slash mark (solidus), *e.g.,* naphthalene/acetic acid. In describing solvent mixtures, it is best to omit the dash if two-word names are involved. For example, the phrase "crystallization from a 1:1 mixture of benzene and ethyl acetate" is better than "crystallization from 1:1 benzene—ethyl acetate".

## Relationship of Locants to the Principal Function

Unless a locant in a single-word name is contained within parentheses or brackets, it refers to the parent compound that bears the principal function. Thus, both 2 and 1 in 2-nitro-1-butanol refer to locations of groups on a butane skeleton. 1-Chloro-3-propylbenzene is 1-ClC$_6$H$_4$CH$_2$-CH$_2$CH$_3$-3 whereas (1-chloropropyl)benzene is C$_6$H$_5$CHClCH$_2$CH$_3$. In the latter name the parentheses specify that the 1 refers to a position in propyl, not in benzene. Primed numbers or letters are not used within parentheses unless the complexity of the group within the parentheses demands it. In a comparable example, 1-ClC$_6$H$_4$CHClCH$_2$CH$_3$-4 is called 1-chloro-4-(1-chloropropyl)benzene and not 1-chloro-4-(1'-chloropropyl)-benzene. Primed numbers or letters are used as necessary, of course, to designate locations in a parent compound, as in $N$-methyl-$N'$-phenylurea for CH$_3$NHCONHC$_6$H$_5$.

The structure 4-BrC$_6$H$_4$CH$_2$CH$_2$CH$_2$COOH is named 4-(4-bromophenyl)butanoic acid or 4-(4-bromophenyl)butyric acid. Similarly, 4-BrC$_6$H$_4$CH$_2$COOH is correctly named (4-bromophenyl)acetic acid whereas the name 4-bromophenylacetic acid is objectionable on technical grounds because the 4 now refers to acetic (even though acetic lacks this number of carbon atoms) rather than to phenyl. The name $p$-bromo-phenylacetic acid would be less open to this objection since $p$- is never used for aliphatic compounds. Strictly speaking, a familiar example of a "single-word name" is open to this kind of objection. Instead of 2,4-dinitrophenylhydrazine, the name should be (2,4-dinitrophenyl)-hydrazine. Omission of parentheses in this name is almost universally the custom, however, because no ambiguity results.

Another illustration is the name for structure 5. The name 2-naphthyl-acetic acid is somewhat objectionable since it could also

5                                    6

designate structure 6, in which the position of the —CH$_2$COOH group is unspecified. (The 2 designates a position in acetic acid when parentheses are omitted.) Structure 5 would be accurately named (2-naphthyl)acetic acid by use of a substitutional prefix, or it could be given the conjunctive name 2-naphthaleneacetic acid.

*Esters*

Ethyl benzoate, butyl hexanoate and methyl benzenesulfonate are good names for simple esters. Errors are met frequently, however, in naming families of esters of specific acids. A good name for CH$_3$COOR where R is any alkyl group is an acetic ester, just as CH$_3$COOH is acetic acid, or CH$_3$COOM (M = a positive ion) is an acetic salt. It is not good style or logic to speak of CH$_3$COOR as an "acetate ester." It is an acetic derivative, not an acetate derivative. It is an acetate, to be sure, but not an "acetate ester."

An acetoacetic ester, CH$_3$COCH$_2$COOR, is an example where one almost never encounters error. The term malonic esters, CH$_2$(COOR)$_2$, is another illustration. In view of the almost universally correct usage of these two terms it is remarkable how frequently one encounters errors in the generalized names for esters of other acids. Thus, one finds "nitrate ester" instead of nitric ester, "phosphate ester" instead of phosphoric ester, "benzenesulfonate ester" instead of benzenesulfonic ester, "sulfate ester" instead of sulfuric ester, "stearate ester" instead of stearic ester, etc.

### Other Acid Derivatives

Names of salts of acids are derived in the same way as those of esters. Thus, a salt of benzoic acid may be called a benzoate but not a benzoate salt. If the general term salt is to appear in the name, the correct form is benzoic salt. A quaternary ammonium salt of hydrochloric acid may be called a chloride but not a chloride salt. A barbiturate may be referred to as a barbituric salt, or a picrate as a picric salt.

Thus, such terms as acetic acid ester or acetic acid anhydride or 1,2-cyclohexanedicarboxylic acid anhydride do not represent good usage because the proton of the acid is no longer present. Likewise, structure 7 should not be called an

$$C_2H_5-CH-CH_2-CH_2$$

(structure 7)

(structure 8, with N—COOR and N—COOR)

**7**                                    **8**

acid lactone. The situation is similar for amides. One should avoid the expression "acid amide" although there is no objection to the periphrase amide of an acid. The terms salicylamide or salicylic amide are acceptable, but "salicylic acid amide" is not. Structure 8 should not be referred to as "hexahydro-1,2-pyridazinedicarboxylic acid esters" since the word "acid" should not be included in the name. There are acid esters, of course, just as there are acid salts. One is ethyl hydrogen oxalate, comparable with sodium hydrogen oxalate.

### Alkyl Groups

Terms such as "2-butyl" for $CH_3CH_2(CH_3)CH-$ and "5-(diethyl-amino)-2-pentyl" for $(C_2H_5)_2NCH_2CH_2CH_2(CH_3)CH-$ represent deviations from rules since the position of attachment of an alkyl group is always numbered 1. A correct name for the first is *sec*-butyl or 1-methylpropyl; the name for the second is 4-(diethylamino)-1-methylbutyl.

In the Geneva, IUC, and IUPAC systems an alkyl group is assumed to be unbranched unless specified otherwise. The 1965 IUPAC Rules utilize *sec*, *tert*, iso, and neo in certain instances but never *n*- for *normal*. Thus, the name 2,2-dipropylpentanoic acid is correct, whereas 2,2-di-*n*-propylpentanoic acid is to be avoided.

### Hybrid Names

One of the most common errors encountered is the use of names derived partly from one system of nomenclature and partly from another. Isopropyl alcohol is a correct name in one system and 2-propanol is its Geneva-IUPAC equivalent. "Isopropanol" is neither but is an objectionable hybrid name. "Isobutene" is another example; the name should be either isobutylene or methylpropene. "*tert*-Butanol" is wrong; the compound should be named 2-methyl-2-propanol or *tert*-butyl alcohol. "*n*-Butanol" is sometimes wrongly used to mean 1-butanol.

The expression "*n*-butenes" is similarly faulty because it suggests that there are other kinds of butenes. Thus, this term implies that $(CH_3)_2C{=}CH_2$ is a butene, which it is not; the compound may be regarded as a "butylene" since it has four carbons, but it is correctly named 2-methylpropene.

One sometimes sees wrongly constructed names such as "3-pyrrole-aldehyde." When an aldehyde group attached to a ring must be designated as the principal function, it is represented by the suffix *carbaldehyde* or *carboxaldehyde*. Thus, the compound referred to would be better called 3-pyrrolecarbaldehyde or 3-pyrrolecarboxaldehyde.

### Carbinol and Carbinyl

"Carbinol," a simple-nucleus parent name, is no longer acceptable. It was once widely used to name secondary and tertiary alcohols. "Triphenylcarbinol," for example, is now named triphenylmethanol by IUPAC rules. Since the use of "carbinol" was never recognized in the older Geneva and IUC rules and since it is specifically ruled against in the present IUPAC rules, the same proscription should apply to the use of "carbinyl" as a substituting-group name. There is really no need for the term since methyl is better. Instead of "diphenylcarbinyl," one should use diphenylmethyl.

### Acid Tautomers of Nitroalkanes

Such compounds may be named with the prefix *aci-* as in *aci*-nitro-phenylmethane for the structure $C_6H_5CH{=}N(O){-}OH$. Its anion is the resonance hybrid 9. Two good names for the sodium salt would be sodium *aci*-nitrophenylmethane and sodium nitrophenylmethanide. The name

$$C_6H_5\overset{-}{C}HNO_2 \longleftrightarrow C_6H_5CH{=}N(O)\overset{-}{O}$$

9

"sodium phenylmethanenitronate" is indefensible although occasionally one sees these *aci* forms referred to as "nitronic acids."

### Operational Prefixes

The prefixes *cyclo, deoxy, oxa,* and others differ from substitutive prefixes in that they denote an operation on a parent structure to form a new parent structure. The prefix in a fusion name, such as *benz* in benz[a]anthracene, also is an operational prefix.

An operational prefix should be attached directly to the parent name it modifies and not placed among the substitutional prefixes. *Deoxy* is no exception although some rules of nomenclature recommend that it be listed alphabetically with the substitutive prefixes. An example of a poor name from a recent paper is methyl 2-amino-2,3-dideoxy-3-mercapto-D-*altro*-pyranoside. A more defensible name would be methyl 2-amino-3-mercapto-D-2,3-dideoxy-*altro*-pyranoside, as shown by the following discussion.

Substitutive prefixes may be detached from the compound's name and still leave a sensible name. For example, one may remove *chloro* from chloronitrobenzene and retain the name nitrobenzene. If operational prefixes are mixed alphabetically with substitutive prefixes and if one then removes the operational prefix only, absurdities may result. Consider as an illustration β-D-2-deoxyribose (**10**), which differs from β-D-ribose in

10                                11

that no oxygen is on carbon 2, but two hydrogens instead. It is a parent compound comparable with cyclohexane or benz[a]anthracene. (This feature makes β-D-2-deoxyribose more defensible as a name than 2-deoxy-β-D-ribose.)   The name 2,2-dimethyl-β-D-2-deoxyribose would clearly represent **11**, in which both of the hydrogens have been replaced by methyl groups, whereas the name "2-deoxy-2,2-dimethyl-β-D-ribose" (alphabetic order of deoxy and methyl) would not make sense if "2-deoxy" were removed.

## Misuse of Substitutive Prefixes

Errors may result if a substitutive prefix is used as though it were part of the root name. Two examples indicate the nature of the problem. The structure $CH_2=CHCH_2COOH$ may be called either vinylacetic acid or 3-butenoic acid. Diketene (**12**) is a structural derivative of this acid that has sometimes been named "vinylaceto-β-lactone." This name is poorly conceived, however, just as the name "β-bromovinylacetic acid" is poorly conceived for $CH_2=CBrCH_2COOH$. The parent compound, acetic acid, has no β-carbon, and one should not be assigned to it by way of the vinyl

$$H_2C=\!\!\!\overset{|}{\underset{|}{C}}\!\!-\!\!O$$
$$H_2C-\!\!CO$$

=CHCOOH

COOH

12                              13                        14

prefix. A good name for diketene as a lactone would be 3-hydroxy-3-butenoic lactone.

Structure **13** is correctly named cyclohexylideneacetic acid, but structure **14** should not be called α,β-epoxycyclohexylideneacetic acid for two reasons: (a) *epoxy*, a substitutive prefix, should not be used to designate addition, and (b) the prefix *epoxy* should not be used to cite a group jointly held by the parent compound and a substituent group. A good name for **14** would be α,1-epoxycyclohexaneacetic acid. A spiro name would be even better.

### Noncapitalization of Structural Prefixes

Many structural prefixes, attached to a name by a hyphen and usually appearing in different type in print, are never capitalized at the beginning of a sentence whether abbreviated or written out in full. The list includes *ortho-*, *meta-*, *para-*, *syn-*, *anti-*, *cis-*, *trans-*, and such abbreviations as *n-*, *sec-*, *tert-*, *sym-*, *o-*, *m-*, *p-*, and others. Correct usage at the beginning of a sentence is:

> *n*-Butyl bromide is . . .          *not N*-butyl bromide is . . .
> *m*-Xylene is . . .                 *not M*-xylene is . . .
> *meta*-Xylene is . . .              *not Meta*-xylene is . . .

Similarly, a prefix that is a capital letter does not serve as the initial capital letter at the beginning of a sentence:

> *N*-Bromosuccinimide is . . .      *not N*-bromosuccinimide is . . .

It makes a difference, of course, if the adjective form is used rather than the prefix form for structural terms. If used, the adjective would not be attached by a hyphen or italicized, and it *would* be capitalized at the beginning of a sentence—*e.g.*, "Normal alkyl bromides are . . .," or "Trans addition is. . . ."

### Conclusion

Present textbooks, journal articles, and chemical catalogs show a striking improvement in organic nomenclature over their counterparts of a few decades ago, but there is still a long road to travel before any perfect state is reached. The secret of proper naming lies in the ability to analyze and to classify a structure correctly before trying to name it. Once classified, the compound should then be named in accordance with rules governing that class. All the various functional groups in a structure must be considered, and from among them a principal function must be selected. If there is no good reason otherwise, the principal function should be the one highest in the order of precedence of functions (Table 8.1); other functions are cited as prefixes.

Sometimes it is helpful to practice naming a simpler structure in the same family before starting on the complex structure at hand. Two examples, **15** and **16**, provide illustrations. Since structure **15** is both a cyclopentene and an ester, it should be named as an ester. The knowledge

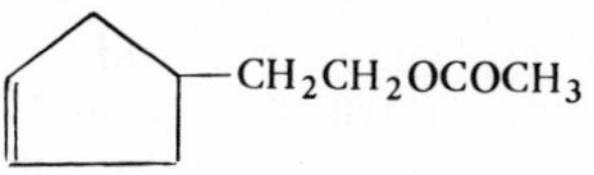

15 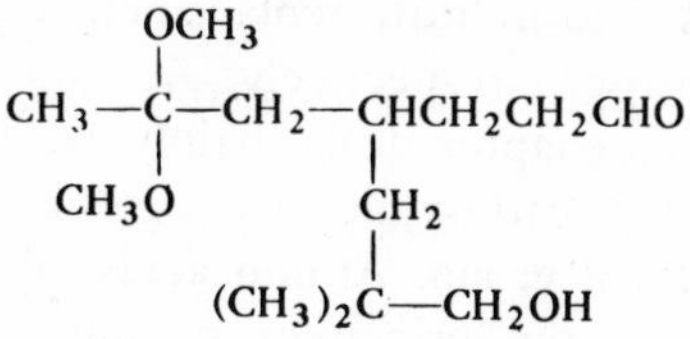                                          16

that $CH_3CH_2OCOCH_3$ is named ethyl acetate provides a parallel style for naming **15** as 2-(3-cyclopentenyl)ethyl acetate.

Structure **16** has many substituents, but one should not be awed by them. Classification is simple enough into aldehyde, alcohol, and acetal functions. Of these, aldehyde takes precedence, and it is readily established that the structure is simply a tetrasubstituted heptanal. A good name, therefore, is 4-(2,2-dimethoxypropyl)-7-hydroxy-6,6-dimethylheptanal.

## Literature Cited

1. C. D. Hurd, *J. Chem. Educ.* (1961) **38**, 43.

# 16

## Acids (Carboxylic, Thiocarboxylic, and Imidic)

Compounds containing one or more carboxy groups (COOH) are called carboxylic acids. The class names **thiocarboxylic acid** and **imidic acid** are used to indicate replacement of oxygen in the COOH group by sulfur and nitrogen, respectively.

This chapter deals mainly with compounds in which a carboxylic or modified carboxylic acid group is the principal and only kind of functional group. **Amino acids,** although they may be named systematically by the principles recommended below, constitute a subclass of sufficient importance to justify a specialized nomenclature (*see* Chapter 30).

*Recommended Nomenclature Practice*

**Carboxylic Acids.** Acyclic **monocarboxylic acids** are named substitutively by adding the suffix *oic acid* to the name of the corresponding hydrocarbon, the final *e* of the hydrocarbon name being elided.

| | |
|---|---|
| $CH_3CH_2CH_2CH_2COOH$ | Pentanoic acid |
| $CH_3CH=CHCOOH$ | 2-Butenoic acid |

Acyclic **dicarboxylic acids** are named similarly, using the suffix *dioic acid* but retaining the final *e* of the hydrocarbon name.

| | |
|---|---|
| $HOOCCH_2CH_2CH_2CH_2COOH$ | Hexanedioic acid |

Cyclic carboxylic acids are named substitutively by adding the suffixes *carboxylic acid, dicarboxylic acid, tricarboxylic acid,* etc. to the name of the ring system to which the COOH group(s) are attached.

Cyclohexanecarboxylic acid

2,7-Naphthalenedicarboxylic acid

This type of substitutive nomenclature is also used for **acyclic polycarboxylic acids** in which more than two COOH groups are attached to the same straight chain since it permits like treatment of the maximum number of COOH groups.

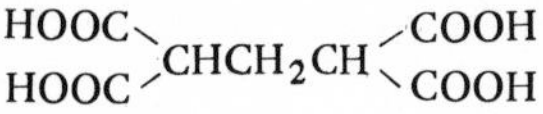

1,1,3,3-Propanetetracarboxylic acid

*not* 2,4-Dicarboxypentanedioic acid

When not all the COOH groups in a polycarboxylic acid can be treated by substitutive nomenclature as recommended above or when another functional group standing higher in the Order of Precedence of Functions (Table 8.1) is present, the terms *carboxy, dicarboxy, tricarboxy,* etc. are used as prefix names.

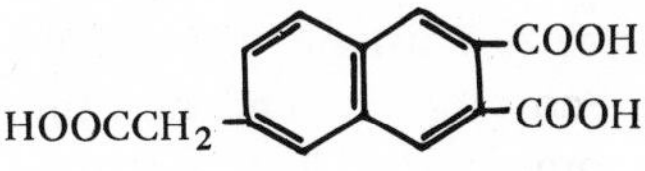

6-(Carboxymethyl)-2,3-naphthalenedicarboxylic acid

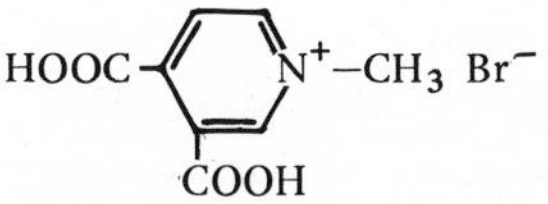

3,4-Dicarboxy-1-methylpyridinium bromide

Several of the more common carboxylic acids have acceptable trivial names. Some of these appear in Table 16.1. For a more complete list of recognized trivial names for carboxylic acids, *see* the 1965 IUPAC Definitive Rules, Section C-404.1, Table VI.

**Thiocarboxylic Acids.** Carboxylic acids in which one or more oxygen atoms of one or more COOH groups have been replaced by S are named according to the above principles, using the suffixes *thioic acid, dithioic acid, carbothioic acid, and carbodithioic acid.* To avoid ambiguity, the multiplying prefixes *bis, tris, tetrakis,* etc. are used with these suffixes. To distinguish between CSOH and COSH, the terms *O*-acid and *S*-acid should be used.

When a thiocarboxylic acid group must be named as a substituent, the terms *thiocarboxy* and *dithiocarboxy* are used as prefix names to designate COSH⇌CSOH and CSSH, respectively. When necessary, distinction between COSH and CSOH should be made with the prefix names *mercaptocarbonyl* and *hydroxy(thiocarbonyl); see* Table 16.1, examples 34-42.

**Imidic Acids.** Carboxylic acids in which the carbonyl oxygen atom of the COOH group has been replaced by =NH are named similarly to their oxygen analogs, using the suffixes *imidic acid* and *carboximidic acid.*

When the C(=NH)OH group must be named as a substituent, the term *imidocarboxy* is used as a prefix name.

For typical names of imidic acids *see* examples 43-46 (Table 16.1).

*Discussion*

Since substitutive nomenclature is the preferred method for naming carboxylic acids, as well as their sulfur and nitrogen analogs, substitutive names appear first for each of the examples listed in Table 16.1. It is worth noting that two kinds of substitutive names are applied to acyclic carboxylic acids—one using the suffix *oic acid,* and the other, *carboxylic acid.* Since *oic acid* designates the groups =O and OH both attached to the same carbon atom, this combined attachment must always be to a terminal carbon in a chain; hence, locants are not usually needed for COOH groups named in this manner (*see* examples 18-22). On the other hand, when the suffix *carboxylic acid* is used, locants are usually required (*see* example 30 ).

Although number locants should always be used in substitutive names, Greek letters have traditionally been used in trivial names of acyclic carboxylic acids to designate positions of substituents. Since 1947, however, *Chemical Abstracts* has abandoned such use of Greek letter locants, except in conjunctive names (*see* example 32). This change has resulted in trivial names such as 2-chloropropionic acid, which is now preferred to α-chloropropionic acid. It must be remembered that when Greek letters are used as locants, the lettering begins with a carbon atom *adjacent* to a COOH group, whereas when number locants are used, the numbering begins with the carbon *of* a COOH group. Unfortunately, when the change to number locants first took place, there was some usage in print whereby numbering was started at the α-carbon atom. This can, of course, lead to confusion if one is using the older literature and is not aware of the existence of such outdated names.

In the Subject Indexes of *Chemical Abstracts,* conjunctive names (*see* Chapter 9) are used as preferred entries for acids having one or more COOH groups attached to a chain connected to a ring system—e.g., 2-naphthaleneacetic acid for 2-naphthylacetic acid. This practice, also used with analogous compounds of other functional classes, has as its purpose the placement of this type of acid in the Subject Indexes at the name of the ring system rather than at the name of the acyclic acid—*i.e.,* at "Naphthalene" rather than at "Acetic acid."

Many trivial names for carboxylic acids have appeared in print over the years, but the present trend in scientific publications is to abandon these in favor of systematic names. Table 16.1 shows that except for familiarity (limited to a few cases) and compactness, trivial names have no advantages.

In naming the three isomeric benzenedicarboxylic acids, the trivial names "*o*-phthalic acid," "*m*-phthalic acid," and "*p*-phthalic acid" should not be used. By definition, phthalic acid is the 1,2-isomer, and the prefixes *o, m,* and *p* are not applicable. The trivial names isophthalic acid and terephthalic acid are acceptable for the 1,3- and 1,4-isomers, respectively.

Finally, pseudosystematic names such as "undecylenic acid," "heptylic acid," and "nonoic acid," which are incorrectly formed, should not be used at all.

The prefixes *thiolo* and *thiono* have found some usage for differentiating between the tautomeric forms of the thiocarboxy group, COSH⇄ CSOH, as in "thioloacetic acid" and "thionoacetic acid." Use of the suffixes *thiolic acid* and *thionic acid* has also been suggested for the same purpose —*e.g.,* "hexanethiolic acid" for $C_5H_{11}COSH$. Neither practice has received official sanction. When necessary, distinction between such tautomers should be made as recommended above by prefixing $S$ or $O$ to the word *acid—e.g.,* hexanethioic $S$-acid or hexanethioic $O$-acid.

It is to be noted that the C(=NH)OH group is tautomeric with the $CONH_2$ group—*i.e.,* an imidic acid may be regarded as the "enol" form of the corresponding amide. Although free imidic acids are generally unstable, their esters are well known, and names for the acids are therefore needed.

Ortho acids, the hypothetically hydrated forms of carboxylic acids— *i.e.,* $RC(OH)_3$ —are best named as *triols* (*see* Chapter 19).

In addition to the above-mentioned prefix names designating the several kinds of acid groups, names for substituting groups formed conceptually by removal of OH from COOH, CSOH, or C(=NH)OH have found wide usage: *acetyl* for $CH_3CO$, *benzoyl* for $C_6H_5CO$, etc. Such acyl prefix names are formed by changing the corresponding acid name suffix *ic* to *yl* or *oyl.* For other oxygen-containing acyl groups, as well as their sulfur and nitrogen analogs, systematic names should be used: *propanoyl* for $C_2H_5CO$; *butanoyl* for $n$-$C_3H_7CO$; *ethanethioyl* for $CH_3CS$; *propanimidoyl* for $C_2H_5C(=NH)$; etc. (*see also* Chapter 17).

## Table 16.1. Examples of Acceptable Usage

| | | |
|---|---|---|
| 1. | HCOOH | Methanoic acid<br>Formic acid |
| 2. | $CH_3COOH$ | Ethanoic acid<br>Acetic acid |

3.  $CH_3CH_2COOH$

Propanoic acid
Propionic acid

4.  $CH_3CH_2CH_2COOH$

Butanoic acid
Butyric acid

5.  $CH_3CHCOOH$
         $|$
         $CH_3$

2-Methylpropanoic acid
Isobutyric acid

6.  $CH_3CH_2CH_2CH_2COOH$

Pentanoic acid
Valeric acid

7.  $CH_3CHCH_2COOH$
         $|$
         $CH_3$

3-Methylbutanoic acid
Isovaleric acid

8.  $CH_3CH_2CH_2CH_2CH_2COOH$

Hexanoic acid

9.  $CH_3(CH_2)_{10}COOH$

Dodecanoic acid
Lauric acid

10. $CH_3(CH_2)_{12}COOH$

Tetradecanoic acid
Myristic acid

11. $CH_3(CH_2)_{14}COOH$

Hexadecanoic acid
Palmitic acid

12. $CH_3(CH_2)_{16}COOH$

Octadecanoic acid
Stearic acid

13. $CH_2=CHCOOH$

Propenoic acid
Acrylic acid

14. $HC\equiv CCOOH$

Propynoic acid
Propiolic acid

15. $CH_3CH=CHCOOH$

2-Butenoic acid
Crotonic acid (*trans*)
Isocrotonic acid (*cis*)

16. $CH_2=CCOOH$
           $|$
           $CH_3$

2-Methylpropenoic acid
Methacrylic acid

17. $CH_3(CH_2)_7CH=CH(CH_2)_7COOH$

9-Octadecenoic acid
Oleic acid (*cis*)

18. $HOOCCOOH$

Ethanedioic acid
Oxalic acid

19. $HOOCCH_2COOH$

Propanedioic acid
Malonic acid

20.  $HOOCCH_2CH_2COOH$

Butanedioic acid
Succinic acid

21.  $HOOCCH_2CH_2CH_2COOH$

Pentanedioic acid
Glutaric acid

22.  $HOOC(CH_2)_4COOH$

Hexanedioic acid
Adipic acid

23.  $C_6H_5COOH$

Benzenecarboxylic acid
Benzoic acid

24. 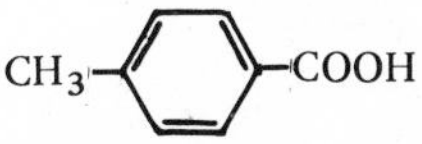  $CH_3$—⟨benzene ring⟩—COOH

4-Methylbenzenecarboxylic acid
*p*-Toluic acid

25. 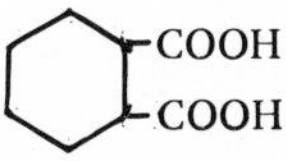  ⟨cyclohexane ring⟩—COOH / —COOH

1,2-Cyclohexanedicarboxylic acid

26. 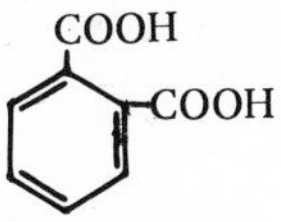  ⟨benzene ring⟩ COOH / —COOH

1,2-Benzenedicarboxylic acid
Phthalic acid

27. 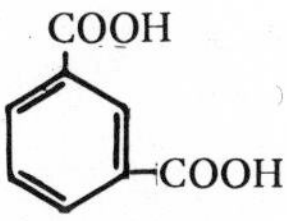  ⟨benzene ring⟩ COOH / —COOH

1,3-Benzenedicarboxylic acid
Isophthalic acid

28. 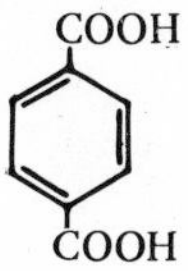  ⟨benzene ring⟩ COOH / COOH

1,4-Benzenedicarboxylic acid
Terephthalic acid

29. 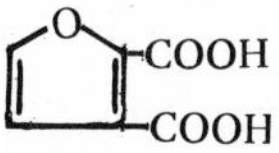  ⟨furan ring⟩—COOH / —COOH

Oxacyclopenta-2,4-diene-2,3-
  dicarboxylic acid
2,3-Furandicarboxylic acid

30.  $HOOCCH_2CHCH_2COOH$
              $|$
            $COOH$

1,2,3-Propanetricarboxylic acid

31. 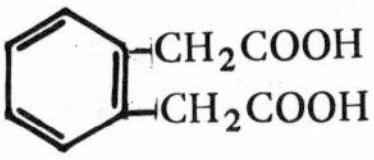  ⟨benzene ring⟩—$CH_2COOH$ / —$CH_2COOH$

1,2-Phenylenediethanoic acid
1,2-Benzenediacetic acid

32.

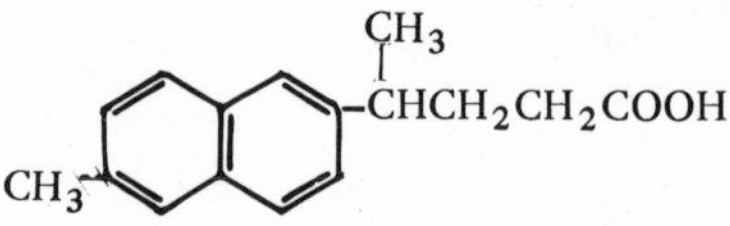

$$CH_3$$

(naphthalene structure with) $-CHCH_2CH_2COOH$

$CH_3-$

4-(6-Methyl-2-naphthyl)pentanoic acid
γ,6-Dimethyl-2-naphthalenebutyric acid

33.    $HOOCCH_2CH_2\overset{+}{\underset{CH_3}{\overset{CH_3}{N}}}CH_3$   $Cl^-$

(2-Carboxyethyl)trimethylammonium
chloride

34.    $CH_3C\left\{\begin{matrix}O\\S\end{matrix}\right\}H$

Ethanethioic acid
*not* Thioacetic acid

35.    $CH_3COSH$

Ethanethioic *S*-acid

36.    $CH_3CSOH$

Ethanethioic *O*-acid

37.    $C_6H_5COSH$

Benzenecarbothioic *S*-acid

38.    $CH_3CH_2CH_2CSSH$

Butanedithioic acid

39.    $HOSCCH_2CH_2COSH$

Butanebis(thioic) *O,S*-acid

40.    $HSSCCH_2CH_2CSSH$

Butanebis(dithioic) acid

41. 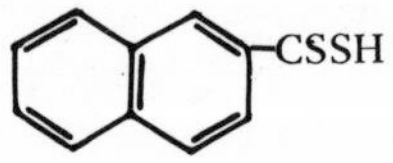    (naphthalene structure with) $-CSSH$

2-Naphthalenecarbodithioic acid

42. 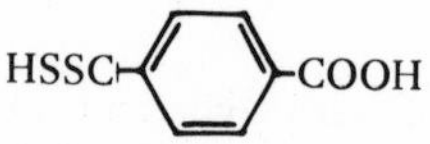    $HSSC-$ (benzene ring) $-COOH$

4-(Dithiocarboxy)benzenecarboxylic acid
4-(Dithiocarboxy)benzoic acid

43.    $CH_3CH_2CH_2C(=NH)OH$

Butanimidic acid
Butyrimidic acid

44.    $HO(HN=)C(CH_2)_5C(=NH)OH$

Heptanediimidic acid

45.

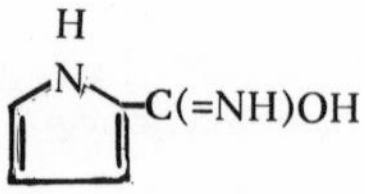

$$H$$

(pyrrole structure with) $-C(=NH)OH$

2-Pyrrolecarboximidic acid

46.

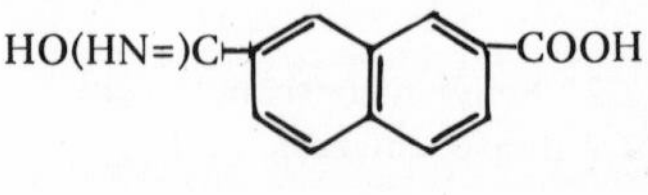

$HO(HN=)C-$ (naphthalene structure with) $-COOH$

7-(Imidocarboxy)-2-naphthalenecarboxylic
acid
7-(Imidocarboxy)-2-naphthoic acid

# Acid Derivatives: Anhydrides, Halides, Esters, and Salts

Chapter 16 has dealt with the nomenclature of compounds containing the carboxylic acid functional group, COOH. Structural modification of this function generates several classes of compounds loosely and collectively called **acid derivatives**. Each of these classes has functional status in its own right and hence confers its own family name on compounds in which it is the principal function. This chapter is concerned with the naming of some of these groups of compounds; others, such as amides, amidines, and nitriles, are discussed in Chapters 21, 22, and 31, respectively.

Both organic and inorganic acids having an acidic hydrogen atom attached to oxygen or sulfur give rise to the classes of organic derivatives covered in this chapter — namely, anhydrides formed by the conceptual elimination of the elements of water between two acid groups; **esters** and **salts**, formed by the replacement of the acidic hydrogen atom by an organic group or metallic atom; and **acid halides**, in which the hydroxy group of the acid group has been replaced by a halogen atom. In the carboxylic acid series, these classes are typified by the structures $(CH_3CO)_2O$, $CH_3COOCH_3$, $CH_3COONa$, and $CH_3COCl$, respectively. Related structures, such as lactones and inner salts, as well as some sulfur and nitrogen analogs are also considered.

### Recommended Nomenclature Practice

**Symmetrical anhydrides** of unsubstituted monocarboxylic acids—i.e., anhydrides derived from two molecules of the same acid—are named by replacing the word *acid* with *anhydride;* the prefix *bis* is used if the parent acid is substituted.

| | |
|---|---|
| $(CH_3CH_2CH_2CO)_2O$ | Butanoic anhydride |
| $(C_6H_{11}CO)_2O$ | Cyclohexanecarboxylic anhydride |
| $(BrCH_2CH_2CO)_2O$ | Bis(3-bromopropanoic) anhydride |

Unsymmetrical (mixed) anhydrides derived from two different mono-carboxylic acids are given three-word names, the first two words (in alphabetical order) representing the two acids, and the third word being *anhydride.*

$C_6H_{11}CO-O-COCH_2CH_3$      Cyclohexanecarboxylic propanoic anhydride

This type of nomenclature also applies to organic anhydrides involving noncarboxylic acids, such as phosphonic or sulfinic acids, and to mixed organic-inorganic anhydrides.

$CH_3CH_2CH_2CO-O-P(O)(CH_3)_2$    Butanoic dimethylphosphinic anhydride

$C_6H_{11}CO-O-SO_2NH_2$      Cyclohexanecarboxylic sulfamic anhydride

Cyclic anhydrides derived from single molecules of polycarboxylic acids are heterocyclic structures but are usually named as simple anhydrides—*e.g.,* 1,2-benzenedicarboxylic anhydride. Open-chain anhydrides derived from polycarboxylic acids or from polybasic inorganic acids are named like partial esters.

$CH_3CH_2CO-O-COCH_2CH_2COOH$   Propanoyl hydrogen butanedioate

$C_6H_{11}CO-O-SO_2OH$      Cyclohexanecarbonyl hydrogen sulfate

When sulfur is present in an anhydride structure, the name is formed in the same way as with the oxygen analogs. A sulfur atom connecting two acid-derived groups is denoted by *thioanhydride.*

$CH_3CS-O-COC_6H_{11}$      Cyclohexanecarboxylic ethanethioic anhydride

$CH_3CH_2CO-S-CSCH_3$      Ethanethioic propanoic thioanhydride

$(C_6H_5CS)_2S$      Benzenecarbodithioic thioanhydride

Acid halides (also called acyl halides), members of a family which also includes acid azides and other less common analogs, are named radico-functionally, the acyl group being cited as a separate word. Names of acyl groups are formed by replacing the ending *ic acid* with the suffix *oyl* or *yl* (*see* 1965 IUPAC Rule C-404.1). Acyl halides derived from acids having names ending in *carboxylic acid* are named using *carbonyl halide* as the suffix.

$CH_3CH_2COBr$      Propanoyl bromide

| | |
|---|---|
| $C_6H_{11}COI$ | Cyclohexanecarbonyl iodide |
| $CH_3CH_2CH_2CH_2CON_3$ | Pentanoyl azide |

Analogous derivatives of **thio acids** and **imidic acids** are named in the same manner.

| | |
|---|---|
| $CH_3CSCl$ | Ethanethioyl chloride |
| $CH_3CH_2C(=NH)Cl$ | Propanimidoyl chloride |

**Acid halides** derived from polycarboxylic acids are named similarly if all of the acid groups have been transformed to acid halide functions. Partial acid halides of polycarboxylic acids are named by using a prefix name such as *chloromethanoyl* (or the more familiar form *chloroformyl*).

| | |
|---|---|
| $ClCOCH_2CH_2CH_2COCl$ | Pentanedioyl dichloride |
| $BrCOCH_2CH_2COOH$ | 3-(Bromomethanoyl)propanoic acid |

**Esters** and **salts** of most types of acids, both organic and inorganic, are named by replacing the endings *ic acid* and *ous acid* by *ate* and *ite*, respectively, the acid stem being preceded by the name of a cation, organic group, or hydrogen (if one or more acidic hydrogens of a polybasic acid has not been replaced), in that order, as separate words.

| | |
|---|---|
| $CH_3CH_2CH_2COOCH_3$ | Methyl butanoate |
| $CH_3CH_2COOLi$ | Lithium propanoate |
| $CH_3CSOCH_2CH_3$ | *O*-Ethyl ethanethioate |
| $CH_3CH_2OCOCH_2COOH$ | Ethyl hydrogen propanedioate |
| $HOOCCOONa$ | Sodium hydrogen ethanedioate |
| $NaOS(O)OCH_3$ | Sodium methyl sulfite |

When named as substituents, ester groups are indicated by terms based on the prefix *carbonyl* denoting the bivalent group $-CO-$. Partial esters of polycarboxylic acids in which all the acid groups can not be included in the suffix name are treated in this manner.

| | |
|---|---|
| (cyclohexane ring)$-COOCH_3$ $-CH_2CH_2COOH$ | 3-[2-(Methoxycarbonyl)cyclohexyl]- propanoic acid |

*Carboxylato* is the corresponding prefix for the anionic carboxylic group, but a periphrase may be used if desired.

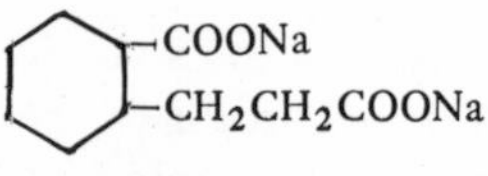

Disodium 2-(2-carboxylatoethyl)-
cyclohexanecarboxylate

—COONa
—CH$_2$CH$_2$COONa

Disodium salt of 2-(2-carboxyethyl)-
cyclohexanecarboxylic acid

Occasionally, the suffix *ic acid* or *oic acid* does not occur in a trivial name of an acid, as in the case of certain α-amino acids, and a periphrase is therefore required. (The preferred systematic name is shown first.)

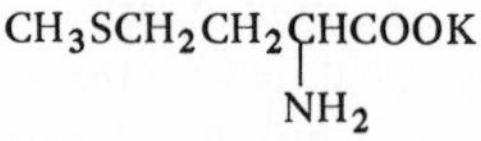

CH$_3$SCH$_2$CH$_2$CHCOOK
            |
            NH$_2$

Potassium 2-amino-4-(methylthio)-
butanoate
Potassium salt of methionine

Lactones (intramolecular esters of hydroxy carboxylic acids) are best named as heterocyclic compounds. However, simple lactones may be named by replacing the word *acid* in the name of the acid with the term *lactone,* along with any necessary locant.

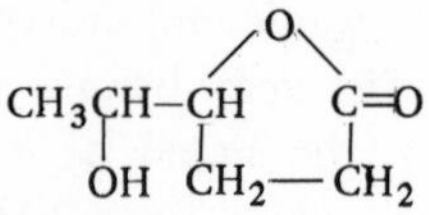

CH$_3$CH—CH    C=O
    |     |      |
    OH   CH$_2$—CH$_2$

5-(1-Hydroxyethyl)oxacyclopentan-2-one
4,5-Dihydroxyhexanoic 1,4-lactone

When a ring sulfur atom occurs in an intramolecular ester, the structure is considered to have been derived from a mercapto-substituted acid. However, heterocyclic names are preferred.

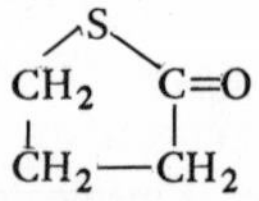

CH$_2$    C=O
 |        |
CH$_2$—CH$_2$

Thiacyclopentan-2-one
4-Mercaptobutanoic lactone

Cyclic intramolecular esters of hydroxy acids are also named as heterocyclic compounds. The class name lactide for these compounds is acceptable.

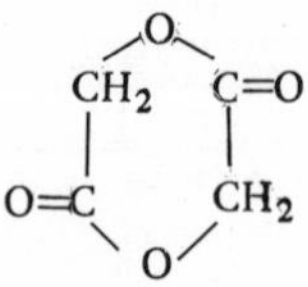

CH$_2$    C=O
 |        |
O=C      CH$_2$
    \   /
     O

1,4-Dioxacyclohexane-2,5-dione

Triesters of the hypothetical ortho acids are named systematically as ethers (*see* Chapter 26).

C$_6$H$_5$C(OCH$_3$)$_3$

(Trimethoxymethyl)benzene
*not* Trimethyl orthobenzoate

## Discussion

Anhydrides are treated briefly in IUC Rule 32 with the statement that they "will retain their present mode of designation according to the names of the corresponding acids." This is satisfactory for simple structures, but for substituted anhydrides *Chemical Abstracts* has found it best to name the parent acid and modify it with the word "anhydride," as in "Acetic acid, methoxy-, anhydride." Mixed anhydrides are usually listed twice—*i.e.,* at the name of each parent acid. Thus, acetic formic anhydride is listed as "Acetic acid, anhydride with formic acid" and as "Formic acid, anhydride with acetic acid."

The word "acid" and "cyclic" in names such as "acetic acid anhydride" and "phthalic acid cyclic anhydride" are redundant and should be omitted. Cyclic anhydrides, although not often given such names by past and present usage, are preferably named as heterocyclic compounds for consistency with the treatment accorded other cyclic acid derivatives.

Sulfur analogs of anhydrides have been named in a variety of ways. *Chemical Abstracts* indexes $CH_3CS-S-COCH_2CH_3$ at both "Acetic acid, dithio-" and "Propionic acid, thio-." Although compounds having a sulfur atom connecting two acyl groups may be regarded as diacyl sulfides, names of this type should be abandoned in favor of the recommended thioanhydride terminology. The term "anhydrosulfide" should not be used in place of thioanhydride.

Diacylamines (imides) are discussed in Chapter 21. Prefix names for acyl groups in which =NH replaces =O are formed systematically by adding *imidoyl* to the acid stem name—*e.g.,* methanimidoyl for $HC(=NH)-$.

Acid halides present few problems, provided the name of the acyl group is known. Special note should be taken of certain amino acids having trivial names ending in *ine* or *an,* for which acyl group names are formed by changing the ending to *yl—e.g.,* valyl and tryptophyl. Acyl cyanides, RCOCN, are better viewed as α-oxo nitriles. In forming a prefix name for the acid chloride group ClCO—, "chlorocarbonyl" might be considered an alternative to chloromethanoyl since alkoxycarbonyl is the preferred prefix form for the ester group ROCO—, but names of the halocarbonyl type have not been officially recognized. Phosgene and thiophosgene are acceptable trivial names for $COCl_2$ (carbonyl dichloride) and $CSCl_2$ (thiocarbonyl dichloride), respectively.

With esters, questions arise when attempts are made to choose among ethyl hydrogen adipate, ethyl adipate, monoethyl adipate, and ethyl adipic acid for the name of the half-ester of adipic acid; the issue is avoided in *Chemical Abstracts* indexes by using "Adipic acid, monoethyl ester." The first type of name leaves no room for uncertainty and is the best choice; the second type of name is sometimes used (incorrectly, except perhaps in indexes) for the diethyl ester and is therefore ambiguous. Monoethyl adipate is unambiguous but requires care to avoid omission of the little-used prefix *mono;* ethyl adipic acid can be confused easily with ethyladipic acid and lacks the characteristic *ate* ending of ester names. The preferred *systematic* name for the compound in question is ethyl hydrogen hexanedioate.

Esters of polyalcohols with simple acids are readily named without resorting to names such as "ethylene glycol diacetate" for $CH_3COOCH_2-CH_2OCOCH_3$, which is systematically named ethylene diethanoate; the half-ester is 2-hydroxyethyl ethanoate. The same applies to esters of glycerol, for which regularly formed names are recommended. Esters of glycerol and simple acids have long been known by the class name glycerides, and trivial names have been formed from the acid names by replacing the ending *ic acid* with *in—e.g.,* 1-monostearin for $CH_3(CH_2)_{16}COOCH_2CH(OH)CH_2OH$. Mixed glycerides are listed in *Chemical Abstracts* indexes at each of the glyceride names. Although names such as tripalmitin enjoy considerable usage, it is recommended that the regularly formed ester names be used whenever possible. Urethanes are another special ester class, and as a class name the term urethane is acceptable. (Prior to Vol. 66 (1967) *Chemical Abstracts* used the spelling "urethan.") Individual urethanes, however, should be named as esters of carbamic acid, $H_2NCOOH$, since little system exists in urethane-type nomenclature. Imidic esters, $RC(=NH)OR'$, are named regularly as esters of imidic acids—*e.g.,* ethyl methanimidate. The term "imido ether" should not be used, either as a class name or in naming individual compounds.

Lactones pose some of the same problems as do anhydrides. Names such as 4-hydroxybutyric acid $\gamma$-lactone contain two redundancies: "acid" and "$\gamma$." Since the modern trend is away from the use of Greek letters as locants, numbers should be used, and then only when necessary, as in 3,4-dihydroxybutyric 1,4-lactone. The additional trend toward heterocyclic nomenclature for all but the simplest cyclic structures may remove the problem altogether. The trivial names $\beta$-propiolactone, $\gamma$-butyrolactone, $\gamma$-valerolactone, and $\delta$-valerolactone are acceptable, but their use is not encouraged. Another system alternative to that of simple heterocyclic treatment is to use the suffix *olide* to denote the lactone function connecting two points in a hydrocarbon chain, or *carbolactone* denoting the same function forming a bridge in a ring system (*see* IUPAC Rule C-472). Although the trivial names coumarin, isocoumarin, and phthalide are acceptable, the corresponding systematic heterocyclic names are preferable for general use.

## Table 17.1. Examples of Acceptable Usage

1. $HCO-O-CHO$

Methanoic anhydride
Formic anhydride

2. $CH_3CO-O-COCH_3$

Ethanoic anhydride
Acetic anhydride

3. $CH_3CH_2CO-O-COCH_2CH_3$

Propanoic anhydride
Propionic anhydride

4. $C_6H_5CO-O-COCH_3$

Benzenecarboxylic ethanoic anhydride
Acetic benzoic anhydride

5. $ClCH_2CO-O-COCH_2Cl$

Bis(chloroethanoic) anhydride
Bis(chloroacetic) anhydride

6. $CH_3CH_2CS-O-CSCH_2CH_3$

Propanethioic anhydride

7. $CH_3C(=NH)-O-C(=NH)CH_3$

Ethanimidic anhydride
Acetimidic anhydride

8.

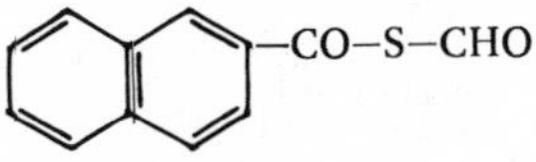

—CO-S-CHO

Methanoic 2-naphthalenecarboxylic
   thioanhydride
Formic 2-naphthoic thioanhydride

9.

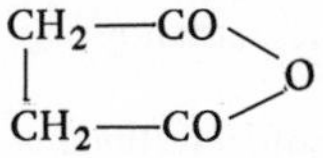

Oxacyclopentane-2,5-dione
Butanedioic anhydride
Succinic anhydride

10.

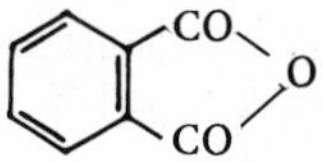

2-Oxaindan-1,3-dione
1,2-Benzenedicarboxylic anhydride
Phthalic anhydride

11.

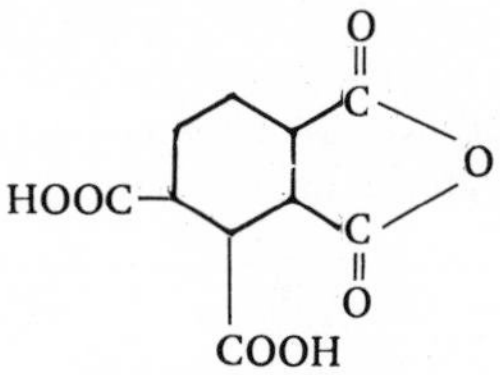

Hexahydro-1,3-dioxo-2-oxaindan-
   -4,5-dicarboxylic acid
1,2,3,4-Cyclohexanetetracarboxylic acid
   1,2-anhydride

12.

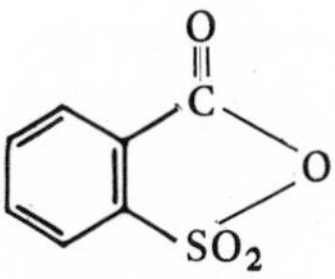

2-Oxa-1-thiaindan-3-one 1,1-dioxide
2-Sulfobenzenecarboxylic anhydride
2-Sulfobenzoic anhydride

13.

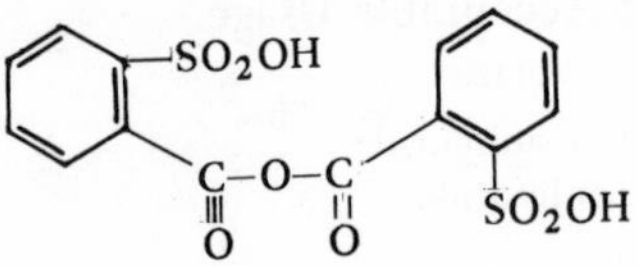

2,2'-(Oxydicarbonyl)dibenzenesulfonic acid
Bis(2-sulfobenzenecarboxylic) anhydride
Bis (2-sulfobenzoic) anhydride

14. $CH_3CH_2CH_2COCl$

Butanoyl chloride
Butyryl chloride

15.

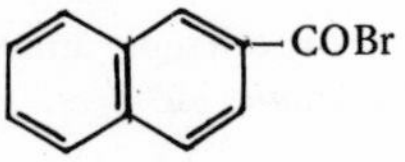

2-Naphthalenecarbonyl bromide
2-Naphthoyl bromide

16.

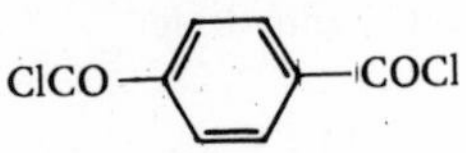

1,4-Benzenedicarbonyl dichloride
Terephthaloyl dichloride

17.

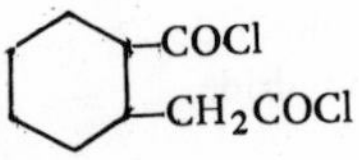

2-[(Chloromethanoyl)methyl]cyclohexane-
carbonyl chloride
2-[(Chloroformyl)methyl]cyclohexane-
carbonyl chloride

18.  $CH_2C(=NH)N_3$
     $CH_2C(=NH)N_3$

Butanediimidoyl diazide

19.

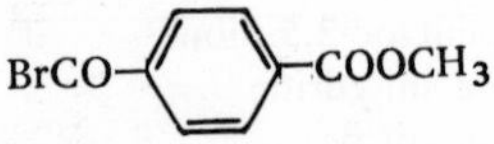

Methyl 4-(bromomethanoyl)benzenecarboxyl-
ate
Methyl 4-(bromoformyl)benzoate

20.

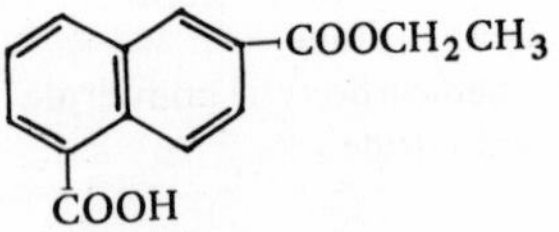

6-Ethyl hydrogen 1,6-naphthalenedicarboxylate
6-(Ethoxycarbonyl)-1-naphthoic acid

21.

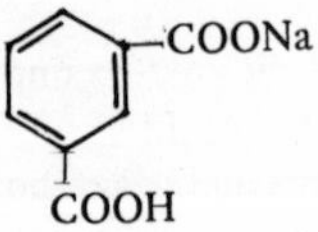

Sodium hydrogen 1,3-benzenedicarboxylate
Sodium hydrogen isophthalate

22.  $CH_3CH_2CH_2COOCH_2$
     $CH_3CH_2CH_2COOCH$
     $CH_3CH_2CH_2COOCH_2$

1,2,3-Propanetriyl tributanoate
*not* Tributyrin

23. $CH_3CH_2CH_2COOCH_2$
$CH_3(CH_2)_{15}COOCH$
$CH_3COOCH_2$

2-(Butanoyloxy)-1-[(ethanoyloxy)methyl]-ethyl heptadecanoate

24. $CH_2COOCH_2$
$CH_2COOCH_2$

1,4-Dioxacyclooctane-5,8-dione
Ethylene butanedioate
Ethylene succinate

25.

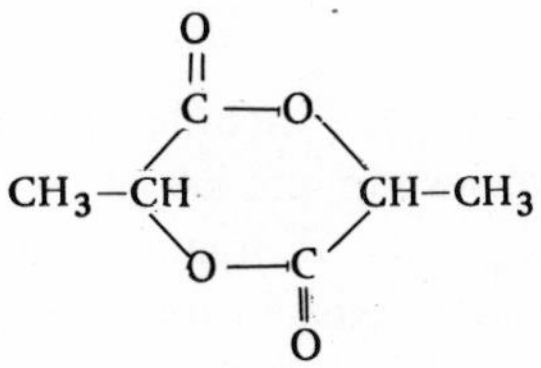

3,6-Dimethyl-1,4-dioxacyclohexane-2,5-dione
3,6-Dimethyl-1,4-dioxane-2,5-dione
*not* Dilactide

26. $C_6H_5NHCOOCH_2CH_3$

Ethyl phenylcarbamate

27. $N_3CSSCH_3$

Methyl azidomethanedithioate

# 18

# Acids (Sulfonic, Sulfinic, Sulfenic) and Derivatives

The organic acids of sulfur discussed in this chapter include those structures having a hydrocarbon or heterocyclic system and a hydroxy group (OH) attached to the same sulfur atom. Three series of organosulfur acids are considered since the central sulfur atom may exist in oxidation states of 2, 4, or 6. In the case of the two higher-valent series of acids, analogous structures in which oxygen bound only to sulfur is replaced by sulfur are also discussed. Typical functional derivatives (halides, esters, and amides) of the organosulfur acids are formed by replacing the hydroxy group with a halogen, alkoxy, or amino group. Removal of the elements of water between the hydroxy groups of any two acids leads to an anhydride structure.

**Selenium and tellurium analogs** of the structures discussed in this chapter are generally named in the same way as the corresponding sulfur compounds (*see* Discussion section).

### Recommended Nomenclature Practice

Names of **organosulfur acids** are analogous to those of cyclic carboxylic acids (*see* Chapter 16). Suffix names (Table 8.1) for the functional groups $-SO_2OH$, $-S(O)OH$, and $-SOH$ are *sulfonic acid, sulfinic acid,* and *sulfenic acid,* respectively. These suffixes are used substitutively with names of hydrocarbon and heterocyclic systems.

| | |
|---|---|
| $C_6H_5SO_2OH$ | Benzenesulfonic acid |
| $CH_3CH_2CH_2CH_2S(O)OH$ | 1-Butanesulfinic acid |
| $C_6H_5CH_2SOH$ | Phenylmethanesulfenic acid |
| 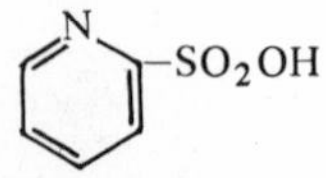 $-SO_2OH$ | 2-Pyridinesulfonic acid |

The corresponding acid groups in which one or more oxygen atoms bound to the central sulfur atom have been replaced by a sulfur atom are named by prefixing the term *thio, dithio,* or *trithio,* as appropriate, to the suffix name. When an unambiguous description of structure is desired, the term *S*-acid (for —SH) or *O*-acid (for —OH) may be used.

| | |
|---|---|
| $C_6H_5(S_3O)H$ | Benzenedithiosulfonic acid |
| $C_6H_5S(O)(S)SH$ | Benzenedithiosulfonic *S*-acid |
| $C_6H_5S(S)_2OH$ | Benzenedithiosulfonic *O*-acid |

**Esters** of organosulfur acids are named radicofunctionally by replacing the ending *ic acid* of the acid name with the suffix *ate* and placing the name of the esterifying group first, as a separate word. If known, the position of the esterifying group in esters of thio sulfur acids is shown by prefixing the capital *S* or *O* (in italics) to the group name. **Salts** are named in the same way as esters.

| | |
|---|---|
| $CH_3CH_2S(O)OCH_3$ | Methyl ethanesulfinate |
| $C_6H_5SO_2SCH_2CH_3$ | *S*-Ethyl benzenethiosulfonate |
| $CH_3SO_2ONa$ | Sodium methanesulfonate |

Cyclic intramolecular esters of hydroxy sulfonic acids (sultones) are preferably named as heterocyclic systems (Table 18.1, Example 15); linear intermolecular esters of hydroxy sulfonic acids are named as esters (Example 16). Structures of the type RSSR are named as disulfides. (*See* Chapter 39).

Functional derivatives of organosulfur acids in which the acidic hydroxy group is replaced by a halogen or a pseudohalogen are named radicofunctionally, the acid-derived group being cited first, as a separate word. Names of such groups (called acyl groups) are formed by replacing the ending *ic acid* of the acid name with the suffix *yl.*

| | |
|---|---|
| $C_6H_5SO_2Br$ | Benzenesulfonyl bromide |
| $CH_3CH_2S(O)CN$ | Ethanesulfinyl cyanide |
| $CH_2[S(O)(S)Cl]_2$ | Methanebis(thiosulfonyl chloride) |

For **organosulfur amides**, where an acyl group is joined to an —NH$_2$, —NHR, or —NR$_2$ group, substitutive nomenclature is used, employing the suffix names *sulfonamide, sulfinamide,* and *sulfenamide.*

| | |
|---|---|
| $C_6H_5SO_2NH_2$ | Benzenesulfonamide |

$CH_3CH_2SNH_2$                                  Ethanesulfenamide

$CH_3SO_2NHCH_2CH_3$                      N-Ethylmethanesulfonamide

Cyclic intramolecular organosulfur amides (sultams) are preferably named as heterocyclic systems (Example 17). Amides in which only the nitrogen atom forms part of a ring are named as derivatives of the heterocyclic system (Example 18).

Symmetrical anhydrides of organosulfur acids are named radicofunctionally by substituting the term *anhydride* for the ending *acid*. The first word of the name of each acid and the term *anhydride*, separated by spaces, are used in naming unsymmetrical anhydrides; *thioanhydride* is used when the connecting atom between the acid groups is sulfur.

$C_6H_5SO_2OSO_2C_6H_5$              Benzenesulfonic anhydride

$CH_3SO_2OS(O)C_6H_5$                Benzenesulfinic methanesulfonic anhydride

$CH_3SO_2SSO_2CH_2CH_3$            Ethanesulfonic methanesulfonic thioanhydride

Cyclic mixed anhydrides between $-COOH$ and $-SO_3H$ may be named by replacing *acid* with *anhydride*, but heterocyclic names are preferable.

2-Oxa-1-thiaindan-3-one 1,1-dioxide
2-Sulfobenzoic anhydride

Radicofunctional treatment of **organosulfur acyl groups** in naming functional derivatives of organosulfur acids such as acid halides and their analogs has been noted above. However, such names for acyl groups are not used substitutively to denote that a sulfur-containing group has replaced a hydrogen atom in a parent hydrocarbon structure. Instead, regular **substitutive names** are formed from the names of the bivalent groups $-SO_2-$, $-S(O)-$, and $-S-$. These groups are named *sulfonyl*, *sulfinyl*, and *thio*, respectively (*see* Table 11.1).

$C_6H_5SO_2-$⟨benzene ring⟩$-COOH$      4-(Phenylsulfonyl)benzoic acid
*not* 4-(Benzenesulfonyl)benzoic acid

Unsubstituted acid groups are given the prefix names *sulfo*, for $-SO_2OH$; *sulfino*, for $-S(O)OH$; and *sulfeno*, for $-SOH$. Names for the corresponding sulfur amide groups are *sulfamoyl*, for $H_2NSO_2-$: *sulfinamoyl*, for $H_2NS(O)-$; and *sulfenamoyl*, for $H_2NS-$.

$$HOSO_2-\langle\ \rangle-COOH$$          4-Sulfobenzoic acid

Prefix names of organosulfur amide groups in which the free bond is on the nitrogen atom are formed from the name of the amide by replacing the final *e* with *o*.

$$C_2H_5SO_2NHCH_2CH_2COOH$$          3-Ethanesulfonamidopropanoic acid

## Discussion

Although used in the past, terms such as "thiolo" and "thiono" should be abandoned in favor of *S-* and *O-* prefixes to denote the location of sulfur and oxygen in acid and ester names where structural ambiguity is possible. When the location of sulfur atoms relative to an acidic hydrogen—*e.g.*, in $CH_3(S_2O_2)H$—need not be specified (as is usually the case), *thio* is sufficient—*e.g.*, methanethiosulfonic acid. For indexing, *Chemical Abstracts* treats *thio*, *dithio*, etc. in this type of name as a separable modifier; thus, the compound used as an example would be indexed at Methanesulfonic acid, thio-.

There is little precedent for naming organosulfur acids in which the NH group is attached to the central sulfur atom. However, by analogy to the names of carboximidic acids (Chapter 16), $CH_3(CH_2)_5S(NH)(O)OH$, for example, may be named 1-hexanesulfonimidic acid.

Esters and salts are ordinarily indexed by *Chemical Abstracts* at the corresponding acid and under these circumstances are described by a periphrase—*e.g.*, "Benzenesulfonic acid, ethyl ester". Occasionally, the uninverted form of a periphrase may be the most satisfactory way to name a structure that contains more than one acid function, for example, $4-NaOSO_2C_6H_4COOCH_3$: "sodium salt of methyl 4-sulfobenzoate." In general, however, periphrases should be avoided.

Intra- and intermolecular condensation products involving organosulfur acids have sometimes presented nomenclature problems. Inclusion of the word "acid" in such names as "benzenesulfonic acid anhydride" is considered to be erroneous; anhydrides and sultones are, however, usually indexed by *Chemical Abstracts* at the names of the acids from which they are derived. Until 1972 cyclic anhydrides and imides were named as such and indexed at the name of the acid; now they are to be found under the heterocyclic names. For example, derivatives of 2-sulfobenzoic imide (commonly called saccharin) will be found in the Subject Index under 1,2-Benzisothiazolin-3-one. Intermolecular anhydrides of hydroxy sulfur acids are best named as ring compounds or systematically as esters when there is no ring; the term "sulfonylide" is no longer used.

Organic selenium and tellurium acids and their derivatives are generally named analogously to the corresponding sulfur compounds, with the stems *selen* or *tellur* replacing *sulf* in the name, as in benzeneselenonic acid.

A few sulfonic acids have been named with more than one function cited in the same ending. Typically, these are compounds such as 1-phenol-4-sulfonic acid, for which the systematic name, 4-hydroxy-benzenesulfonic acid, is preferred. Certain other trivial names of organo-sulfur acids have also been used: ethionic acid, for $HOSO_2CH_2CH_2SO_2$-OH; methionic acid, for $CH_2(SO_2OH)_2$; isethionic acid, for $HO(CH_2)_2$-$SO_2OH$; taurine, for $H_2N(CH_2)_2SO_2OH$; and naphthionic acid, for $4\text{-}H_2NC_{10}H_6SO_3H\text{-}1$ are examples. Continued used of such names is to be discouraged. However, both sulfanilic acid and metanilic acid, along with their derivative names, are well-established trivial names; the names of several sulfanilamide derivatives (the "sulfa drugs") appear to be equally well entrenched. A few shortened prefix forms of acid group names such as mesyl (for methylsulfonyl or methanesulfonyl) and tosyl (for 4-tolylsulfonyl or 4-toluenesulfonyl) are common but should be used sparingly; the term "tosylate" should be avoided. Other shortened names of this type should not be used.

### Table 18.1. Examples of Acceptable Usage

| | | |
|---|---|---|
| 1.. | $CH_3CHCH_2CHCH_3$ with $Cl$ and $SO_2OH$ | 4-Chloro-2-pentanesulfonic acid |
| 2. | $CH_3S$—⟨benzene ring⟩—$S(O)(S)OCH_3$ | *O*-Methyl 4-(methylthio)benzenethiosulfonate |
| 3. | $CH_3CH_2$—⟨benzene ring⟩—$SO_2SH$ | 4-Ethylbenzenethiosulfonic *S*-acid |
| 4. | $CH_3CH_2S$—⟨benzene ring⟩—$SO_2OH$ | 4-(Ethylthio)benzenesulfonic acid |
| 5. | ⟨benzene ring⟩—$S(O)OH$, —$S(O)OH$ | 1,2-Benzenedisulfinic acid |

6.

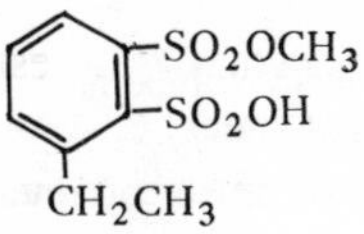

1-Methyl hydrogen 3-ethyl-1,2-benzene-
disulfonate

7.  $NaOSCH_2CH_2CH_2SO_2OCH_3$

Sodium 3-(methoxysulfonyl)-1-propanesulfenate

8. 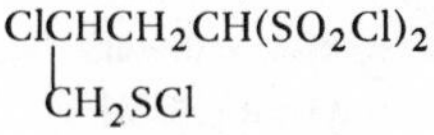  $ClCHCH_2CH(SO_2Cl)_2$
       |
       $CH_2SCl$

3-Chloro-4-(chlorothio)-1,1-butane-
disulfonyl dichloride

9.

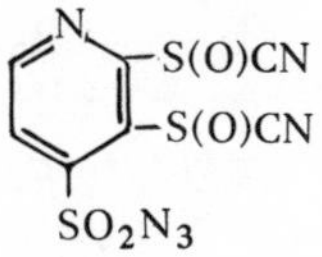

2,3-Bis(cyanosulfinyl)-4-pyridinesulfonyl azide

10. 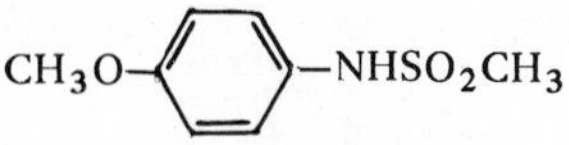  $CH_3O$—⟨benzene⟩—$NHSO_2CH_3$

N-(4-Methoxyphenyl)methanesulfonamide

11. 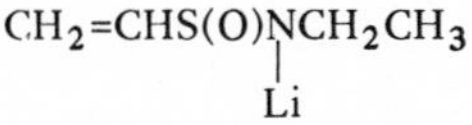  $CH_2=CHS(O)NCH_2CH_3$
              |
              $Li$

N-Lithio-N-ethylethenesulfinamide

12.  $C_6H_5SO_2NHSO_2CH_3$

N-(Methylsulfonyl)benzenesulfonamide

13. 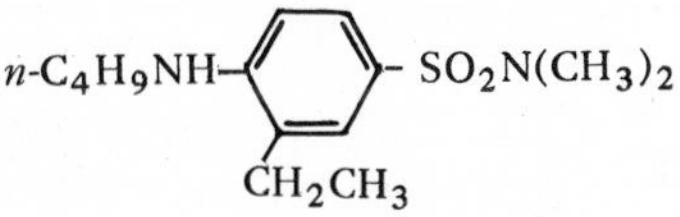  $n$-$C_4H_9NH$—⟨benzene⟩—$SO_2N(CH_3)_2$
                      |
                   $CH_2CH_3$

4-(Butylamino)-3-ethyl-N,N-dimethylbenzene-
    sulfonamide
$N^4$-Butyl-3-ethyl-$N^1$,$N^1$-dimethylsulfanilamide

14. 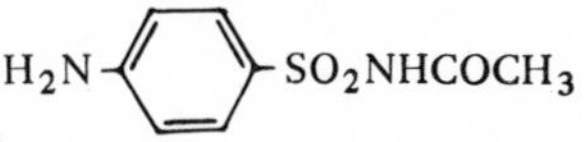  $H_2N$—⟨benzene⟩—$SO_2NHCOCH_3$

N-Acetyl-4-aminobenzenesulfonamide
$N^1$-Acetylsulfanilamide
N-Sulfanilylacetamide

15.

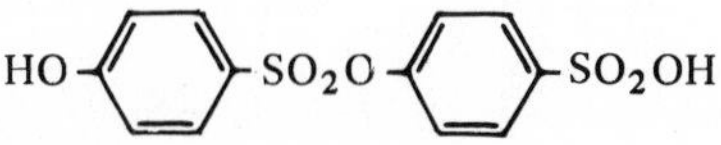

$HO$—⟨benzene⟩—$SO_2O$—⟨benzene⟩—$SO_2OH$

4-(4-Hydroxyphenylsulfonyloxy)-
    benzenesulfonic acid
4-Sulfophenyl  4-hydroxybenzenesulfonate

16.

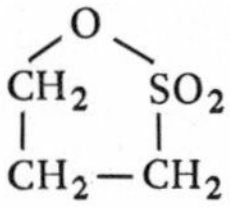

1-Oxa-2-thiacyclopentane 2,2-dioxide
1,2-Oxathiane  2,2-dioxide

17.

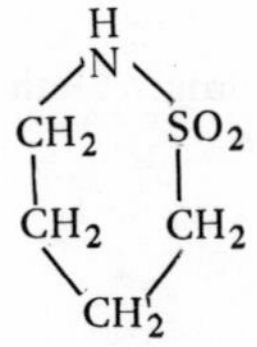

$$
\begin{array}{c}
\text{H} \\
\text{N} \\
\diagup \quad \diagdown \\
\text{CH}_2 \qquad \text{SO}_2 \\
| \qquad\qquad | \\
\text{CH}_2 \qquad \text{CH}_2 \\
\diagdown \qquad \diagup \\
\text{CH}_2
\end{array}
$$

1-Thia-2-azacyclohexane
  1,1-dioxide
Tetrahydro-2*H*-1,2-thiazine
  1,1-dioxide

18. 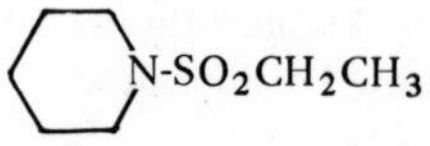     N-SO$_2$CH$_2$CH$_3$

1-(Ethylsulfonyl)azacyclohexane
1-(Ethylsulfonyl)piperidine

# Alcohols and Thioalcohols

**A**lcohols are compounds containing one or more hydroxy groups (OH), not part of another functional group, each attached to a carbon atom in a chain or nonaromatic ring. Compounds having one or more hydroxy groups attached to an aromatic ring are classified as phenols (*see* Chapter 35).

**Polyhydric alcohols** are those containing two or more OH groups; **dihydric alcohols** are often called "glycols." The class name **polyol** includes both polyhydric alcohols and polyhydric phenols, as do the class names **diol, triol,** etc. Certain polyhydric alcohols closely related to the sugars have been assigned special names authorized by the official ACS Rules of Carbohydrate Nomenclature (*see* Chapter 30.)

**Thiols** are compounds containing one or more mercapto groups (SH) each attached to a carbon atom in a chain or ring system regardless of saturation; hence, the generic name **thiol** includes the sulfur analogs of both alcohols and phenols. Within the broad class of thiols, the subclass names **thiophenol** and **thioalcohol** are used to distinguish between aromatic and nonaromatic structures carrying SH groups. The class name "mercaptan," which has found wide use in the older literature to denote thiols, is no longer acceptable (*see* Discussion).

*Recommended Nomenclature Practice*

### Alcohols

**Monohydric alcohols** are named substitutively by adding the suffix *ol* to the name of the corresponding hydrocarbon, the final *e* of the hydrocarbon name being elided.

$$CH_3CH_2CH_2CH_2OH \qquad \text{1-Butanol}$$

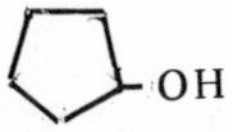

Cyclopentanol

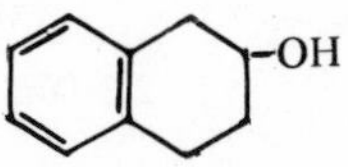

1,2,3,4-Tetrahydro-2-naphthalenol

**Polyhydric alcohols** are named substitutively by adding the appropriate suffix *diol, triol, tetrol, pentol,* etc. to the name of the corresponding hydrocarbon or heterocyclic system.

$HOCH_2CH_2CH_2CH_2CH_2CH_2OH$          1,6-Hexanediol

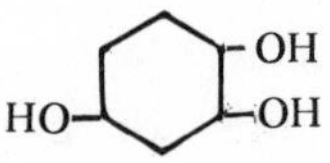

1,2,4-Cyclohexanetriol

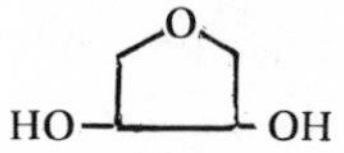

Oxacyclopentane-3,4-diol

When one or more OH groups must be designated as substituents, the prefix names *hydroxy, dihydroxy, trihydroxy,* etc. are used.

$HOCH_2CH_2CH_2COOH$          4-Hydroxybutanoic acid

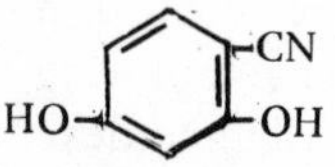

2,4-Dihydroxybenzonitrile

A few trivial names for alcohols are still acceptable. Three of these (Examples 11, 17, 18) are shown in Table 19.1. For a more complete list of recognized trivial names, *see* the 1965 IUPAC Rules.

### Thioalcohols

**Thioalcohols** are named substitutively according to the above principles using the suffixes *thiol, dithiol, trithiol,* etc.

$CH_3CH_2SH$          Ethanethiol

$HSCH_2CH_2CH_2CH_2SH$          1,4-Butanedithiol

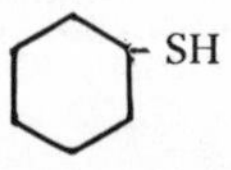

Cyclohexanethiol

The prefix name for the SH group is *mercapto.*

HS-CHCOOH
|
HS-CHCOOH          2,3-Dimercaptobutanedioic acid

## Salts

Ionic compounds in which a metal replaces the H of an alcoholic OH or SH group are named as **salts** of the corresponding alcohol or thioalcohol. The names of anions are formed by changing the parent suffixes *ol* and *thiol* to *olate* and *thiolate,* respectively. Commonly used abbreviated names ending in *oxide* are also recognized in the 1965 IUPAC Rules.

$CH_3CH_2CH_2ONa$      Sodium 1-propanolate
Sodium propoxide

$(CH_3O)_2Ca$      Calcium dimethanolate
Calcium dimethoxide

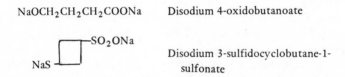      Dipotassium 1,4-cyclohexanedithiolate

When the anionic group must be named as a substituent, the prefix names *oxido* and *sulfido* are used.

$NaOCH_2CH_2CH_2COONa$      Disodium 4-oxidobutanoate

Disodium 3-sulfidocyclobutane-1-sulfonate

*Discussion*

Substitutive names, which are listed first in each case in Table 19.1, represent the simplest and most broadly useful system of designating alcohols and thioalcohols and are therefore generally preferred. In present-day usage the trend is away from radicofunctional names ending in the word *alcohol,* although such names are still used widely, especially for the most common members of the class (*see* Examples 1—8, 11, and 20.)

The radicofunctional method of naming thioalcohols using the class name "mercaptan"—*e.g.,* "methyl mercaptan" for $CH_3SH$, has been officially abandoned (*see* the 1965 IUPAC Rules) and should be avoided entirely. Similarly, "mercaptide" as a radicofunctional class name for salts of thiols, as in "sodium ethyl mercaptide" for $CH_3CH_2SNa$, has been dropped in favor of the suffix name *thiolate.*

Radicofunctional names for dihydric alcohols employing the class name *glycol* have found considerable use, but except for ethylene glycol and propylene glycol (Examples 15 and 16), this type of name should be abandoned. As noted above in the introductory section, "glycol" has also been used in a generic sense to designate the class of dihydric alcohols;

however, the more systematic term alkanediol is now preferred. The use of "glycol" as an abbreviated trivial form meaning 1,2-ethanediol is clearly poor practice and should be abandoned.

Alcohols and thioalcohols having one or more OH or SH groups attached to a chain connected to a ring system are given conjunctive names as the preferred entries in the Subject Indexes of *Chemical Abstracts* (*see* second names of Examples 23-25, also Chapter 9), but substitutive names are preferable for general use. In the special case of methyl-substituted benzene derivatives, $\alpha$, $\alpha'$, $\alpha''$, etc., have been used as locants for OH or SH attached to the methyl groups (*see* Examples 20 and 21), but names requiring such locants are no longer recommended.

Acyclic alcohols and thioalcohols containing several hetero atoms in the principal chain may be named substitutively, by means of replacement nomenclature (*see* Example 28, also Chapter 7).

As defined in the introductory section of this chapter, the class names alcohol and thioalcohol do not denote structures where OH or SH is attached directly to an atom other than carbon; hence, the suffix names *ol* and *thiol* are not applicable to such structures except for silanols (*see* Chapter 38). The most familiar compounds of this type, however, are those in which OH or SH is attached to a nitrogen atom of a heterocyclic ring. In these cases, the OH or SH group is treated as a substituent of the ring system (*see* Example 31).

It is desirable to distinguish among *hydroxy* as the prefix name for the OH group (*see* Examples 26 and 27), *hydroxyl* for the ·OH free radical, and *hydroxide* for the OH⁻ anion (*see also* Chapter 28).

The frequently encountered names "isopropanol," "isobutanol," "*sec*-butanol," "*tert*-butanol," and the like are incorrectly formed and should not be used. Correct radicofunctional names for these alcohols are shown as alternative names in Examples 4 and 6—8.

Use of "carbinol" as a synonym for methanol, and the prefix name "carbinyl" instead of *methyl*, in naming substituted methanols is disappearing and should be abandoned entirely (*see* Examples 20 and 22). The same can be said of acid-derived names such as "stearyl alcohol" and "phthalyl alcohol," which are incorrectly formed, and of "methylol" as a prefix name for $-CH_2OH$ (Example 32).

Halohydrin and cyanohydrin names (Examples 29 and 30), although still used occasionally, should be abandoned in favor of systematic names, as should ortho acid names (Example 33).

## Table 19.1. Examples of Acceptable Usage

1. $CH_3OH$

    Methanol
    Methyl alcohol

2. $CH_3CH_2OH$

    Ethanol
    Ethyl alcohol

3. $CH_3CH_2CH_2OH$

    1-Propanol
    Propyl alcohol

4. $CH_3\overset{\displaystyle |}{\underset{\displaystyle OH}{C}}HCH_3$

    2-Propanol
    Isopropyl alcohol
    *not*  Isopropanol

5. $CH_3CH_2CH_2CH_2OH$

    1-Butanol
    Butyl alcohol

6. $\underset{\displaystyle CH_3CHCH_2OH}{\overset{\displaystyle CH_3}{|}}$

    2-Methyl-1-propanol
    Isobutyl alcohol
    *not*  Isobutanol

7. $CH_3CH_2\overset{\displaystyle |}{\underset{\displaystyle OH}{C}}HCH_3$

    2-Butanol
    *sec*-Butyl alcohol
    *not*  *sec*-Butanol

8. $CH_3\overset{\displaystyle CH_3}{\underset{\displaystyle OH}{C}}-CH_3$

    2-Methyl-2-propanol
    *tert*-Butyl alcohol
    *not*  *tert*-Butanol

9. $CH_3CH_2CH_2CH_2CH_2OH$

    1-Pentanol

10. $CH_3CH_2CH_2CH_2CH_2CH_2OH$

    1-Hexanol

11. $CH_2=CHCH_2OH$

    2-Propen-1-ol
    Allyl alcohol

12. $(C_6H_5)_2C=\overset{\displaystyle |}{\underset{\displaystyle OH}{C}}C_6H_5$

    Triphenylethenol

13.

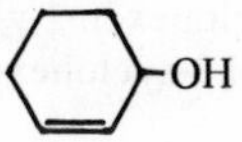

    2-Cyclohexen-1-ol

14.

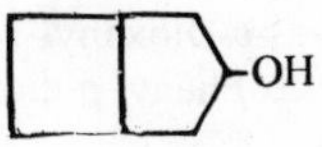

    Bicyclo[3.2.0] heptan-3-ol

15.  $HOCH_2CH_2OH$                     1,2-Ethanediol
                                        Ethylene glycol

16.  $CH_3CHCH_2OH$                     1,2-Propanediol
         $OH$                           Propylene glycol

17.  $HOCH_2CHCH_2OH$                   1,2,3-Propanetriol
          $OH$                          Glycerol
                                   *not*  Glycerine

         $CH_2OH$
18.  $HOCH_2CCH_2OH$                    2,2-Bis(hydroxymethyl)-1,3-propanediol
          $CH_2OH$                      Pentaerythritol

19. 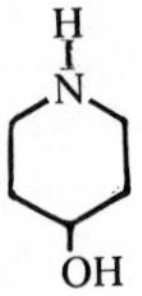
                                        4-Piperidinol

20.  $C_6H_5CH_2OH$                     Phenylmethanol
                                        Benzyl alcohol
                                   *not*  Phenylcarbinol
                                   *not*  α-Toluenol

21.  $C_6H_5CH_2SH$                     Phenylmethanethiol
                                   *not*  α-Toluenethiol
                                   *not*  Benzyl mercaptan

22.  $(C_6H_5)_3COH$                    Triphenylmethanol
                                   *not*  Triphenylcarbinol
                                   *not*  Trityl alcohol

23.

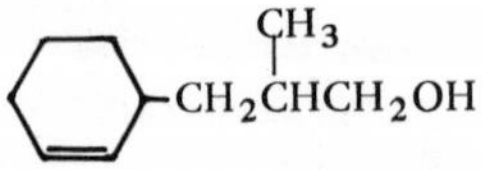

                            3-(2-Cyclohexen-1-yl)-2-methyl-1-propanol
                            β-Methyl-2-cyclohexene-1-propanol

24. 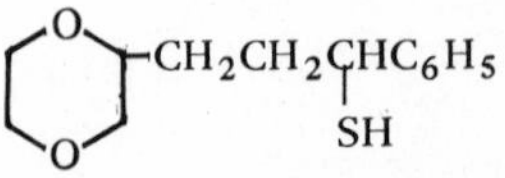
                            3-p-Dioxanyl-1-phenyl-1-propanethiol
                            α-Phenyl-p-dioxanepropanethiol

25.

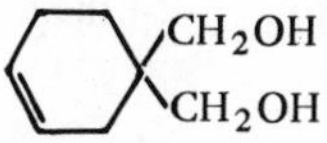

$CH_2OH$
$CH_2OH$

3-Cyclohexen-1-ylidenedimethanol
3-Cyclohexene-1,1-dimethanol

26. $HOCH_2CH_2CHCH_2OH$
$CH_2OH$

2-(Hydroxymethyl)-1,4-butanediol

27. $HOCH_2CH_2CH_2N^+(CH_3)_3$ $Cl^-$

(3-Hydroxypropyl)trimethylammonium
chloride

28. $HOCH_2CH_2OCH_2CH_2OCH_2CH_2OH$

3,6-Dioxaoctane-1,8-diol
2,2'-(Ethylenedioxy)diethanol
*not* Triethylene glycol

29. $CH_3CHCH_2OH$
$Cl$

2-Chloro-1-propanol
*not* Propylene chlorohydrin

30.
$CH_3$
$CH_3CCN$
$OH$

2-Hydroxy-2-methylpropanenitrile
*not* Acetone cyanohydrin

31. 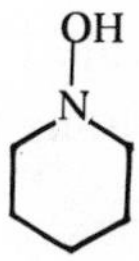
$OH$
$N$

1-Hydroxyazacyclohexane
1-Hydroxypiperidine
*not* 1-Piperidinol

32. 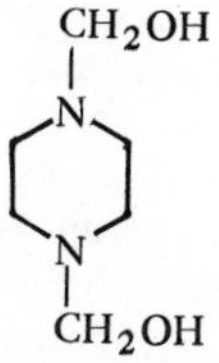
$CH_2OH$
$N$
$N$
$CH_2OH$

1,4-Diazacyclohexane-1,4-diyldimethanol
1,4-Piperazinediyldimethanol
*not* 1,4-Dimethylolpiperazine

33. $CH_3C(OH)_3$

1,1,1-Ethanetriol
*not* Orthoacetic acid

# 20

# Aldehydes

C ompounds containing the carbonyl group ($>$C=O) attached only to carbon or hydrogen have been divided traditionally into two functional classes called aldehydes and ketones. This distinction has been made because of certain characteristic differences in the reactivity of the C=O group depending upon whether or not it is attached to hydrogen. Thus, aldehydes are carbonyl compounds containing one or more —CH=O groups (commonly written CHO), and ketones (*see* Chapter 29) are carbonyl compounds in which the C=O group is attached to two carbon atoms. **Thioaldehydes** contain the —CH=S group (commonly written CHS).

Some oxygen and sulfur derivatives of aldehydes are also treated in this chapter. These include the **acetals** and **hemiacetals**, which contain the groups —CH(OR)$_2$ and —CH(OH)(OR), respectively, and their analogs in which one or more oxygen atoms have been replaced by sulfur.

Nitrogen derivatives of aldehydes, such as oximes and hydrazones, are discussed in Chapter 32.

### Recommended Nomenclature Practice

Acyclic monoaldehydes are named substitutively by adding the suffix *al* to the name of the corresponding hydrocarbon, the final *e* of the hydrocarbon name being elided.

CH$_3$CH$_2$CH$_2$CH$_2$CH$_2$CHO                    Hexanal

CH$_2$=CHCH$_2$CH$_2$CHO                    4-Pentenal

Acyclic dialdehydes are named similarly by using the suffix *dial* but retaining the final *e* of the hydrocarbon name.

OHCCH$_2$CH$_2$CH$_2$CH$_2$CH$_2$CHO                    Heptanedial

Acyclic polyaldehydes in which more than two CHO groups are attached to the same straight chain are named substitutively by using the suffixes *tricarbaldehyde, tetracarbaldehyde,* etc. with the name of the chain.

$$OHCCH_2CHCH_2CHO$$
$$|$$
$$CHO$$

1,2,3-Propanetricarbaldehyde

Cyclic aldehydes in which one or more CHO groups are attached to a ring system are named substitutively by adding the suffixes *carbaldehyde, dicarbaldehyde, tricarbaldehyde,* etc. to the name of the ring system.

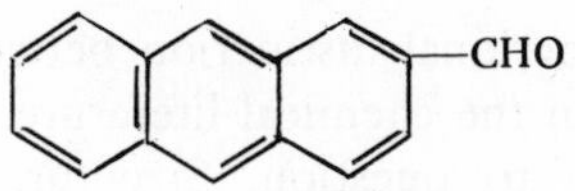

2-Anthracenecarbaldehyde

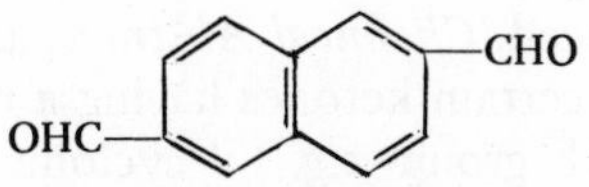

2,6-Naphthalenedicarbaldehyde

When all of the CHO groups in a polyaldehyde cannot be treated as recommended above or when another functional group standing higher in the Order of Precedence of Functions (Table 8.1) is present, the terms *methanoyl, dimethanoyl, trimethanoyl,* etc. (or the more familiar *formyl, diformyl, triformyl,* etc.) are used as **prefix names.**

$$CH_2CH_2CH_2CHO$$
$$|$$
$$OHCCH_2CHCH_2CH_2CHO$$

4-(Methanoylmethyl)octanedial

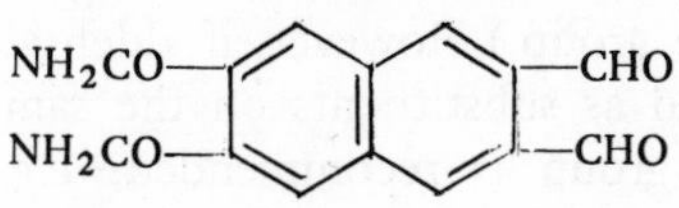

6,7-Dimethanoyl-2,3-naphthalene-
dicarboxamide

Trivial names for the more common aldehydes based on trivial names of the corresponding carboxylic acids (*see* Chapter 16) have been used widely. Such names are formed by changing the suffix *ic acid* or *oic acid* to *aldehyde,* as in acetaldehyde for $CH_3CHO$ and benzaldehyde for $C_6H_5CHO$. When polyaldehydes are named in this manner, it is implied that all the COOH groups have been converted to CHO groups—*e.g.,* succinaldehyde for $OHCCH_2CH_2CHO$. Some acceptable trivial names of aldehydes are included in Table 20.1.

Thioaldehydes are named analogously to aldehydes by using the suffixes *thial* and *carbothialdehyde* in place of *al* and *carbaldehyde;* the term *methanethioyl* (or the more familiar *thioformyl*) is the prefix name for the CHS group. For typical names of thioaldehydes *see* Examples 22-24 in Table 20.1.

Acetals are preferably named systematically as ethers—*e.g.*, 1,1-dimethoxyethane for $CH_3 CH(OCH_3)_2$ —and **hemiacetals** as alcohols—*e.g.*, 1-methoxyethanol for $CH_3 CH(OH)(OCH_3)$. **Thioacetals and thiohemiacetals** containing one or more sulfur atoms are named as thioethers and thiols or alcohols, respectively; cyclic structures are treated as heterocyclic systems (*see* Examples 27-30).

## Discussion

Although, as pointed out above, the functional distinction between aldehydes and ketones is firmly established in the chemical literature, its value in systematic nomenclature is open to question. Structurally, aldehydes bear the same relationship to ketones as do primary alcohols to secondary alcohols and could be named using the suffix *1-one* rather than *al,—e.g.*, 1-propanone for $CH_3 CH_2 CHO$. Indeed, *Chemical Abstracts* uses this approach in forming parent names for certain ketones having a ring system directly attached to the carbonyl group—*e.g.*, 1-cyclohexyl-1-butanone for $C_6 H_{11} COCH_2 CH_2 CH_3$. However, the *1-one* type of nomenclature has not as yet been officially recognized or used for aldehydes.

Although the identity of the aldehyde function is customarily preserved by use of the suffixes *al* and *carbaldehyde,* as recommended above, the situation is different when the CHO group is named as a prefix. For a substituting CHO group attached to a chain, either *methanoyl* (or *formyl*) or *oxo* (denoting =O only) may be used as a prefix, according to the 1965 IUPAC Rules—*e.g.*, 4-methanoylbutanoic acid or 5-oxopentanoic acid for $OHCCH_2 CH_2 CH_2 COOH$. The former alternative is the choice of *Chemical Abstracts* and seems preferable since it maintains the identity of the aldehyde function. (The prefix *oxo* does not, of course, distinguish between an aldehyde group and a ketone group.) However, if aldehyde and ketone groups are both to be named as substituents on the same chain, use of *oxo* for both types of group is recommended,—*e.g.*, 3,5-dioxopentanoic acid for $OHCCH_2 COCH_2 COOH$.

The IUPAC-recommended suffix *carbaldehyde* for CHO has not yet found extensive use; *Chemical Abstracts* still uses the older form, *carboxaldehyde.* However, *carbaldehyde* is recommended here because strictly speaking both the *ox* and *aldehyde* portions of *carboxaldehyde* denote an oxygen atom, resulting in unnecessary and possibly misleading redundancy.

Acyclic aldehydes in which CHO groups are separated from a ring system by one or more carbon atoms are given conjunctive names (*see* Chapter 9) in many instances by *Chemical Abstracts, e.g.,* 2-pyridine-propionaldehyde for 3-(2-pyridyl)propionaldehyde, but substitutive names are preferred for general use.

Trivial names for aldehydes based on those trivial names of carboxylic acids listed as acceptable in Chapter 16 are, of course, implicitly acceptable themselves. However, except for these, and a few other very common trivial names included in Table 20.1, trivial names of aldehydes should be abandoned in favor of systematic names.

In forming acid-derived trivial names for aldehydes, two points warrant emphasis: the acid name must be trivial, not systematic, and both the ending *ic* and the word *acid* are replaced by the suffix *aldehyde*. Thus, names such as "ethanaldehyde" for acetaldehyde and "oleic aldehyde" for olealdehyde are incorrect and should not be used. Also, radicofunctional names such as "*n*-octyl aldehyde," which have found some use, should be abandoned.

Trivial aldehydic acid names such as malonaldehydic acid for $OHCCH_2COOH$ are recognized in the 1965 IUPAC Rules and until 1972 were used by *Chemical Abstracts,* but the corresponding systematic names are to be preferred.

The prefix *thio* should not be used in naming individual thioaldehydes, as in "thioformaldehyde, " "thiobenzaldehyde," and the like. As in other classes of compounds the use of *thio* to denote replacement of oxygen by sulfur can often lead to ambiguous names. The class name thioaldehyde is, of course, acceptable, as are the class names monothioacetal, dithioacetal, monothiohemiacetal, and dithiohemiacetal.

Such older names for acetals as acetaldehyde diethyl acetal for $CH_3CH(OC_2H_5)_2$ are recognized in the 1965 IUPAC Rules, but ether-type systematic names are preferred. The trivial names methylal for $H_2C(OCH_3)_2$ and acetal for $CH_3CH(OCH_2CH_3)_2$ should no longer be used.

## Table 20.1  Examples of Acceptable Usage

| 1. | HCHO | Methanal |
| | | Formaldehyde |
| 2. | $CH_3CHO$ | Ethanal |
| | | Acetaldehyde |
| 3. | $CH_3CH_2CHO$ | Propanal |
| | | Propionaldehyde |
| 4. | $CH_3CH_2CH_2CHO$ | Butanal |
| | | Butyraldehyde |
| 5. | $CH_3\underset{\underset{CH_3}{\vert}}{CH}CHO$ | 2-Methylpropanal |
| | | Isobutyraldehyde |
| 6. | $CH_3CH_2CH_2CH_2CHO$ | Pentanal |
| | | Valeraldehyde |

7. $CH_3CHCH_2CHO$
   |
   $CH_3$

3-Methylbutanal
Isovaleraldehyde

8. $CH_3CH_2CH_2CH_2CH_2CHO$

Hexanal

9. $Cl_3CCHO$

Trichloroethanal
Trichloroacetaldehyde
Chloral

10. $CH_2=CHCHO$

Propenal
Acrylaldehyde
Acrolein

11. $OHCCHO$

Ethanedial
Oxalaldehyde
Glyoxal

12. $OHCCH_2CHO$

Propanedial
Malonaldehyde

13. $OHCCH_2CH_2CH_2CH_2CHO$

Hexanedial
Adipaldehyde

14. $C_6H_5CHO$

Benzenecarbaldehyde
Benzaldehyde

15. $CH_3$—⟨benzene ring⟩—CHO

4-Methylbenzenecarbaldehyde
p-Tolualdehyde

16. ⟨cyclohexane ring⟩—CHO

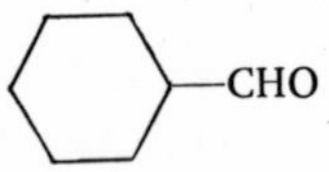

Cyclohexanecarbaldehyde

17. ⟨benzene ring⟩—CHO / —CHO

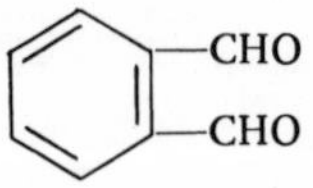

1,2-Benzenedicarbaldehyde
Phthalaldehyde

18. $OHCCHCHO$
    |
    $CHO$

Methanetricarbaldehyde

19. ⟨furan ring⟩—CHO

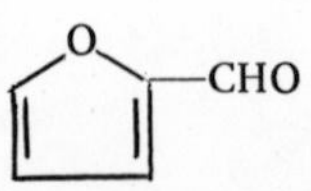

Oxacyclopenta-2,4-diene-2-carbaldehyde
2-Furancarbaldehyde
2-Furaldehyde
Furfural

20.   OHCCH$_2$—⟨benzene ring⟩—CH$_2$CHO     1,4-Phenylenediethanal
                                                                      1,4-Benzenediacetaldehyde

21. 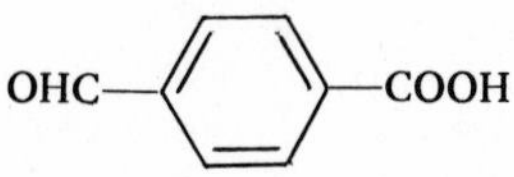  OHC—⟨benzene ring⟩—COOH     4-Methanoylbenzenecarboxylic acid
                                                          4-Formylbenzoic acid

22.  CH$_3$CH$_2$CHS                              Propanethial
                                                           *not* Thiopropionaldehyde

23. 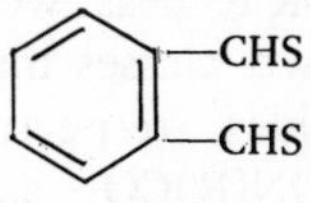  ⟨benzene ring⟩—CHS / —CHS     1,2-Benzenedicarbothialdehyde
                                                              *not* Dithiophthalaldehyde

24.  SHCCH$_2$CH$_2$COOH     3-Methanethioylpropanoic acid
                                                 3-(Thioformyl)propionic acid
                                                 *not* 4-Thiosuccinaldehydic acid

25.  CH$_3$CH$_2$CH$_2$CH(OCH$_3$)$_2$     1,1-Dimethoxybutane

26.  CH$_3$OCHCH$_2$CH$_2$CHOCH$_3$
           |                    |                    1,4-Dimethoxy-1,4-butanediol
          OH                 OH

27.  CH$_3$CH$_2$CH(SCH$_2$CH$_3$)$_2$     1,1-Bis(ethylthio)propane

28.  CH$_3$CH$_2$CH$_2$CHSCH$_3$
                           |                        1-(Methylthio)-1-butanethiol
                          SH

29.  CH$_3$CH$_2$CHSCH$_2$CH$_3$
                   |                            1-(Ethylthio)-1-propanol
                  OH

30. 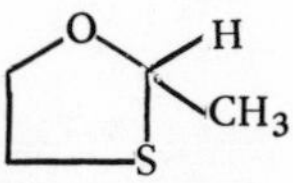  ⟨oxathiacyclopentane ring⟩ O, H, CH$_3$, S     2-Methyl-1-oxa-3-thiacyclopentane

# 21

# Amides and Imides

**D**erivatives of carboxylic acids in which the OH portion of the COOH group has been replaced by $NH_2$ (as such or substituted) are called amides or, more accurately, **carboxamides**. Such compounds have the generic structures $RCONH_2$, RCONHR', and RCONR'R", as well as RCONHCOR," and RCON(COR')COR". The latter two classes may be referred to as **secondary** amides and **tertiary** amides, respectively. Compounds containing the group —CONHCO— or —CON(R)CO— as part of a ring are called **imides** or, more accurately, **carboximides**.

Broadly speaking, the class terms amide and imide also denote the corresponding nitrogen derivatives of noncarboxylic acids such as those discussed in Chapters 18, 36, and 40—*e.g.,* sulfonamides and amides of phosphorus acids, as well as ionic derivatives of ammonia such as sodium amide ($NaNH_2$).

Sulfur analogs of carboxamides and dicarboximides, called **thioamides** and **thioimides**, are included in this chapter.

Naming of **peptides**, a special subclass made up of amides derived from $\alpha$-amino carboxylic acids, is discussed in Chapter 30.

### *Recommended Nomenclature Practice*

Compounds of the type $RCONH_2$ are preferably named by changing the ending *oic acid* in the corresponding systematic alkanoic or alkane-dioic acid name to *amide,* or the ending *carboxylic acid,* where that is used in the acid name, to *carboxamide.*

$$CH_3CH_2CH_2\overset{\overset{\displaystyle CH_3}{|}}{C}HCONH_2 \qquad \text{2-Methylpentanamide}$$

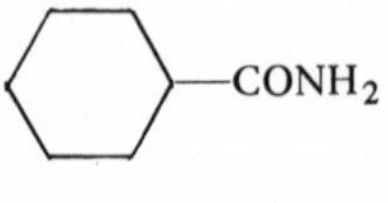

—$CONH_2$     Cyclohexanecarboxamide

$H_2NCOCH_2CH_2CH_2CH_2CONH_2$     Hexanediamide

166

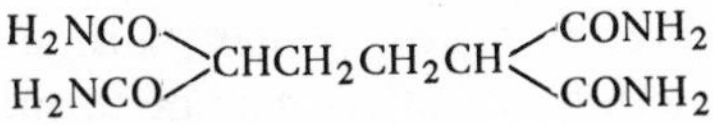

$$\underset{H_2NCO}{\overset{H_2NCO}{\diagdown}}CHCH_2CH_2CH\underset{CONH_2}{\overset{CONH_2}{\diagup}}$$

1,1,4,4-Butanetetracarboxamide

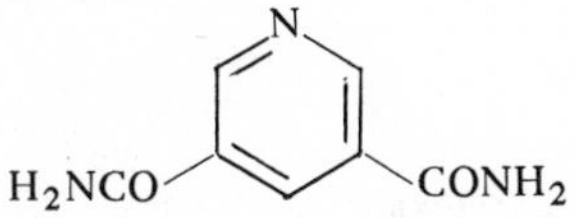

Azabenzene-3,5-dicarboxamide

Compounds of the types **RCONHR'** and **RCONR'R''** are named as N-substituted amides, excepting when the nitrogen atom is part of a ring; in that case they are named as substituted heterocyclic compounds.

$$\underset{CH_3CH_2CH_2CHCONCH=CH_2}{\overset{\overset{CH_3}{|}\quad\overset{CH_3}{|}}{\phantom{.}}}$$

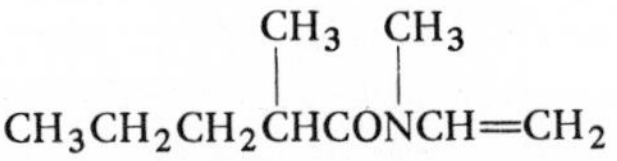

N-Ethenyl-N,2-dimethylpentanamide

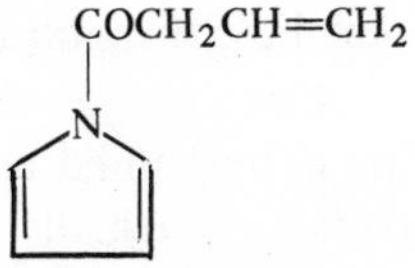

1-(3-Butenoyl)azacyclopenta-2,4-diene

Likewise, compounds of the types $(RCO)_2NH$ and $(RCO)_3N$, whether all R groups are alike or different, are best treated as N-substituted amides. For unsymmetrical structures the largest or most complex acyl group is preferably expressed in the parent name..

$$CH_3CH_2CONHCOCH_2CH_3$$

N-Propanoylpropanamide

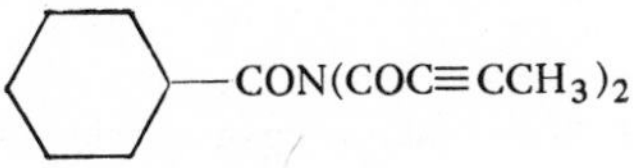

N,N-Di-2-butynoylcyclohexane-
carboxamide

Amides in which the $-CONH-$ group is part of a ring, often called lactams, are best named as heterocyclic compounds—*i.e.*, as ketones.

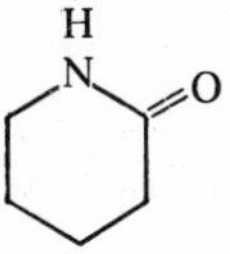

Azacyclohexan-2-one

Similar treatment is generally to be preferred for imides, although systematic names expressing the imide function as a suffix may be formed by replacing the *edioic* acid ending of the corresponding dibasic acid name with *imide* or by changing the ending *dicarboxylic acid* to *dicarboximide*.

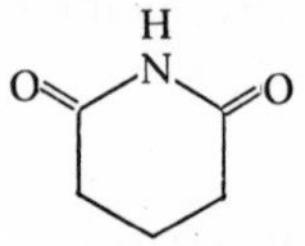

Azacyclohexane-2,6-dione
Pentanimide

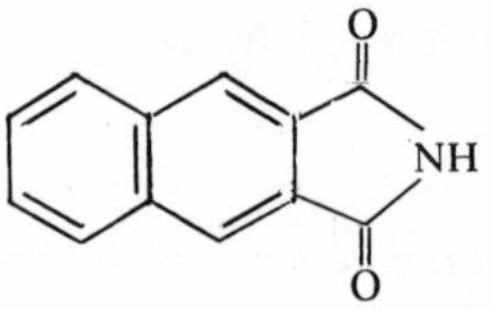

2-Azabenz[*f*]indan-1,3-dione
2,3-Naphthalenedicarboximide

Thioamides are named analogously to amides by using the suffixes *thioamide* and *carbothioamide* (*see* Examples 15-18 in Table 21.1).

Thioimides should be named as heterocyclic compounds (*see* Example 19).

When the amide or imide function must be named as a substituting group, one of two types of **prefix names** is usually applicable. If attachment is through the carbon atom(s), the prefixes *carbamoyl* for $H_2NCO-$ and *iminodicarbonyl* for $-CONHCO-$ are used; if attachment is through the nitrogen atom, the prefix name is formed by changing the final *e* of the amide or imide name to *o* although for imide groups heterocyclic names are to be preferred.

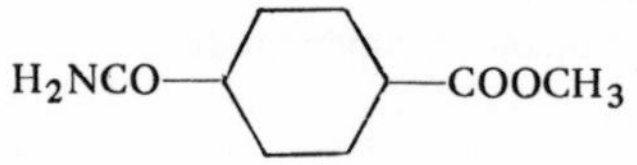

Methyl 4-carbamoylcyclohexane-
    carboxylate

$$CH_2{=}CHCH_2CONHCH_2CH_2COOH$$

3-(3-Butenamido)propanoic acid

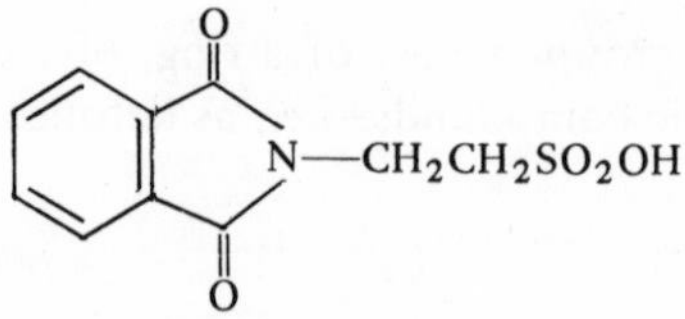

2-(1,3-Dioxo-2-azaindan-2-yl)ethane-
    sulfonic acid
2-(1,2-Benzenedicarboximido)ethane-
    sulfonic acid

Alternatively, substituting groups of the types $(RCO)_2N-$ and $RCON{<}$ may be given diacylamino and acylimino names, respectively.

$(CH_3CO)_2NCH_2CH_2COOH$ 

3-(*N*-Ethanoylethanamido)propanoic acid
3-(Diethanoylamino)propanoic acid

$COC_6H_5$

$HOSO_2$——N——$SO_2OH$    4,4'-(Benzoylimino)dibenzenesulfonic
acid

The prefix name for the $H_2NCS-$ group is *thiocarbamoyl*.

Trivial names may be used for amides and imides in instances where the corresponding acid has an acceptable trivial name (*see* Chapter 16). As with systematic names, the ending *ic acid* or *oic acid* is changed to *amide* or *imide*. Monoamides of dibasic acids having acceptable trivial names may be named by changing the *ic acid* ending to *amic acid,* except for oxalic and carbonic acids, where the contracted forms oxamic acid and carbamic acid are used. Examples of acceptable trivial names for amides and imides are included in Table 21.1.

As with other functional classes, conjunctive names (*see* Chapter 9) are used by *Chemical Abstracts* for acyclic amides in which $CONH_2$ groups are separated from a ring system by one or more carbon atoms—*e.g.*, 2-naphthaleneacetamide, but substitutive names are generally preferable.

*Discussion*

The trivial name urea is well established for the simplest acyclic diamide, $H_2NCONH_2$, as is biuret for $H_2NCONHCONH_2$ and triuret for $H_2NCONHCONHCONH_2$. When these names are used as parent names, a special numbering sequence including both the nitrogen and oxygen atoms is used to designate location of substituents (*see* Example 32). The prefix names *ureido* for the substituting group $H_2NCONH-$ and *ureylene* for $-NHCONH-$ are acceptable as alternatives to the corresponding systematic names. The group $H_2NCON<$ is named systematically as *carbamoylimino*. The names "carbamide" for $NH_2CONH_2$, "carbanilide" for $C_6H_5NHCONHC_6H_5$, and "carbanilic acid" for $C_6H_5NHCOOH$ should be abandoned.

The tautomeric form of urea, which has the imidic acid structure $H_2NC(=NH)OH$ and is known chiefly in the form of derivatives, is commonly called isourea (1965 IUPAC Rules) or pseudourea.

The sulfur analogs of urea and isourea have the trivial names thiourea and isothiourea or 2-thiopseudourea.

In naming amides of amino acids having officially recognized trivial names ending in *ine*, the final *e* of the amino acid name is changed to *amide*—*e.g.*, alaninamide for $CH_3CH(NH_2)CONH_2$.

*N*-Acylated derivatives of hydroxylamine and hydrazine, which may be regarded as *N*-hydroxy and *N*-amino amides, respectively, are discussed in Chapter 32.

Acyl derivatives of some aromatic and heterocyclic amines having trivial names have often been named as derivatives of the amines—*i.e.*, as anilides, *p*-toluidides, piperidides, morpholides, etc. Such usage is declining and desirably so; the 1965 IUPAC Rules authorize the anilide names only.

The terms secondary amide and tertiary amide are defined in the 1965 IUPAC Rules as denoting compounds having two and three acyl groups, respectively, attached to a nitrogen atom; these terms do not, therefore, refer, as supposed by some chemists, to amides of the types RCONHR and RCONR'R''. For symmetrical secondary and tertiary amides, names such as diacetamide for $(CH_3CO)_2NH$ and triacetamide for $(CH_3CO)_3N$ are recognized in the 1965 IUPAC Rules.

According to the 1965 IUPAC Rules and *Chemical Abstracts* current practice the class name imides is applied only to *cyclic* secondary amides and thus does not include acyclic compounds having the structure RCONHCOR. This has been a point of confusion in the past.

Amides in which one —CONH— linkage is part of a ring may be called lactams as a class. However, individually they should be named systematically as heterocyclic compounds.

The diamides formed by cyclization of dipeptides are also preferably named as heterocyclic compounds—*i.e.*, as derivatives of 1,4-diazacyclohexane-2,5-dione. Amides containing the ureylene group as part of a ring are also best named as heterocyclic compounds.

Table 21.1. Examples of Acceptable Usage

| | |
|---|---|
| 1.  $HCONH_2$ | Methanamide<br>Formamide |
| 2.  $CH_3CONH_2$ | Ethanamide<br>Acetamide |
| 3.  $CH_2{=}CHCH_2CONH_2$ | 3-Butenamide |
| 4.  $C_6H_5CONH_2$ | Benzenecarboxamide<br>Benzamide |
| 5.  $CH_3CONHC_6H_5$ | *N*-Phenylethanamide<br>*N*-Phenylacetamide<br>Acetanilide |
| 6.  $H_2NCONH_2$ | 1-Aminomethanamide<br>Urea<br>*not*  Carbamide |

7. H$_2$NCOCONH$_2$

Ethanediamide
Oxamide (*not* Oxalamide)

8. H$_2$NCOCH$_2$CH$_2$CONH$_2$

Butanediamide
Succinamide

9. H$_2$NCOCH$_2$C≡CCH$_2$CONH$_2$

3-Hexynediamide

10.

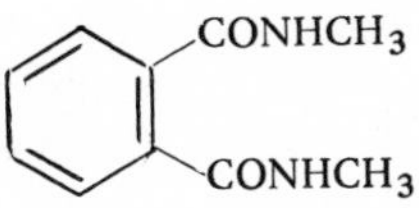

*N,N'*-Dimethyl-1,2-benzenedicarboxamide
*N,N'*-Dimethylphthalamide

11. CH$_3$CON(COCH$_3$)$_2$

*N,N*-Diethanoylethanamide
*N,N*-Diacetylacetamide
Triacetamide

12.

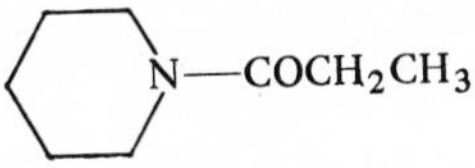

1-Propanoylazacyclohexane
1-Propionylpiperidine

13.

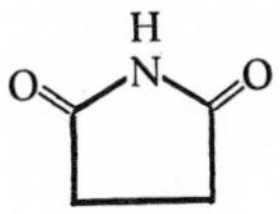

Azacyclopentane-2,5-dione
2,5-Pyrrolidinedione
Butanimide
Succinimide

14.

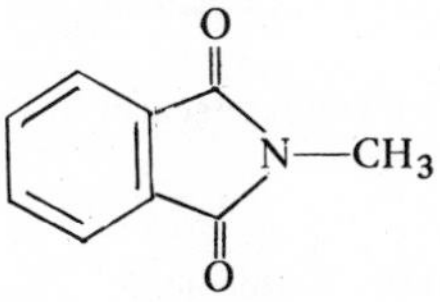

2-Methyl-2-azaindan-1,3-dione
*N*-Methyl-1,2-benzenedicarboximide
*N*-Methylphthalimide

15. CH$_3$CH$_2$CH$_2$CSNH$_2$

Butanethioamide
*not* Thiobutanamide
*not* Thiobutyramide

16. C$_6$H$_5$CSNH$_2$

Benzenecarbothioamide
*not* Thiobenzamide

17. H$_2$NCSNH$_2$

1-Aminomethanethioamide
Thiourea

18.

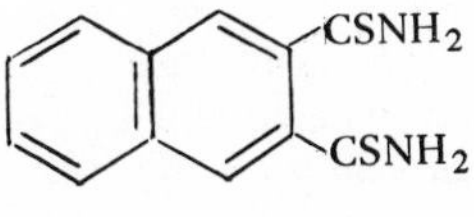

2,3-Naphthalenedicarbothioamide

19. 

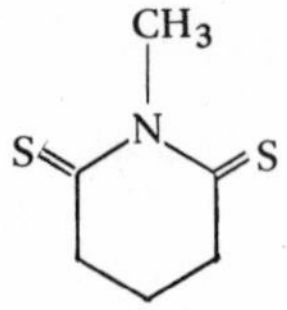

CH₃

1-Methylazacyclohexane-2,6-dithione
1-Methyl-2,6-piperidinedithione

20. 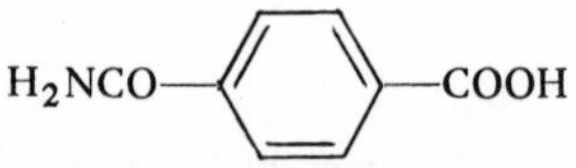 H₂NCO—⟨benzene⟩—COOH

4-Carbamoylbenzenecarboxylic acid
4-Carbamoylbenzoic acid
Terephthalamic acid

21. $H_2NCOOH$

Aminomethanoic acid
Carbamic acid

22. $CH_3CONHCH_2CH_2SO_2Cl$

2-Ethanamidoethanesulfonyl chloride
2-Acetamidoethanesulfonyl chloride

23. $C_6H_5CONHCH_2CH_2COOH$

3-Benzenecarboxamidopropanoic acid
3-Benzamidopropionic acid

24. 

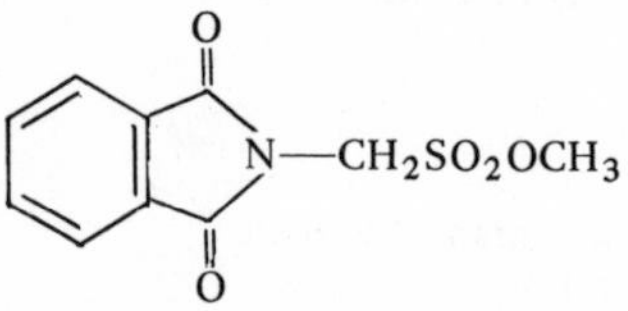

N—CH₂SO₂OCH₃

Methyl (1,3-dioxo-2-azaindan-2-yl)methanesulfonat
Methyl (1,2-benzenedicarboximido)methanesulfon
Methyl phthalimidomethanesulfonate

25. 

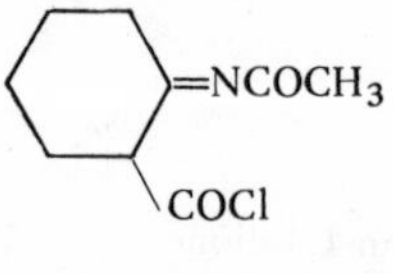

=NCOCH₃
COCl

2-(Ethanoylimino)cyclohexanecarbonyl
  chloride
2-(Acetylimino)cyclohexanecarbonyl
  chloride

26. $H_2NCSCH_2CH_2CH_2COOH$

4-(Thiocarbamoyl)butanoic acid
4-(Thiocarbamoyl)butyric acid

27.

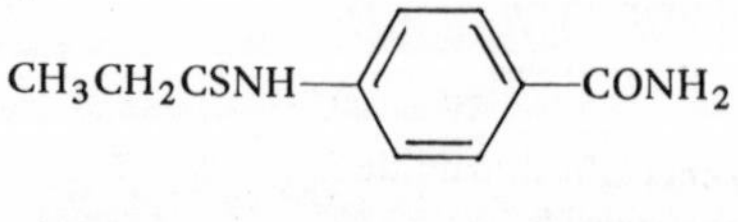

$CH_3CH_2CSNH$—⟨benzene⟩—$CONH_2$

4-(Propanethioamido)benzenecarboxamide
4-(Propanethioamido)benzamide

28. $HOOCCH_2CONHCOCH_2COOH$

(Iminodicarbonyl)diethanoic acid
(Iminodicarbonyl)diacetic acid

29. $H_2NCONHCH_2CH_2CH_2COOH$

4-(1-Aminomethanamido)butanoic acid
4-Ureidobutyric acid

30. $H_2NCONHCH_2OH$

1-Amino-*N*-(hydroxymethyl)methanamide
(Hydroxymethyl)urea
*not* Methylolurea

31. $H_2NCONHCONH_2$

1,1'-Iminodimethanamide
Biuret

32.

$$\underset{CH_3NHCONCONHCH_3}{\overset{\overset{\displaystyle CH_3}{|}}{}}$$

1,1'-(Methylimino)bis(*N*-methylmethanamide)
1,3,5-Trimethylbiuret

33. $H_2NCSNHCONH_2$

1-(1-Aminomethanethioamido)methanamide
Thiobiuret

34. $HN(COOCH_2CH_3)_2$

Diethyl iminodimethanoate
Diethyl iminodiformate

35. $H_2NCH_2CONH_2$

2-Aminoethanamide
2-Aminoacetamide
Glycinamide

# 22

# Amidines

Compounds containing one or more $-C(=NH)NH_2$ groups are called amidines. The amidine function may be regarded as being derived from the carboxylic acid function by simultaneous replacement of OH with $NH_2$ and of $=O$ with $=NH$; alternatively, an amidine may be considered to be the amide of the corresponding carboximidic acid (*see* Chapter 16). Although the ending *amidine* has been used in naming analogous structures derived from sulfinic acids (*see* IUPAC Rule C-641.9), extension of the class name amidine to include such nitrogen derivatives of noncarboxylic acids has not otherwise become established.

*Recommended Nomenclature Practice*

Acyclic and cyclic amidines are named preferably by replacing the ending *oic acid* of the corresponding systematic alkanoic or alkanedioic acid name with the suffix *amidine*, or by changing the ending *carboxylic acid*, where that is used in the acid name, to *carboxamidine*.

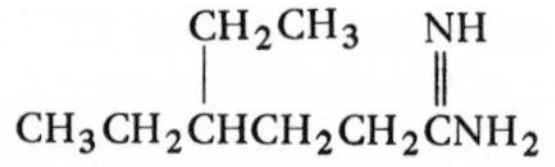

$$\underset{CH_3CH_2\underset{\underset{CH_2CH_3}{|}}{CH}CH_2CH_2\overset{\overset{NH}{\|}}{C}NH_2}{}$$  4-Ethylhexanamidine

$$H_2N\overset{\overset{NH}{\|}}{C}CH_2CH_2CH_2\overset{\overset{NH}{\|}}{C}NH_2$$  Pentanediamidine

$$\underset{\underset{NH}{\|}}{H_2N\overset{\overset{NH}{\|}}{C}}\!\!\diagdown\!\!\underset{H_2N\overset{}{C}}{}\!\!\diagup\!\!CHCH_2CH\!\!\diagdown\!\!\underset{\underset{NH}{\|}}{\overset{\overset{NH}{\|}}{C}NH_2}\!\!\diagup\!\!\underset{CNH_2}{}$$  1,1,3,3-Propanetetracarboxamidine

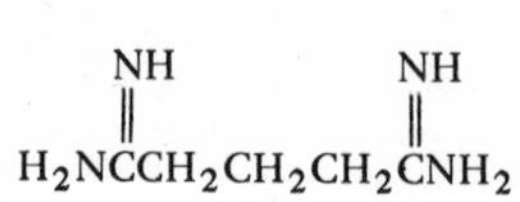

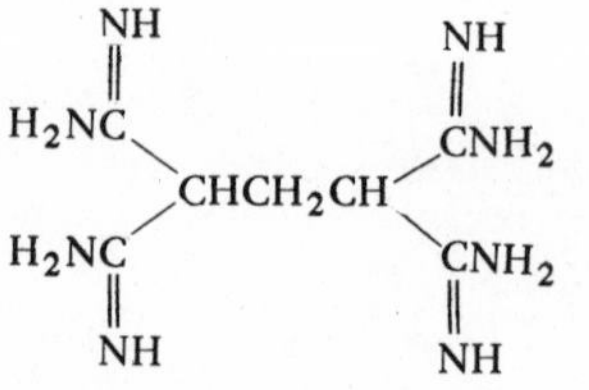

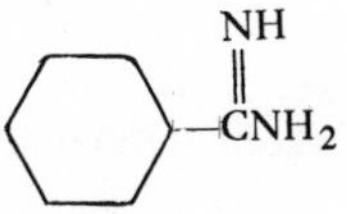

Cyclohexanecarboxamidine

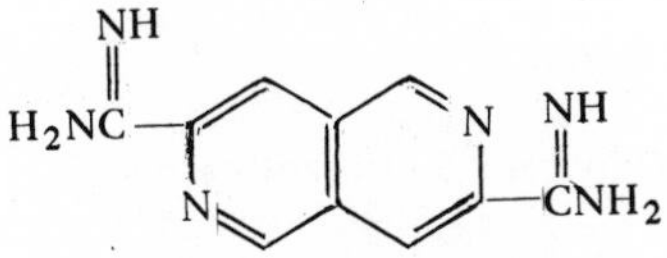

2,6-Diazanaphthalene-3,7-dicarboxamidine

Substituted amidines are named by using either of two types of locant depending on the degree of structural specificity desired in the name. If a distinction between the amide and imide nitrogen atoms is intended, the locants $N^1$ and $N^2$ are used to denote the $NH_2$ and $=NH$ groups, respectively; if no such distinction is to be made, the locants $N$ and $N'$ are used.

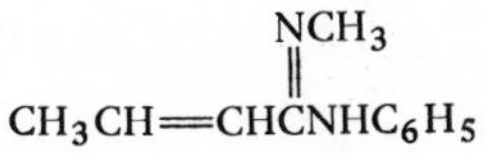

$N^2$-Methyl-$N^1$-phenyl-2-butenamidine

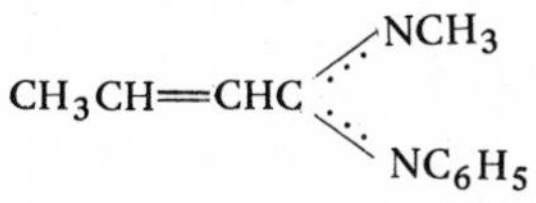

$N$-Methyl-$N'$-phenyl-2-butenamidine

When one or both nitrogen atoms of an amidine group are members of a ring system, the compound is named on the basis of its heterocyclic structure (*see* Chapter 6).

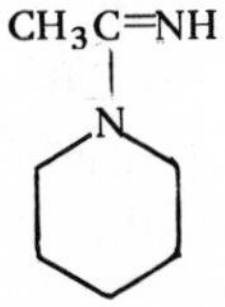

1-Ethanimidoylazacyclohexane

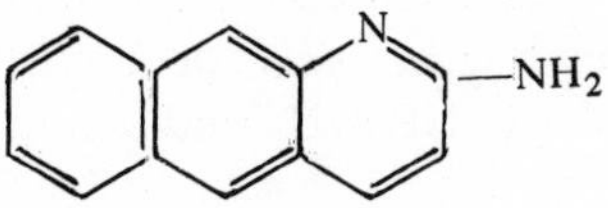

1-Azaanthracen-2-amine

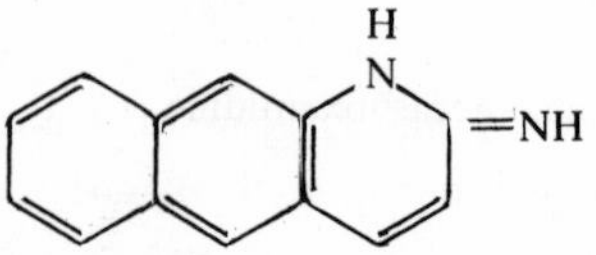

1-Azaanthracen-2(1*H*)-imine

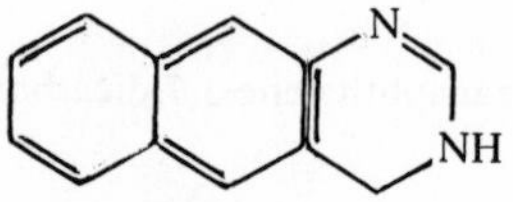

3,4-Dihydro-1,3-diazaanthracene

Compounds of the types [RC(=NH)]$_2$NH and [RC(=NH)]$_3$N, whether or not all R groups are alike or different, are best treated as *N*-substituted amidines. For unsymmetrical structures the largest or most complex acyl group is preferably expressed in the parent name.

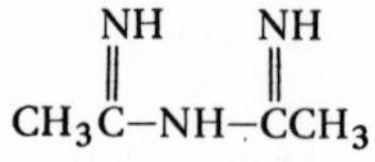

*N*-Ethanimidoylethanamidine

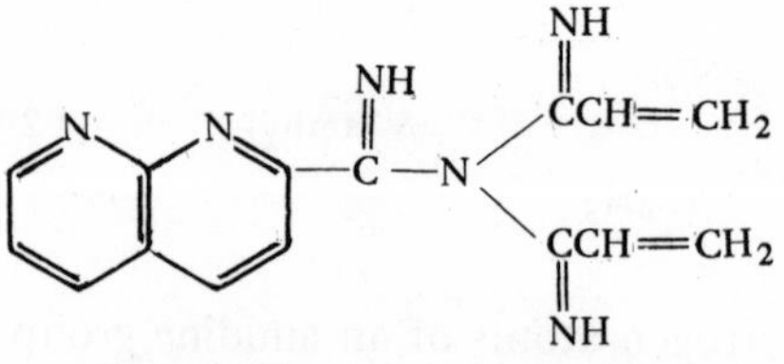

*N,N*-Dipropenimidoyl-1,8-diazanaphthalene-2-carboxamidine

Selection of **prefix names** for the amidine function when it must be treated as a substituting group depends upon the mode of attachment: the prefix *carbamimidoyl* is to be preferred for H$_2$NC(=NH)— whereas names of amidine groups attached through nitrogen are best formed by combining the names of the individual components—*e.g.,* (ethanimidoyl-amino) or [(1-iminoethyl)amino] for CH$_3$C(=NH)NH— and [(1-amino-ethylidene)amino] for CH$_3$C(NH$_2$)=N—.

In cases where the corresponding carboxylic acid has an acceptable trivial name (*see* Chapter 16) trivial names may be used for amidines. Such names are formed by changing the ending *ic acid* or *oic acid* to *amidine—e.g.,* propionamidine for CH$_3$CH$_2$C(=NH)NH$_2$ and benzamidine for C$_6$H$_5$C(=NH)NH$_2$.

*Discussion*

Although in many respects the naming of amidines parallels that of amides, the former do present some added complexities since unlike amides they contain two dissimilar nitrogen atoms each of which may carry substituents. However, as long as one hydrogen atom is present which can shift tautomerically from one nitrogen atom to the other, it is usually not necessary to specify in the name which nitrogen is in the imino form and which is in the amino form. Thus, although the 1965 IUPAC Rules suggest locants ($N^1$ and $N^2$) that provide a means for naming either tautomeric form of a substituted amidine unambiguously, *Chemical Abstracts* does not make this distinction in the index name itself but simply uses the less specific locants $N$ and $N'$.

Unlike class names for amides the class terms secondary amidine and tertiary amidine have neither been used much nor officially recognized for structures in which two or three RC(=NH)— groups, respectively, are attached to a single nitrogen atom. Accordingly, *Chemical Abstracts* does not use names such as "diacetamidine" or "tribenzamidine" and this type of name is not recognized by IUPAC. There is also no separate class name for cyclic amidines having structures analogous to those of imides.

In its indexes *Chemical Abstracts* uses conjunctive names (*see* Chapter 9) for acyclic amidines wherein the amidine group is connected to a ring system through a chain of one or more carbon atoms. Nevertheless, as recommended generally throughout this book, substitutive names are preferred.

Although often seen in the older literature, the term "guanyl" should not be used as a prefix name for the substituting group $H_2N(HN=)C—$. For a number of years *Chemical Abstracts* has used *amidino* rather than "guanyl," but this prefix does not express the carbon atom, nor does a terminal *o* denote attachment through carbon. Actually, by analogy with *amido,* the term *amidino* might be used logically to name amidine substituents attached through nitrogen—*e.g.,* ethanamidino for $CH_3C$-(=NH)NH— and cyclohexanecarboxamidino for $C_6H_{11}C(=NH)NH—$, but such usage has no official status.

Among acceptable trivial names for amidines, the names guanidine for $H_2NC(=NH)NH_2$, biguanide for $H_2NC(=NH)NHC(=NH)NH_2$, isourea (1965 IUPAC Rules) or pseudourea (*Chemical Abstracts*) for $HOC(=NH)$-$NH_2$, and isothiourea or thiopseudourea for $HSC(=NH)NH_2$ have been used widely to the virtual exclusion of systematic names. When these trivial names are used as parent names, special numbering is required to indicate location of substituents (*see* Examples 6, 15, and 16 in Table 22.1).

The trivial prefix name *guanidino* for $H_2NC(=NH)NH—$ as used by *Chemical Abstracts* is acceptable as an alternative to the systematic name *carbamimidoylamino*; the substituting group $(H_2N)_2C=N-$ is named systematically as *diaminomethyleneamino*.

Naming of compounds containing amidine groups carrying OH and/or $NH_2$ substituents attached to either of the nitrogen atoms is discussed in Chapter 32.

## Table 22.1. Examples Of Acceptable Usage

1.

$$\underset{HCNH_2}{\overset{NH}{\|}}$$

Methanamidine
Formamidine

2.

$$\underset{CH_3CNH_2}{\overset{NH}{\|}}$$

Ethanamidine
Acetamidine

3.

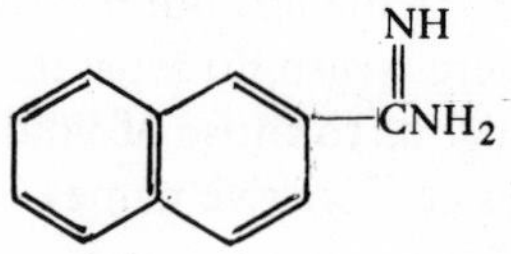

$$\overset{NH}{\overset{\|}{CNH_2}}$$

2-Naphthalenecarboxamidine
2-Naphthamidine

4.

$$C_6H_5CH=CH\underset{}{\overset{NH}{\overset{\|}{C}}}NHCH_3$$

$N^1$-Methyl-3-phenylpropenamidine

5.

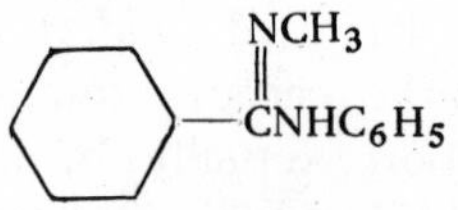

$$\overset{NCH_3}{\overset{\|}{CNHC_6H_5}}$$

$N^2$-Methyl-$N^1$-phenylcyclohexanecarbox-
amidine

6.

$$\underset{\underset{1\quad 3}{H_2N\overset{}{C}NH_2}}{\overset{\overset{2}{NH}}{\|}}$$

1-Aminomethanamidine
Guanidine (numbering shown)

7.

$$\underset{H_2NC-CNH_2}{\overset{HN\quad NH}{\|\quad\|}}$$

Ethanediamidine
Oxamidine (*not* Oxalamidine)

8.

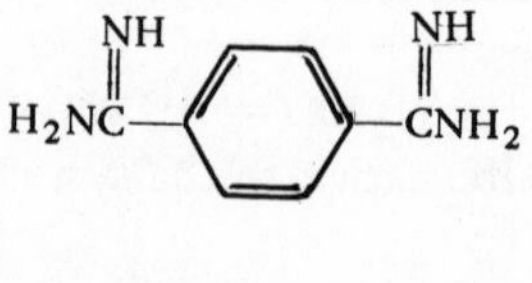

$$H_2N\overset{NH}{\overset{\|}{C}}\!\!-\!\!\!\!\!\!\!\!\!\overset{NH}{\overset{\|}{CNH_2}}$$

1,4-Benzenedicarboxamidine
Terephthalamidine

9.

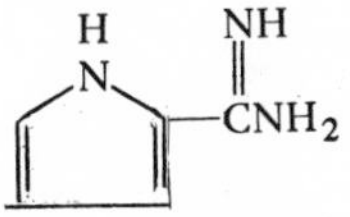

Azacyclopenta-2,4-diene-2-carboxamidine
Pyrrole-2-carboxamidine

10.

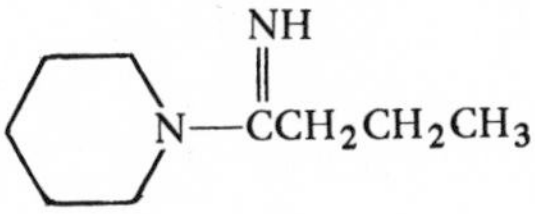

1-Butanimidoylazacyclohexane
1-Butanimidoylpiperidine

11.

$$\underset{NH_2}{\overset{NH}{\parallel}}CCH_2CH_2CH_2COOH$$

4-Carbamimidoylbutanoic acid
4-Carbamimidoylbutyric acid

12.

$$CH_3\overset{NH}{\overset{\parallel}{C}}NHCH_2COOCH_3$$

Methyl (ethanimidoylamino)ethanoate
Methyl (acetimidoylamino)acetate

13.

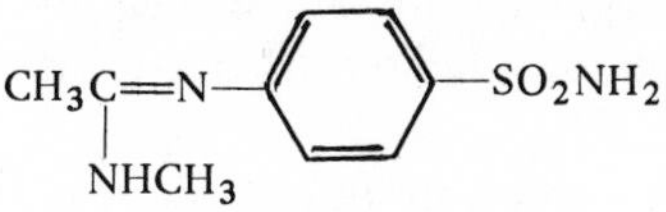

4-[1-Methylamino)ethylideneamino]-
benzenesulfonamide

14.

$$\underset{1\quad\quad 3\quad\quad 5}{NH_2\overset{\overset{2}{NH}}{\overset{\parallel}{C}}NH\overset{\overset{4}{NH}}{\overset{\parallel}{C}}NH_2}$$

1,1'-Iminodimethanamidine
Biguanide (numbering shown)

15.

$$\underset{3\quad 2}{H_2N\overset{\overset{1}{NH}}{\overset{\parallel}{C}}OH}$$

1-Aminomethanimidic acid
Isourea (numbering shown)
Pseudourea[a]

16.

$$\underset{3\quad 2}{H_2N\overset{\overset{1}{NH}}{\overset{\parallel}{C}}SH}$$

1-Aminomethanimidothioic acid
Isothiourea (numbering shown)
2-Thiopseudourea[a]

17.

$$H_2N\overset{NH}{\overset{\parallel}{C}}NHCH_2CH_2CONH_2$$

3-(Carbamimidoylamino)propanamide
3-Guanidinopropionamide

[a]The names pseudourea and 2-thiopseudourea, although not recognized in the 1965 IUPAC Rules, have been used by *Chemical Abstracts*. For both of these names the locants 1, 2, and 3 designate substitution on =NH, O (or S), and -NH$_2$, respectively.

# 23

# Amines and Imines

A n amine may be defined as a derivative of ammonia ($NH_3$) in which one or more hydrogen atoms have been replaced by non-acyl organic groups linked to the nitrogen atom through carbon. In addition, the subclassifying terms **primary, secondary,** and **tertiary** are used to distinguish among amines conceptually derived by the replacement of one, two, or three such hydrogen atoms, respectively. Compounds in which two hydrogen atoms of $NH_3$ have been replaced by a group attached through a single carbon atom with formation of a double bond are called imines; when the remaining hydrogen atom is replaced by a non-acyl organic group the structure may be regarded as either a substituted imine or a substituted amine.

When a nitrogen atom replaces carbon in an organic ring system, the resulting structure is named as a heterocyclic compound (*see* Chapter 6); for nomenclature purposes it is not formally classed as an amine or imine.

### Recommended Nomenclature Practice

Primary monoamines are preferably named substitutively by adding the suffix *amine* to the name of the chain or ring system to which the $NH_2$ group is attached, with elision of the terminal *e* in names wherein it occurs.

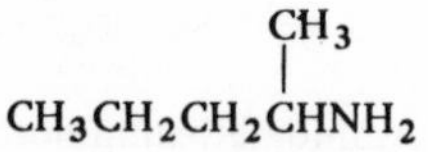

2-Pentanamine

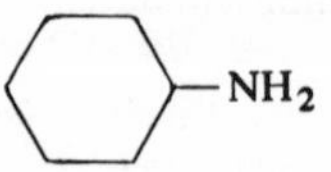

Cyclohexanamine

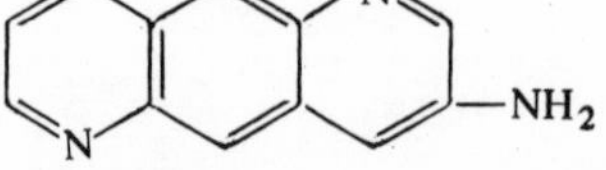

1,5-Diazaanthracen-3-amine

Primary **polyamines** are best named substitutively, by using the suffixes *diamine, triamine, tetramine, pentamine,* etc.

180

$$\underset{\phantom{x}}{\overset{\overset{\displaystyle CH_3}{|}}{H_2NCH_2CH_2CHNH_2}}$$

1,3-Butanediamine

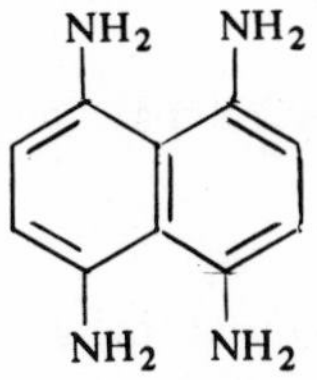

1,4,5,8-Naphthalenetetramine

   **Secondary and tertiary acyclic amines** having no more than two NH or NR groups within a single straight chain otherwise composed of carbon atoms are preferably named substitutively as derivatives of the highest-ranking primary amine structure present. The parent primary amine is selected by applying the following criteria in the order listed: (1) largest number of amine groups expressed in the suffix, (2) largest number of unsaturated linkages, (3) longest parent chain, and (4) lowest locants for amine groups expressed in the suffix. The locants $N^1$, $N^2$, $N^3$, etc. (or $N$ and $N'$ when no ambiguity would result) are used to designate substituents attached to $NH_2$ groups of the parent structure.

$CH_3CH_2CH_2NHCH_3$                      *N*-Methyl-1-propanamine

$$\underset{\phantom{x}}{\overset{\phantom{x}}{CH_3CH_2CH_2NHCH_2\underset{\underset{\displaystyle NHCH_3}{|}}{CH}CH_2NHCH_3}}$$

$N^1$, $N^2$-Dimethyl-$N^3$-propyl-1,2,3-propanetriamine

$$\underset{\phantom{x}}{\overset{\overset{\displaystyle CH_3}{|}}{CH_2{=}CHCH_2NCH_2C{\equiv}CH}}$$

*N*-Methyl-*N*-(2-propynyl)-2-propen-1-amine

   As alternatives to alkanamine-type names for **symmetrical acyclic second-ary and tertiary monoamines**, substitutive names based on *amine*, denoting the parent compound $NH_3$, are acceptable.

$CH_3CH_2NHCH_2CH_3$                      *N*-Ethylethanamine
                                         Diethylamine

$(ClCH_2CH_2)_3N$                         2-Chloro-*N*,*N*-bis(2-chloroethyl)-
                                         ethanamine
                                         Tris(2-chloroethyl)amine

**Acyclic polyamines** having more than two NH or NR groups within a single straight chain are best named by using replacement nomenclature (*see* Chapter 7).

$$CH_3NHCH_2CH_2NCH_2CH_2NHCH_3$$

with CH₃ branch: 

CH₃
|
CH₃NHCH₂CH₂NCH₂CH₂NHCH₃          5-Methyl-2,5,8-triazanonane

Cyclic amines in which one or more nitrogen atoms are each directly connected to two or more ring systems are preferably named substitutively as derivatives of the highest-ranking primary amine structure present. The parent primary amine is selected by applying the following criteria in the order listed: (1) largest number of amine groups expressed in the suffix, and (2) seniority of parent ring system as specified by IUPAC Rule C-14.1 and the Introduction to the Subject Index, *Chem. Abstr.* (1967) 66, ¶ 29-33 and (3) largest number of *N*-substituents. In the ranks of ring systems, all heterocycles are preferred to all carbocycles, and nitrogen-containing heterocycles are preferred to those not containing nitrogen; consideration is next given to (*a*) number of rings, (*b*) type of ring fusion, in the order spiro, bridged, and non-bridged, and (*c*) total number of atoms in the ring system. As recommended above for acyclic secondary and tertiary amines, the locants $N^1, N^2, N^3$, etc. are used when necessary to avoid ambiguity.

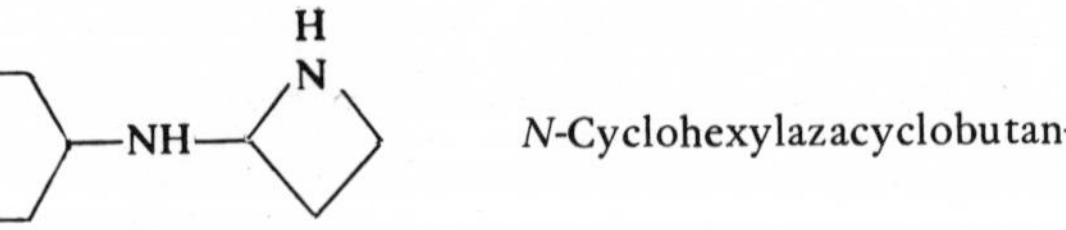

$N^1$-(4-Aminophenyl)-$N^4$-phenyl-1,4- benzenediamine

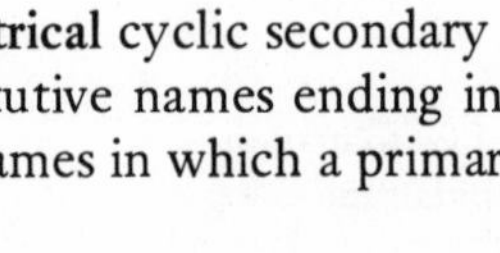

*N*-Cyclohexylazacyclobutan-2-amine

For **symmetrical** cyclic secondary and tertiary monoamines the commonly used substitutive names ending in *amine* are acceptable alternatives to the preferred names in which a primary amine is the parent compound.

*N*-Phenylbenzenamine
Diphenylamine

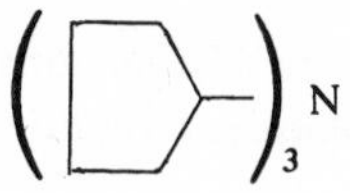

$$\left(\left[\square\right]-\right)_3 N \qquad \begin{array}{l} N,N\text{-Dicyclopentylcyclopentanamine} \\ \text{Tricyclopentylamine} \end{array}$$

Cyclic-acyclic amines are best named substitutively as derivatives of the senior primary amine structure present. In selecting the parent amine, the chain or ring system carrying the most amine groups is preferred; if there is a further choice, all heterocyclic systems are preferred to all cyclic and acyclic hydrocarbon systems; otherwise, the largest hydrocarbon is chosen. Additional seniority criteria are provided in the Introduction to the Subject Index, *Chem. Abstr.* (1967) *66*, ¶ 29-33.

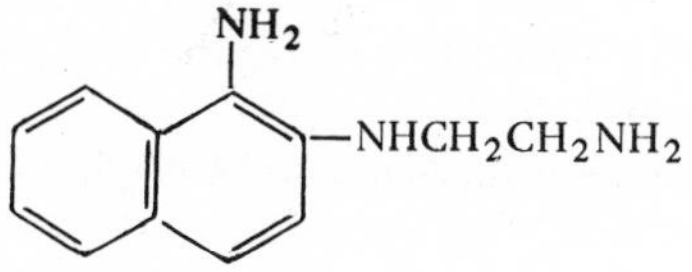

$$\text{—NHCH}_2\text{CH}_2\text{NH}_2 \qquad N^2\text{-(2-Aminoethyl)-1,2-naphthalenediamine}$$

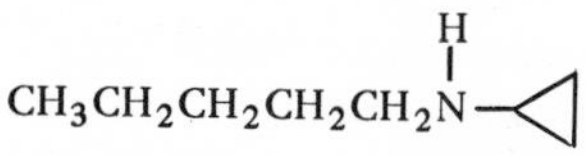

$$\text{CH}_3\text{CH}_2\text{CH}_2\text{CH}_2\text{CH}_2\overset{\overset{\text{H}}{|}}{\text{N}}\text{—}\triangleleft \qquad N\text{-Pentylcyclopropanamine}$$

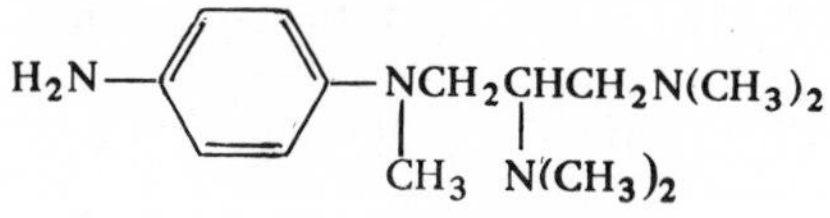

$$\text{H}_2\text{N—}\left[\bigcirc\right]\text{—NCH}_2\underset{\underset{\text{N(CH}_3)_2}{|}}{\overset{\overset{}{}}{\text{CHCH}_2}}\text{N(CH}_3)_2$$

$$\underset{\text{CH}_3}{|}$$

$N^1$-(4-Aminophenyl)-$N^1,N^2,N^2,N^3,N^3$-pentamethyl-1,2,3-propanetriamine

Straight chains containing more than two NH or NR groups are best named by replacement nomenclature (*see* Chapter 7).

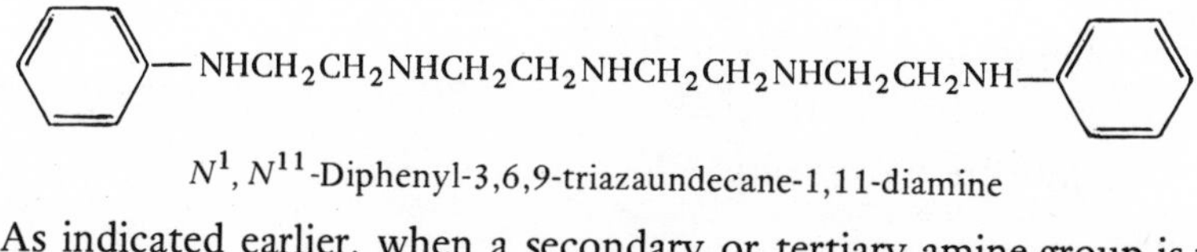

$$\text{—NHCH}_2\text{CH}_2\text{NHCH}_2\text{CH}_2\text{NHCH}_2\text{CH}_2\text{NHCH}_2\text{CH}_2\text{NH—}$$

$N^1, N^{11}$-Diphenyl-3,6,9-triazaundecane-1,11-diamine

As indicated earlier, when a secondary or tertiary amine group is part of a ring system, the latter is named as a heterocycle (*see* Chapter 6). The prefix *aza* denotes the amine function in replacement names for heterocyclic systems.

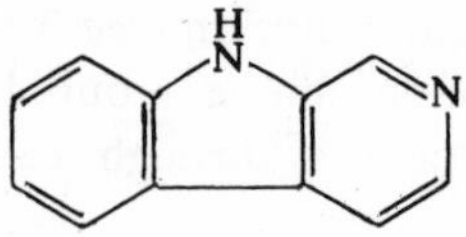

*9H*-2,9-Diazafluorene

As an alternative to the regularly formed heterocycle names for structures in which a nitrogen atom forms a bridge connecting two atoms of a ring system, the bivalent non-detachable prefix *epimino* may be used.

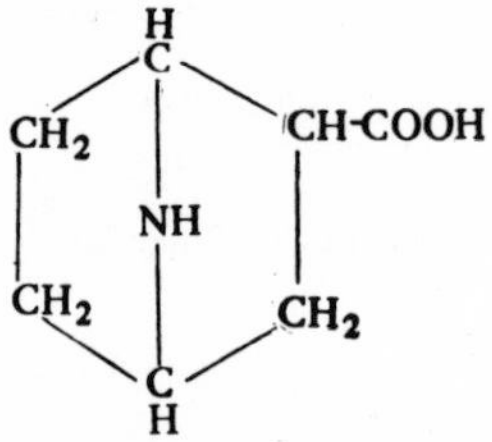

7-Azabicyclo[2.2.1]heptane-2-carboxylic acid
2,5-Epiminocyclohexanecarboxylic acid

When the amine function must be named as a substituent one of the prefix names *amino, imino,* or *nitrilo* is used, depending upon the number of groups (one, two, or three) to which the nitrogen atom is attached.

$(CH_3)_2NCH_2CH_2CH_2OH$          3-(Dimethylamino)-1-propanol

$HOOCCH_2NHCH_2COOH$          Iminodiethanoic acid

$N(CH_2CH_2CN)_3$          3,3',3''-Nitrilotripropanenitrile

The trivial name **aniline** for $C_6H_5NH_2$ is acceptable. Other amines should be named systematically as described above.

**Acyclic and cyclic imines** are preferably named substitutively by adding the suffix *imine* to the name of the chain or ring system to which the =NH group is attached, with elision of the terminal *e* in names where it occurs. **Polyimines** are named similarly by using the customary multiplying prefixes and retaining the terminal *e*.

$CH_3CH_2CH_2CH_2CH{=}NH$          1-Pentanimine

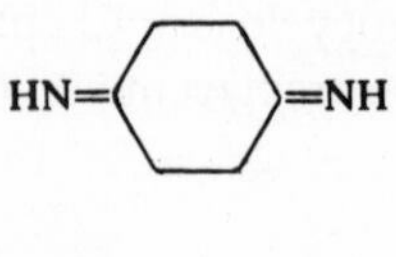

1,4-Cyclohexanediimine

Since the amine function is senior to the imine function (*see* Table 8.1), structures in which the hydrogen atom of an imine group has been replaced by a non-acyl organic group attached through carbon are preferably named as substituted amines.

$=NCH_2CH_2CH_3$     *N*-Cyclohexylidene-1-propanamine

The prefix name for the imine group is *imino*.

HN=     =O     4-Imino-2,5-cyclohexadien-1-one

Amine and imine derivatives having an oxygen atom bound to the nitrogen atom are preferably named as **azane oxides** but the more familiar amine oxide type of name is an acceptable alternative.

$(CH_3CH_2)_3NO$     Triethylazane oxide
                     Triethylamine oxide

$CH_3CH_2CH=N(O) C_6H_5$     Phenylpropylidenazane oxide
                            *N*-Propylidenebenzenamine oxide

## Discussion

Over a period of many years, two methods, both systematic, have been used in naming amines. Although both these methods are substitutive, they differ in the choice of the parent compound. No clear-cut distinction has evolved, but in general monoamines have more often been named by citing the substituting group(s) attached to the parent molecule $NH_3$; in this type of name the latter is called "amine" rather than ammonia. In the alternative kind of name, which has been used predominately for polyamines, the term *amine* is a functional suffix appended to the name of a chain or ring system denoting substitution of H by $NH_2$; secondary and tertiary amines are treated as *N*-substituted primary amines. For reasons of consistency with the treatment of other functional classes, names in which the amine function is expressed as a suffix are now recommended for all amines except those containing a higher-ranking functional group (*see* Table 8.1).

Conjunctive names (*see* Chapter 9) such as 1-naphthalenemethylamine for 1-(1-naphthyl)methanamine, although used by *Chemical Abstracts*, are not recommended, nor is the special treatment accorded nitrogen-containing heterocyclic amines whereby directly attached $NH_2$ groups are named as prefixes rather than suffixes—*e.g.*, "2,6-diaminopyridine" rather than 2,6-pyridinediamine. For general use, regularly formed substitutive names are much to be preferred.

Semisystematic names such as "triethylenetetramine" for $H_2N$-$CH_2CH_2NH$ $CH_2CH_2NH$ $CH_2CH_2NH_2$ should be abandoned in favor of systematic substitutive names, *e.g.*, *N,N'*-bis(2-aminoethyl)-1,2-ethanediamine, or, if desired, replacement names—*e.g.*, 3,6-diazaoctane-1,8-diamine.

If unsymmetrical secondary and tertiary amines are to be named as substitution products of $NH_3$, each group attached to nitrogen should be cited separately, in its entirety. Names such as 1-chloro-2'-fluorodiethylamine, though recognized in the 1965 IUPAC Rules, are poorly formed because they imply that the parent compound is $(C_2H_5)_2NH$ rather than $NH_3$. The name (1-chloroethyl)(2-fluoroethyl)amine is therefore a much better choice.

The use of "imine" as a terminal modifying word denoting conversion of =O to =NH in quinone-type structures, although officially sanctioned, is poor practice and should be discontinued (*see* Example 20).

Organic structures in which a nitrogen atom is directly connected to an atom other than carbon and hydrogen, as in amine oxides, are best named systematically as derivatives of azane, diazane, triazane, etc. (*see* Chapter 32).

By this approach $(CH_3)_3NO$ is called trimethylazane oxide and $(CH_3)_2NOH$ is called dimethylazanol. Such names, though logically formed, have not yet been officially recognized; current usage calls for trimethylamine oxide and *N,N*-dimethylhydroxylamine, and these names are, of course, entirely acceptable.

Naming of oxides of *N*-alkylidenimines as derivatives of the parent compound "nitrone", $CH_2=NH(O)$, should be discontinued (*see* Example 21).

## Table 23.1.   Examples Of Acceptable Usage

| 1. | $CH_3NH_2$ | Methanamine<br>Methylamine |
|---|---|---|
| 2. | $CH_3\underset{\underset{NH_2}{\vert}}{CH}CH_2CH_3$ | 2-Butanamine<br>*sec*-Butylamine |
| 3. | $C_6H_5NH_2$ | Benzenamine<br>Phenylamine<br>Aniline |

4.

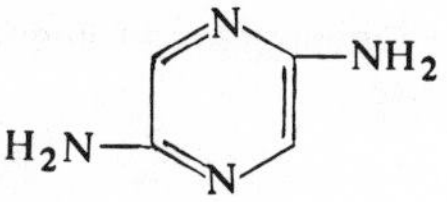

1,4-Diazabenzene-2,5-diamine
2,5-Pyrazinediamine

5. $H_2NCH_2CH_2NHCH_2CH_2NH_2$

*N*-(2-Aminoethyl)-1,2-ethanediamine
Bis(2-aminoethyl)amine
*not* Diethylenetriamine

6.

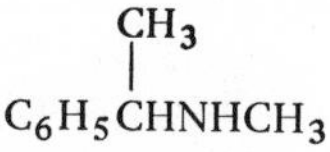

$$\underset{C_6H_5CHNHCH_3}{\overset{CH_3}{|}}$$

*N*-Methyl-1-phenylethanamine

7. $(C_6H_5CH_2)_2NH$

1-Phenyl-*N*-(phenylmethyl)methanamine
Dibenzylamine

8. $(CH_3)_2NCH_2N(CH_3)_2$

*N,N,N,'N'*-Tetramethylmethanediamine

9. $(CH_3CH_2CH_2CH_2)_3N$

*N,N*-Dibutyl-1-butanamine
Tributylamine

10.

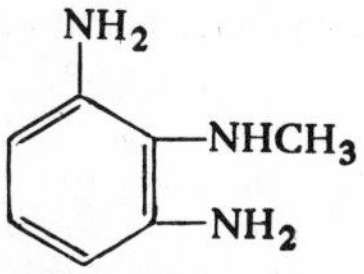

$N^2$-Methyl-1,2,3-benzenetriamine

11. $H_2NCH_2CH_2(NHCH_2CH_2)_2NHCH_2CH_2NH_2$

3,6,9-Triazaundecane-1,11-diamine
*not* Tetraethylenepentamine

12.

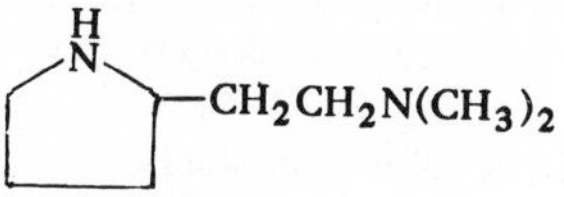

2-(Azacyclopent-2-yl)-*N,N*-dimethyl-
ethanamine

13.

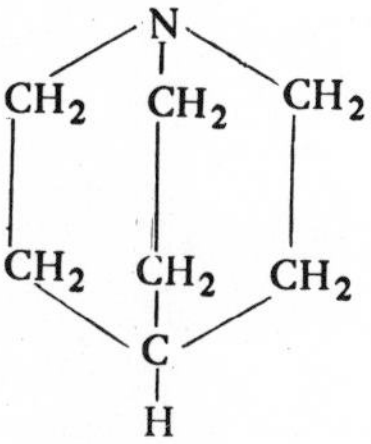

1-Azabicyclo[2.2.2]octane
Quinuclidine

14. $(C_6H_{11})_3N$

*N,N*-Dicyclohexylcyclohexanamine
Tricyclohexylamine

15.   $C_6H_5NHCH_2CH_2CH_2NHCH_3$        *N*-Methyl-*N'*-phenyl-1,3-propanediamine

16. 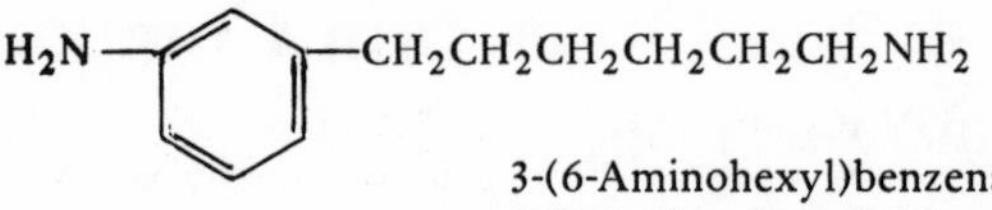   $H_2N$ —⟨benzene ring⟩— $CH_2CH_2CH_2CH_2CH_2CH_2NH_2$

3-(6-Aminohexyl)benzenamine
6-(3-Aminophenyl)-1-hexanamine

17.   $C_6H_5N=CHCH_3$                    *N*-Ethylidenebenzenamine

18.   $\underset{NH\quad\ NH}{CH_3\overset{\parallel}{C}CH_2\overset{\parallel}{C}H_2\overset{\parallel}{C}CH_3}$

2,5-Hexanediimine

19.   $C_6H_5\overset{O}{N}(CH_3)_2$

Dimethylphenylazane oxide
*N,N*-Dimethylbenzenamine oxide

20. 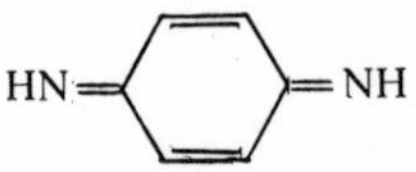   $HN=$⟨cyclohexadiene ring⟩$=NH$

2,5-Cyclohexadiene-1,4-diimine
*not* *p*-Benzoquinone diimine

21.   $CH_3CH=\overset{O}{N}CH_2CH_2OH$

Ethylidene(2-hydroxyethyl)azane oxide
*N*-Ethylidene-2-hydroxyethanamine oxide
*not* *N*-(2-hydroxyethyl)-α-methylnitrone

22.   $C_6H_{11}N=C=NC_6H_{11}$          Dicyclohexylmethanediimine
                                         Dicyclohexylcarbodiimide
                                         (*see* Chapter 32)

# 24

# Ammonium and Other Ium Compounds

Ion-pair structures containing one or more organic cationic groups in which the positively charged atom is a nonmetallic element other than carbon are commonly referred to as **ium compounds**. Members of this family are divided into two main categories: (1) ordinary salts formed by protonation of an atom carrying an unshared pair of electrons. and (2) ionic compounds in which the positively charged atom is not attached to hydrogen. The latter are called **onium compounds**, and within this category the most commonly encountered examples are those in which the cationic atom is nitrogen—namely the **ammonium compounds**, $R_4N^+ X^-$. Since there has been some usage of the term "ammonium" in naming protonated amino groups, as in "dimethylammonium chloride" for $(CH_3)_2NH_2^+ Cl^-$, the class name **quaternary ammonium compounds** has been introduced to prevent ambiguity.

This chapter deals chiefly with the naming of amine salts and quaternary ammonium compounds. The general principles so exemplified may be extended to other ium and onium compounds. **Diazonium compounds** are considered here, but other nitrogen-based cationic structures are discussed in Chapter 32. Ions and ion-pairs in which a carbon atom is positively or negatively charged are treated in Chapter 28; although **carbenium compounds** certainly fall within the broad class of ium compounds, common usage of the latter term has usually implied

**Table 24.1. Names of Organic Onium Ions**
**(Substituent Names Omitted)**

| Formula | Parent Name | Prefix Name |
|---|---|---|
| $R_4N^+$ | Ammonium | Ammonio |
| $R_4P^+$ | Phosphonium | Phosphonio |
| $R_4As^+$ | Arsonium | Arsonio |
| $R_4Sb^+$ | Stibonium | Stibonio |
| $R_3O^+$ | [a] Oxonium | Oxonio |
| $R_3S^+$ | Sulfonium | Sulfonio |
| $R_3Se^+$ | Selenonium | Selenonio |
| $R_2F^+$ | Fluoronium | Fluoronio |
| $R_2Cl^+$ | Chloronium | Chloronio |
| $R_2Br^+$ | Bromonium | Bromonio |
| $R_2I^+$ | Iodonium | Iodonio |

[a]Hydronium is more often used for the unsubstituted cation $H_3O^+$.

189

structures containing organic cationic groups in which the central atom is an element other than carbon.

### Recommended Nomenclature Practice

Organic cationic groups in which the positive charge is considered to be at a central atom (other than carbon) not attached to hydrogen are named substitutively as derivatives of the parent cations listed in Table 24.1, except when the charged atom is part of a ring system.

$(CH_3CH_2)_4N^+$ $Cl^-$          Tetraethylammonium chloride

$C_6H_5\overset{+}{P}(CH_3)_3$ $OH^-$          Trimethylphenylphosphonium hydroxide

$(CH_3)_3S^+$ $Br^-$          Trimethylsulfonium bromide

Cyclic cationic structures are named as derivatives of heterocyclic systems (*see* Chapter 6). When replacement nomenclature is used, the non-detachable prefixes *azonia, phosphonia,* etc. designate cationic hetero atoms. In Hantzsch-Widman and acceptable trivial names, the suffix *ium* is added (with elision of a terminal *e* if present) to the name of the corresponding heterocycle.

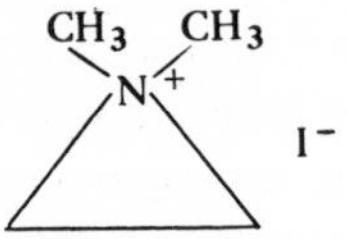

1,1-Dimethylazoniacyclopropane iodide
1,1-Dimethylaziridinium iodide

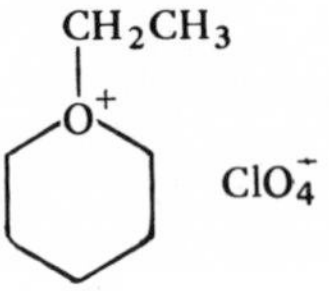

1-Ethyloxoniacyclohexane perchlorate
1-Ethyltetrahydro-2*H*-pyranium perchlorate

Onium compounds, both acyclic and cyclic, in which the compensating negative center resides in the same molecule are preferably named by citing the cationic group as a prefix (*see* Table 24.1) and the anionic group as a suffix.

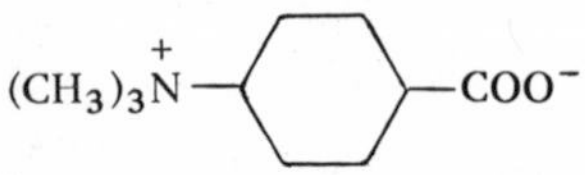

4-(Trimethylammonio)cyclohexane-carboxylate

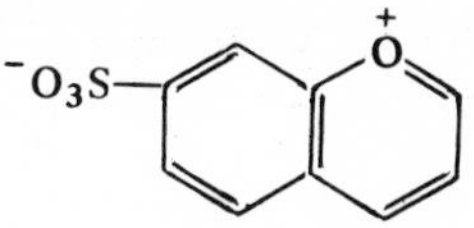

1-Oxonia-7-naphthalenesulfonate

The prefix names shown in Table 24.1 are also used whenever an acyclic onium group must be named as a substituent. This occurs, for example, when more than one such group is present in the same molecule (*see* Discussion).

$$(CH_3)_2\overset{+}{S}CH_2CH_2\overset{+}{P}(CH_3)_3 \quad 2\ Cl^-$$

[2-(Dimethylsulfonio)ethyl] trimethylphosphonium dichloride

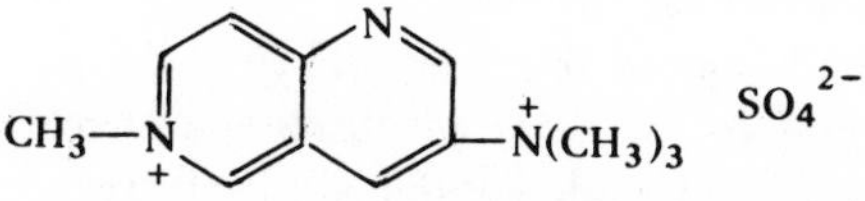

6-Methyl-3-(trimethylammonio)-6-azonia-1-azanaphthalene sulfate

Salts in which a proton has become attached to nitrogen or some other nonmetallic atom are preferably named by adding the suffix *ium* to the name of the unprotonated molecule, with elision of a final *e* when present. The remaining anionic portion is cited as a separate word or words following the name of the cationic moiety.

$CH_3CH_2CH_2CH_2\overset{+}{N}H_3\ Cl^-$      1-Butanaminium chloride

$(CH_3CH_2)_3\overset{+}{N}H\ HSO_4^-$      Triethylaminium hydrogen sulfate

$C_6H_{11}\overset{+}{N}H(CH_3)_2\ Br^-$      *N,N*-Dimethylcyclohexanaminium bromide

$(CH_3)_2\overset{+}{N}HN(CH_3)_2\ I^-$      Tetramethyldiazanium iodide

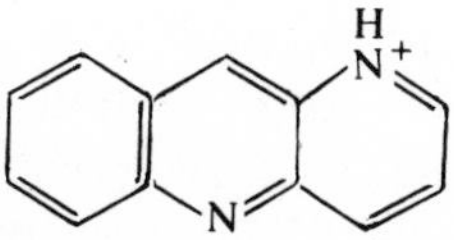

$ClO_4^-$      1,10-Diazaanthracen-1-ium perchlorate

$(CH_3CH_2CH{=}\overset{+}{N}H_2)_2\ SO_4^{2-}$      Bis-1-propaniminium sulfate

$CH_3SSS\overset{+}{S}HCH_3\ BF_4^-$      Dimethyltetrasulfan-1-ium tetrafluoroborate

Compounds having the generic structure $RN_2^+ X^-$ are named by adding the suffix *diazonium* to the name of the hydrocarbon RH, followed by the name of the anion $X^-$.

$C_6H_5N_2^+ Cl^-$                     Benzenediazonium chloride

## Discussion

In the past, although the naming of quaternary ammonium compounds, $R_4N^+ X^-$, has been fairly consistent, the treatment of amine salts having the structure $RNH_3^+ X^-$, $R_2NH_2^+ X^-$, or $R_3NH^+ X^-$ has varied. For salts in which $X^-$ is a halide ion, usage has been heavily in favor of names in which the name of the free base is cited first, followed by the name of the acid with change of the *ic* ending to *ide*—e.g., methylamine hydrochloride. This type of name is acceptable, but systematic names using the suffix *ium* as recommended above are preferred.

Protonated cationic groups have also been given names ending in *onium*—e.g., methylammonium chloride for $CH_3NH_3^+ Cl^-$. This practice should be discontinued so as to reserve the suffix *onium* for fully substituted cationic groups, as shown in Table 24.1.

Use of the suffix *ium*, designating addition of a proton to an atom carrying an unshared pair of electrons, is the preferred method for naming organic cationic groups. When the central atom is a nonmetallic element other than nitrogen, *ium* replaces the terminal *e* of the corresponding hydride name: phosphinium, arsinium, sulfanium, etc. Names of protonated nitrogen groups end in *aminium* or *iminium* depending on whether the free base (hydride) is named as an amine or imine.

If an organic cationic group, either partially or fully substituted, must be named as a substituent, its prefix name is formed by changing the terminal *um* of the parent name to *o* as in dimethylaminio, phenylphosphinio, diethylsulfonio, etc.; however, because organic cations stand near the top of the order of precedence of functions (*see* Table 8.1), such prefix names are not often required.

If a choice must be made between two or among more cationic groups to select a parent compound, the group standing highest in Table 24.1 is chosen; among groups having the same element as their central atoms, cyclic structures are senior to acyclic structures; a further choice, if necessary, is made by applying the rules for seniority among heterocyclic systems (*see* Chapter 6).

In naming cationic structures in which the positive charge is on a hetero atom contained in a ring system, the locant specifying position of the charge is omitted unless ambiguity would otherwise result (*see* Examples 6 and 7, Table 24.2). If more than one species of hetero atom is present, locants are required (*see* Example 8). These principles apply to both systematic and trivial names of cationic heterocyclic groups.

Trivial names of heterocyclic cations ending in *ylium*—e.g., pyrylium, flavylium, chromenylium, and xanthylium—although recognized in the 1965 IUPAC Rules, should be abandoned in favor of regularly formed names (*see* Examples 9 and 10). The same recommendation applies to the names betaine and thetin (*see* Examples 11 and 12); the latter term with a terminal *e* added (*i.e.*, thetine) has been used occasionally in place of *sulfonium* in naming salts having the structure $R_3 S^+ X^-$, but this practice has no official status and should be avoided.

The commonly used method of naming protonated amine salts of hydrohalogen acids in terms of the free base, *e.g.*, aniline hydrobromide for $C_6 H_5 NH_3^+ Br^-$, should not be extended to quaternary ammonium compounds. Names such as pyridine methiodide, although convenient, should no longer be used (*see* Example 6).

### Table 24.2. Examples of Acceptable Usage

1. $(C_6 H_5 CH_2 CH_2)_4 N^+ \ OH^-$    Tetrakis(2-phenylethyl)ammonium hydroxide

2.
$$CH_2 CH_3$$
$$\mid +$$
$$CH_3 CH_2 CH_2 NCH_2 CH_2 OH \quad Cl^-$$
$$\mid$$
$$CH_2 CH_2 OH$$

Ethylbis(2-hydroxyethyl)propylammonium chloride

3. $\left[ C_6 H_5 \overset{+}{P}(CH_3)_3 \right]_2 \ SO_4^{2-}$    Bis(trimethylphenylphosphonium) sulfate

4. $(CH_3 CH_2 CH_2)_2 I^+ \ ClO_4^-$    Dipropyliodonium perchlorate

5. $(CH_3)_3 \overset{+}{N} CH_2 CH_2 \overset{+}{N}(CH_3)_3 \ 2Br^-$    Ethylenebis(trimethylammonium) dibromide

6.

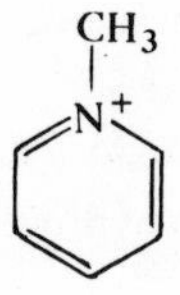

1-Methylazoniabenzene iodide
1-Methylpyridinium iodide
*not* Pyridine methiodide

7.

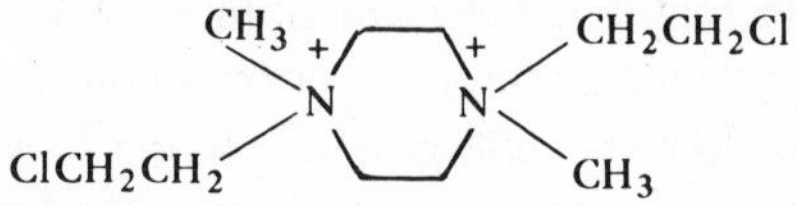

$$\begin{array}{c} CH_3 \\ \overset{+}{N} \end{array} \overset{+}{N} \begin{array}{c} CH_2CH_2Cl \\ \end{array} \qquad 2Cl^-$$

$$ClCH_2CH_2 \qquad \qquad CH_3$$

1,4-Bis(2-chloroethyl)-1,4-dimethyl-1,4-diazoniacyclohexane dichloride
1,4-Bis(2-chloroethyl)-1,4-dimethylpiperazinediium dichloride

8.

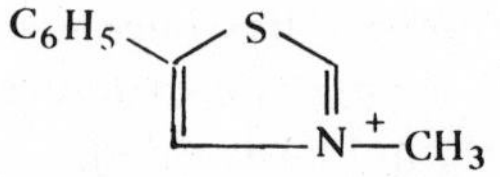

$$C_6H_5 \overset{S}{\underset{N \overset{+}{-} CH_3}{\parallel}} \qquad H_2PO_4^-$$

3-Methyl-5-phenyl-1-thia-3-azoniacyclopenta-2,4-diene dihydrogen
   phosphate
3-Methyl-5-phenyl-1,3-thiazol-3-ium dihydrogen phosphate

9.

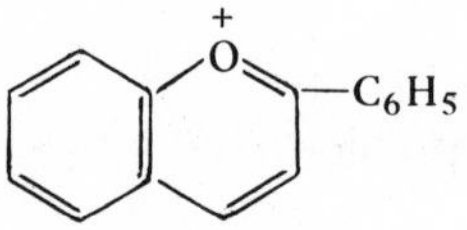

$$\overset{+}{O} - C_6H_5 \qquad Cl^-$$

2-Phenyl-1-oxonianaphthalene chloride
*not* Flavylium chloride

10.

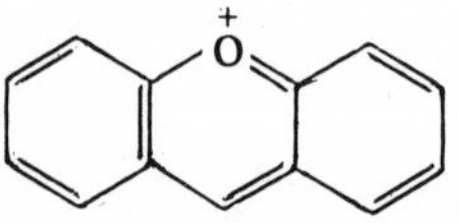

$$\overset{+}{O} \qquad I^-$$

9-Oxoniaanthracene iodide
*not* Xanthylium iodide

11. $(CH_3)_3\overset{+}{N}CH_2COO^-$

(Trimethylammonio)ethanoate
(Trimethylammonio)acetate
*not* Betaine

12. $(CH_3)_2\overset{+}{S}CH_2COO^-$

(Dimethylsulfonio)ethanoate
(Dimethylsulfonio)acetate
*not* Thetin

13. $(CH_3)_3\overset{+}{N}CH_2CH_2OH \quad OH^-$

(2-Hydroxyethyl)trimethylammonium
   hydroxide
*not* Choline

14. $CH_3CH_2CH_2\overset{+}{N}H_3 \quad Br^-$

1-Propanaminium bromide
1-Propanamine hydrobromide
*not* Propylammonium bromide

15. $(C_6H_5)_2\overset{+}{N}H_2 \quad I^-$

*N*-Phenylbenzenaminium iodide
Diphenylaminium iodide
Diphenylamine hydriodide
*not* Diphenylammonium iodide

16.

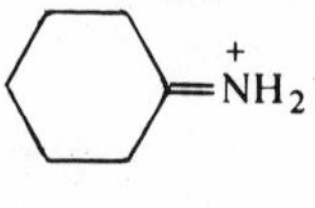

$$\overset{+}{=NH_2} \qquad HSO_4^-$$

Cyclohexaniminium hydrogen sulfate

17. $[(HOCH_2)_3 \overset{+}{P}H]_2$  $PtCl_6{}^{2-}$    Tris(hydroxymethyl)phosphinium
hexachloroplatinate

18. 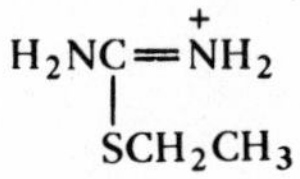 $H_2NC \overset{+}{=\!\!=} NH_2$  $Cl^-$    1-Amino-1-(ethylthio)methaniminium
$\underset{|}{\overset{}{SCH_2CH_3}}$    chloride

# 25

# Boron Compounds

S tructures containing at least one organic group attached to a boron atom through carbon are called **organoboron compounds**; the broader class name **organic boron compounds** also includes structures not containing carbon-to-boron bonds. Except when functional groups are present that take precedence, organic boron compounds are preferably named as derivatives of inorganic parent structures in accordance with official rules adopted by the American Chemical Society (*1*). The ACS Rules are comprehensive and explicit; since they are readily available, only a few principles enabling one to name the more common types of organic derivatives are included in this chapter.

*Recommended Nomenclature Practice*

The compound $BH_3$ is called **borane**; this term is also used to designate boron hydrides, $B_xH_y$, as a class. Hydrides containing more than one boron atom are named by using the customary multiplying prefixes and adding an Arabic numeral within parenthesis as a suffix to denote the number of hydrogen atoms present.

| | |
|---|---|
| $B_5H_9$ | Pentaborane (9) |
| $B_5H_{11}$ | Pentaborane (11) |

Organic derivatives are named substitutively in the usual manner, the highest-ranking function being expressed as a suffix.

| | |
|---|---|
| $CH_3BH_2BH_2CH_3$ | 1,2-Dimethyldiborane (6) |
| $CH_3CH_2B(OCH_3)_2$ | Ethyldimethoxyborane |
| $(CH_3)_2NBHCH_2CH_2CH_3$ | *N,N*-Dimethyl-1-propylboranamine |
| $C_6H_5CH_2BCl_2$ | Dichloro(phenylmethyl)borane |

Compounds having the structures $(RO)_3B$, $(RO)_2BOH$, and $ROB(OH)_2$ are named as esters of **boric acid**, $B(OH)_3$.

$(C_6H_5O)_3B$                  Triphenyl borate

$(CH_3CH_2CH_2O)_2BOH$          Dipropyl hydrogen borate

Organic ring systems containing one or more boron atoms are preferably named by replacement nomenclature as described for heterocyclic systems in Chapter 6.

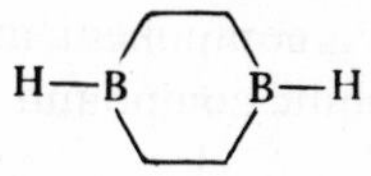

H—B        B—H          1,4-Diboracyclohexane

For three particularly stable **inorganic ring systems** containing boron, ACS Rules recommend retention of well established trivial names. Numbering is fixed as shown.

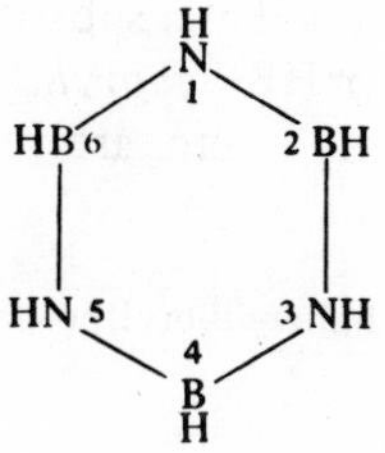

Borazine

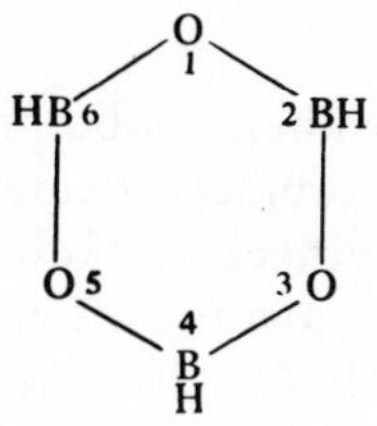

Boroxin

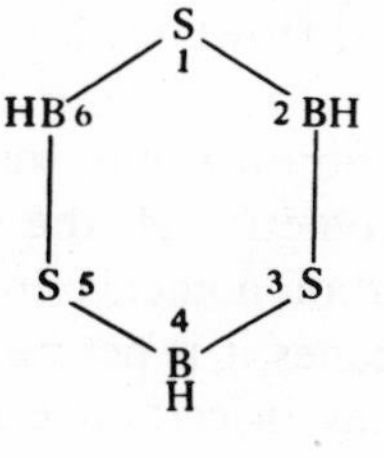

Borthiin

The class of compounds in which one or more carbon atoms replace boron in polyboron hydrides (polyboranes) is called **carbaboranes**. Individual members are named by using the replacement prefixes *carba*, *dicarba*, etc.

$B_{10}C_2H_{12}$                               Dicarbadodecaborane (12)

Structures in which a nucleophilic group has combined with boron by completing the latter's octet of electrons are regarded as *addition compounds* and are named by citing the two component molecules and joining their names with a dash. The nucleophilic compound is listed first.

$CH_3NH_2 \cdot BH_2CH_2CH_3$               Methanamine—ethylborane

**Boron-containing ions** are named as coordination compounds by applying the regular inorganic rules.

$[(CH_3)_4N]\ [BCl_4]$                         Tetramethylammonium
                                              tetrachloroborate

$K[B(C_6H_5)_4]$                               Potassium tetraphenylborate

When a group containing boron must be named as a substituent, prefix names based on *boryl* for $H_2B-$, *borylene* for $HB<$, *borylidyne* for $B<$, *diboran(4)yl* for $B_2H_3-$, *diboran(6)yl* for $B_2H_5-$, etc. are formed in the regular manner.

$(C_6H_5)_2BCH_2CH_2COOCH_3$               Methyl 3-(diphenylboryl)propanoate

*Discussion*

The unusual types of bonding encountered in many boron compounds require that specialized methods be used in forming systematic names for these structures. With this in mind, committees of both organic and inorganic chemists have developed what are now the official ACS Rules for the Nomenclature of Boron Compounds (*1*). In general the Rules adhere to principles previously established and approved by the IUPAC Commissions on Organic and Inorganic Nomenclature; they also take note of practices employed by *Chemical Abstracts.* However, a few comments are in order.

In naming boron hydrides containing 20 boron atoms, the multiplying prefix *icosa* (to emphasize the spatial geometry of the molecule) is preferred to the form *eicosa* used in hydrocarbon nomenclature.

The ACS Rules state that in naming boranes it is permissible to omit the numerical designation of hydrogen atoms in certain cases where no ambiguity would result. However, since this practice requires careful judgement and tends to produce apparent inconsistencies throughout the published literature, it is not recommended here, except for the simple and very common hydride $BH_3$.

Systematic names for the inorganic heterocyclic structures $(BHNH)_3$, $(BHO)_3$, and $(BHS)_3$ formed by replacement nomenclature—i.e., 1,3,5-tri-azacyclohexaborane, 1,3,5-trioxacyclohexaborane, and 1,3,5-trithiacyclo-hexaborane—can be formed, but these have not yet received official recognition. Thus, although systematic names are generally preferred, the three trivial names recommended by the ACS Rules are here regarded as currently preferable. The older name "borazole" for borazine should no longer be used.

There has been considerable use of "carborane," a shortened form of carbaborane, both as a class name and as a trivial name for the hydride $B_{10}C_2H_{12}$. Neither practice is recommended.

In recognition of significant usage among organic chemists, *Chemical Abstracts* has named structures of the types $RB(OH)_2$ and $R_2BOH$ as "boronic acids" and "borinic acids," respectively. As with the corresponding organophosphorus compounds (*see* Chapter 36), this treatment may be extended to include not only the esters $RB(OR)_2$ and $R_2BOR'$ but other formal acid derivatives, such as halides, anhydrides, and amides, as well. However, in naming organoboron compounds containing functional groups attached to boron, the application of principles used in naming organosilicon compounds (*see* Chapter 38) is more appropriate because of the close chemical analogy of boron to silicon; this method leads to the functional parent names boranol for $H_2BOH$ and boranediol for $HB(OH)_2$, boranamine for $H_2BNH_2$ and boranediamine for $HB(NH_2)_2$, but it does not affect the naming of the corresponding halogen or OR derivatives. Except for their use by *Chemical Abstracts*, the aforementioned phosphorus-type names for organoboron compounds have not received official sanction.

For the compound $B(OH)_3$, the 1970 IUPAC Inorganic Rules recognize both the names orthoboric acid and boric acid. Usage has heavily favored the latter. Structures in which halogen or nitrogen have replaced one or more OH groups in boric acid should be named as borane derivatives—*e.g.*, aminoboranediol for $H_2NB(OH)_2$, *not* amidoboric acid.

The prefix names for boron-containing groups recommended in the ACS Rules are logically formed by analogy with their carbon counterparts. The prefix *borono* for $(HO)_2B-$, although used by *Chemical Abstracts*, should be abandoned in favor of *dihydroxyboryl*.

## Table 25.1. Examples of Acceptable Usage

1.     $CH_3CH_2B(CH_3)_2$       Ethyldimethylborane

2.

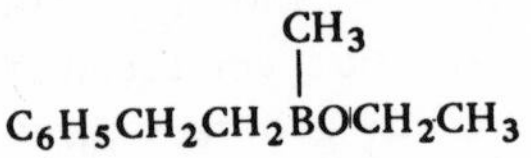

$$\overset{\displaystyle CH_3}{\underset{\displaystyle C_6H_5CH_2CH_2BOCH_2CH_3}{|}}$$

Ethoxy(methyl)(2-phenylethyl)borane

3.     $Cl_2BCH_2CH_2BCl_2$       Ethylenebis(dichloroborane)

4.

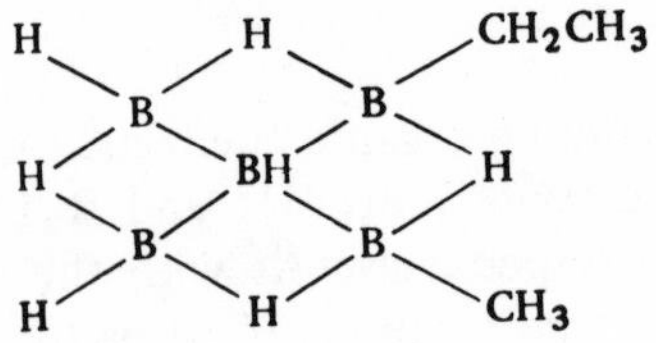

2-Ethyl-3-methylpentaborane (9)

(for structures and numbering of other polyboranes, *see* the ACS Rules)

5.

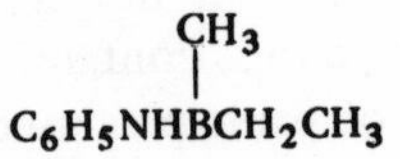

$$\overset{\displaystyle CH_3}{\underset{\displaystyle C_6H_5NHBCH_2CH_3}{|}}$$

1-Ethyl-1-methyl-*N*-phenylboranamine
Ethylmethyl(phenylamino)borane

6.

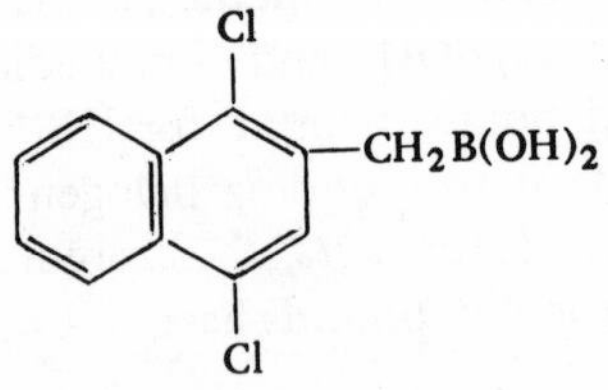

(1,4-Dichloro-2-naphthylmethyl)boranediol
(1,4-Dichloro-2-naphthylmethyl)dihydroxy-borane

7.     $CH_3CH_2CH_2B(OCH_3)_2$       Dimethoxypropylborane

8.     $(CH_3CH_2O)_3B$       Triethyl borate
                       *not* Ethyl borate

9.     $CH_3CH=CHCH_2OB(OH)_2$       2-Butenyl dihydrogen borate

10.     $C_6H_5CH_2CH_2OBCl_2$       Dichloro(2-phenylethoxy)borane

11.     $(CH_3COO)_3B$       Boric ethanoic trianhydride
                            Boric acetic trianhydride

12.        $(CH_3O)_2BB(OCH_2CH_3)_2$     1,1-Diethoxy-2,2-dimethoxydiborane(4)

13.

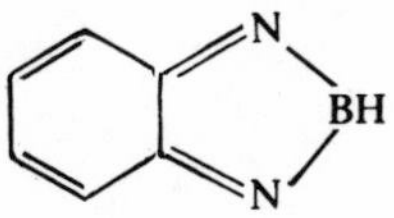

2*H*-1,3-Diaza-2-boraindene
2*H*-Benzo[*d*]-1,3,2-diazaborole

14.

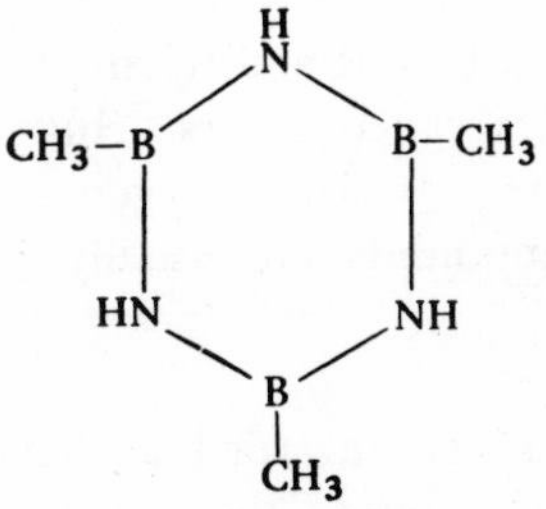

2,4,6-Trimethylborazine

15.        $Na[HB(OCH_3)_3]$             Sodium hydrotrimethoxyborate

16.

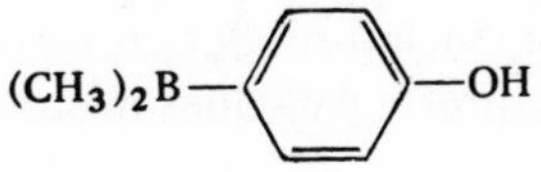

$(CH_3)_2B$—⟨benzene ring⟩—OH

4-(Dimethylboryl)benzenol
4-(Dimethylboryl)phenol

## Literature Cited

1. "The Nomenclature of Boron Compounds," (an ACS report), *Inorg. Chem.* (1968)
7, 1945.

# 26

# Ethers

Compounds having the structure ROR', where R and R' are the same or different acyclic or cyclic groups, are called **ethers**. This class also includes cyclic structures in which one or more oxygen atoms are members of a ring, although such compounds are usually named as heterocyclic systems.

Conceptually, the ether function may be regarded as having been derived by replacement of the hydrogen atom of the OH group in an alcohol or phenol with a hydrocarbon or heterocyclic group. Thus the term **partial ether** denotes a modified polyol structure in which some but not all of the OH groups have been converted to OR.

From another viewpoint an ether group is produced when an oxygen atom replaces a $CH_2$ group in a hydrocarbon chain or ring. Hence, the prefix *oxa* as used in replacement nomenclature denotes the ether function in the same manner in which the prefix *oxy* does in substitutive nomenclature.

Naming of sulfur analogs of ethers, which are called **sulfides**, is discussed in Chapter 39. **Acetals,** which constitute a subclass of ethers derived from aldehydes and ketones, are treated in Chapter 20.

## Recommended Nomenclature Practice

Acyclic ethers having no more than two —O— linkages within a single straight chain otherwise composed of carbon atoms are preferably named substitutively as *oxy* derivatives of one of the all-carbon chains present in the molecule. The parent (senior) chain is selected by applying the following criteria in the order listed: (1) largest number of attached OR groups, (2) highest degree of unsaturation, (3) longest chain, and (4) lowest locants for attached OR groups (*see* 1965 IUPAC Rule C-13.1). Names for acyclic OR groups are formed systematically by combining the prefix name of the R group with *oxy*—e.g., hexyloxy—but the shortened form *alkoxy* is preferred when the alkyl group contains less than five carbon atoms (Table 26.1).

## Table 26.1.Contractions of Names for Acyclic OR Groups

| | |
|---|---|
| $CH_3O-$ | Methoxy |
| $CH_3CH_2O-$ | Ethoxy |
| $CH_3CH_2CH_2O-$ | Propoxy |
| $CH_3CH_2CH_2CH_2O-$ | Butoxy |

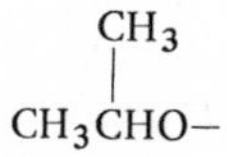

$$\underset{CH_3CHO-}{\overset{CH_3}{|}}$$ Isopropoxy[a]

$$\underset{CH_3CHCH_2O-}{\overset{CH_3}{|}}$$ Isobutoxy[a]

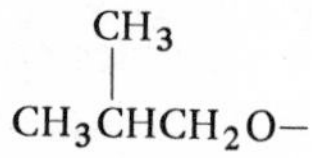

$$\underset{CH_3CH_2CHO-}{\overset{CH_3}{|}}$$ *sec*-Butoxy[a]

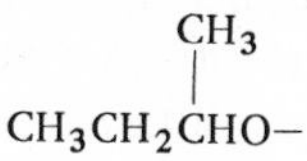

$$\underset{\underset{CH_3}{|}}{\overset{CH_3}{\underset{CH_3CO-}{|}}}$$ *tert*-Butoxy[a]

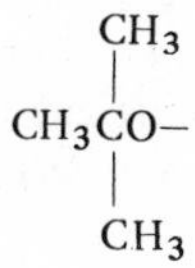

[a] For the unsubstituted group only.

As alternatives to substitutive names for acyclic monoethers of symmetrical structure, radicofunctional names terminating in the word *ether* may be used.

| | |
|---|---|
| $CH_3OCH_2CH_2CH_2CH_3$ | 1-Methoxybutane |
| $CH_3OCH_2CH_2OCH_2CH_2CH_3$ | 1-Methoxy-2-propoxyethane |

$$\underset{CH_2=CHOCH_2CHCH_3}{\overset{CH_3}{|}}$$ Isobutoxyethene

$$\underset{CH_3CH_2CHOCHCH_2CH_3}{\overset{CH_3\ \ \ CH_3}{|\ \ \ \ \ |}}$$ 2-*sec*-Butoxybutane
Di-*sec*-butyl ether

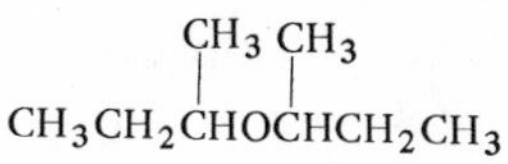

$$OCH_3$$
$$|$$
$$CH_3OCH_2CHCH_2OCH_3$$          1,2,3-Trimethoxypropane

Acyclic polyethers having more than two —O— linkages within a single straight chain are best named by using replacement nomenclature (*see* Chapter 7).

$$CH_3OCH_2CH_2OCH_2CH_2OCH_2CH_3$$     2,5,8-Trioxadecane

Cyclic ethers having one or more —O— linkages connecting two ring systems are preferably named substitutively as *oxy* derivatives of the senior ring system carrying the most OR groups. In choosing the senior ring system as the parent compound, all heterocycles are preferred to all carbocycles, and nitrogen-containing heterocycles are preferred to those not containing nitrogen. If no heterocyclic system is present, the senior carbocyclic system containing the largest number of OR groups is selected as the parent compound. (For more detailed information on seniority of ring systems, *see* 1965 IUPAC Rule C-14.1 and the Introduction to the Subject Index, *Chem. Abstr.* 1967, 66, ¶ 29-33). Names of cyclic OR groups are formed systematically by combining the prefix name of the R group with *oxy—e.g.,*· cyclohexyloxy—except that the contraction *phenoxy* is preferred for $C_6H_5O$-. For **symmetrical** cyclic monoethers, radicofunctional names ending with the word *ether* are acceptable alternatives to the preferred substitutive names.

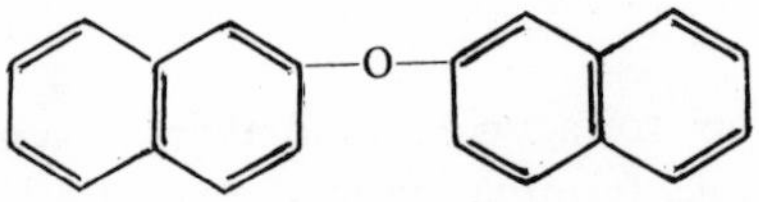

2-(2-Naphthyloxy)naphthalene
Di-2-naphthyl ether

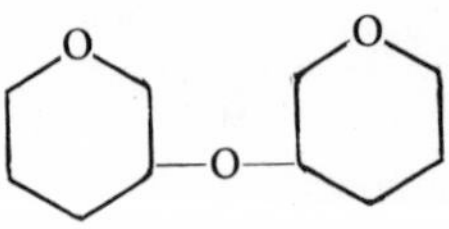

3-(Oxacyclohex-3-yloxy)oxa-
cyclohexane
Bis(tetrahydropyran-3-yl) ether

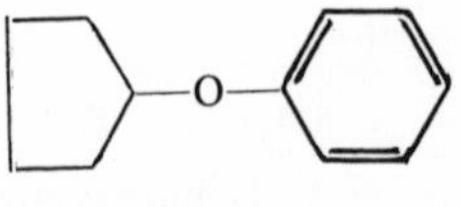

(Cyclopentyloxy)benzene

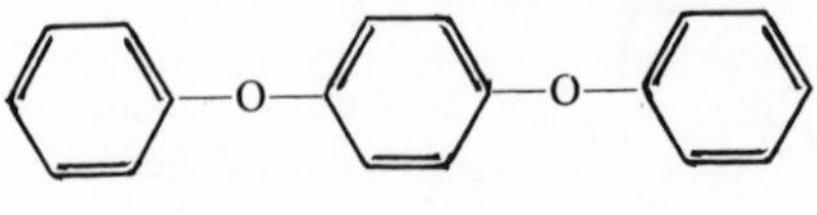

1,4-Diphenoxybenzene

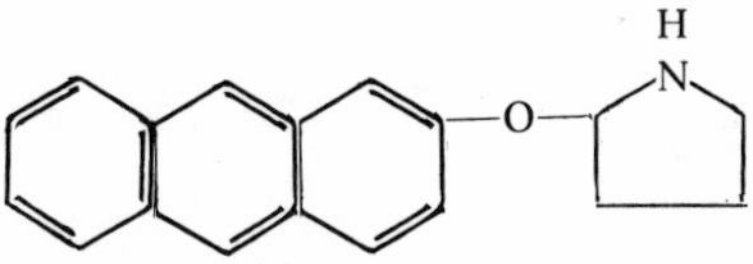

2-(2-Anthryloxy)azacyclopentane

Cyclic-acyclic monoethers are preferably named substitutively as *oxy* derivatives of the senior ring or chain present. When there is a choice, all heterocyclic systems are preferred to all cyclic and acyclic hydrocarbon systems; otherwise the largest hydrocarbon is selected as the parent. Symmetrical structures may also be named radicofunctionally.

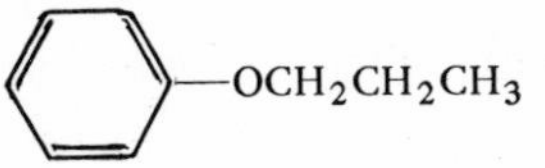

Propoxybenzene

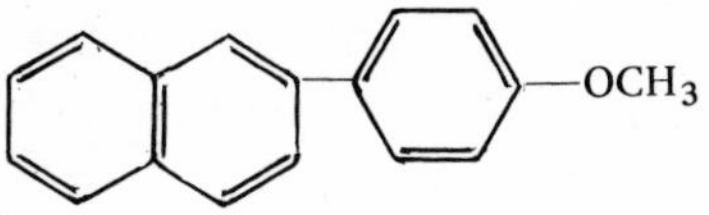

2-(4-Methoxyphenyl)naphthalene

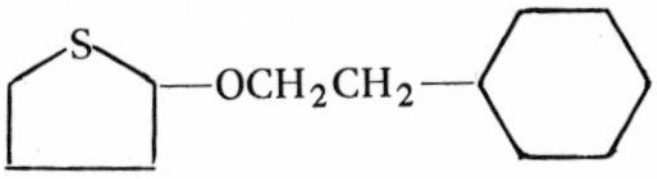

2-(2-Cyclohexylethoxy)thiacyclopentane

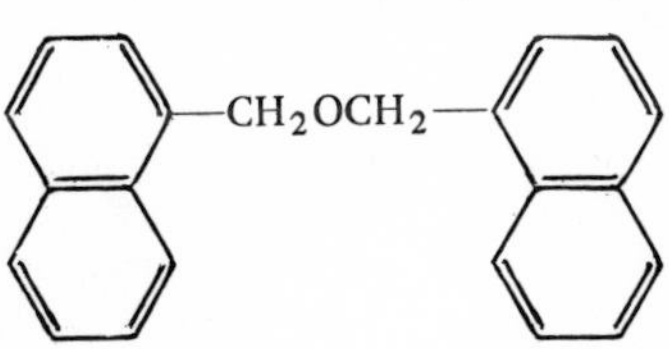

1-(1-Naphthylmethoxymethyl)naphthalene
Bis(1-naphthylmethyl) ether

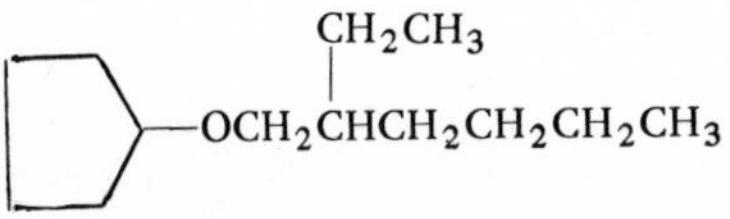

3-[(Cyclopentyloxy)methyl]heptane

Cyclic-acyclic polyethers are named similarly, except that in selecting the parent chain or ring system consideration is first given to the number of attached OR groups.

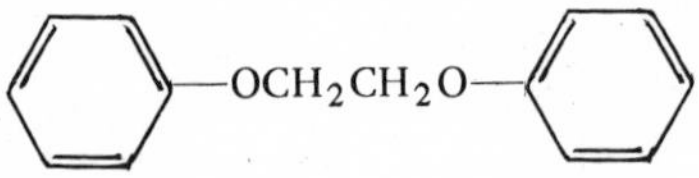

1,2-Diphenoxyethane

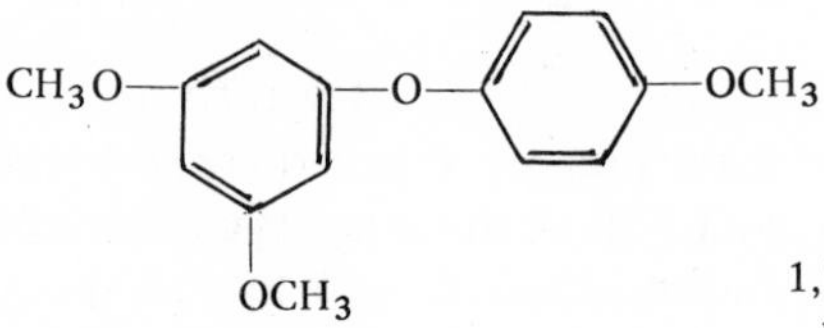

1,3-Dimethoxy-5-(4-methoxyphenoxy)-benzene

Straight chains containing more than two —O— linkages are best named by using replacement nomenclature (*see* Chapter 7).

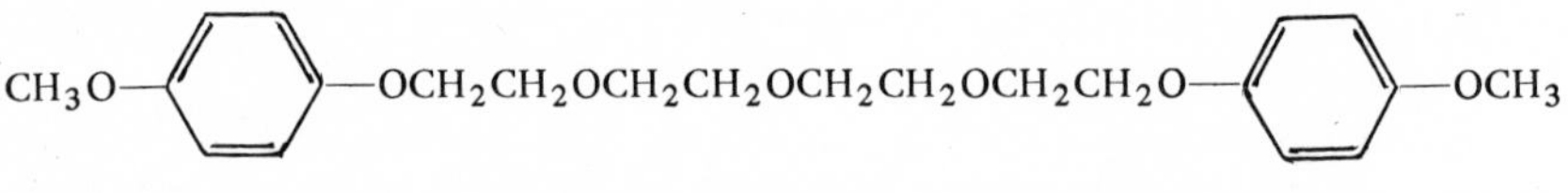

1,11-Bis(4-methoxyphenoxy)-3,6,9-trioxaundecane

Structures that contain one or more —O— linkages in a ring system are preferably named as heterocyclic compounds (*see* Chapter 6).

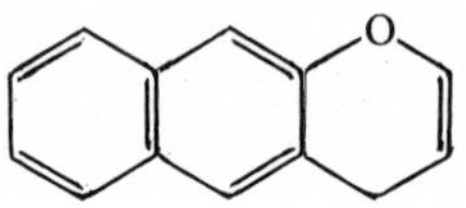

4*H*-1-Oxaanthracene

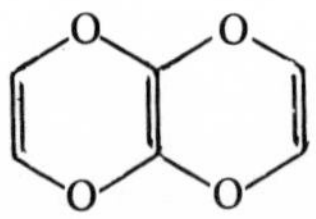

1,4,5,8-Tetraoxanaphthalene

As an alternative to the regularly formed heterocycle names for structures in which O is attached to two adjacent carbon atoms of a chain or ring, the bivalent substitutive prefix *epoxy* may be used.

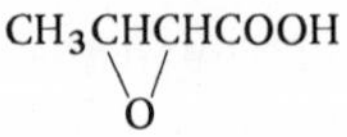

$$CH_3CHCHCOOH$$
$$\diagdown O \diagup$$

3-Methyloxacyclopropane-2-carboxylic acid
2,3-Epoxybutanoic acid

The term *epoxy* is also used as a nondetachable prefix (*see* Chapter 6, discussion section) in naming bridged polycyclic aromatic structures.

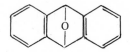 9,10-Dihydro-9,10-epoxyanthracene

## Discussion

The word ether is properly used only as a generic term denoting a functional class of compounds; it should never be used in place of diethyl ether as a specific name for the compound $CH_3CH_2OCH_2CH_3$.

Since, in substitutive nomenclature, the ether function is always cited as a prefix, it is not considered in selecting the principal function when naming compounds of mixed functions substitutively (*see* Chapter 10). For this reason, the ether function does not appear in the second part of Table 8.1; it is included, however, by both IUPAC and *Chemical Abstracts* in their order-of-precedence listings, which apply not only to substitutive but also to radicofunctional nomenclature.

Although radicofunctional names are still widely used in naming ethers, notably by *Chemical Abstracts* until 1972, substitutive nomenclature is preferred for reasons of systematization. However, for symmetrical monoethers, particularly those of simple structure, the more familiar radicofunctional names ending in the word *ether* (not oxide) are acceptable second choices.

Although *Chemical Abstracts* has indexed simple symmetrical ethers under names such as "Ethyl ether" for diethyl ether and "Phenyl ether" for diphenyl ether, omission of the prefix *di* or *bis* is incorrect by modern principles of organic nomenclature and should be avoided.

Additive names such as "ethylene oxide" and "styrene oxide" should be abandoned in favor of the systematic names oxirane and phenyloxirane.

As noted in Chapter 7, replacement nomenclature offers a convenient, systematic means of naming acyclic ethers, particularly when more than two —O— linkages occur within a single straight chain. Although replacement names have been officially recommended and used only for polyethers of relatively complex structure, names such as oxapropane for $CH_3OCH_3$ and 2,5-dioxahexane for $CH_3OCH_2CH_2OCH_3$ are certainly acceptable.

Partial ethers of polyhydroxy compounds are best named substitutively as *oxy* derivatives of the appropriate alcohol or phenol (*see* Chapters 19 and 35). Names such as "diethylene glycol" for $HOCH_2CH_2OCH_2CH_2OH$ and "ethylene glycol monoethyl ether" for $CH_3CH_2OCH_2CH_2OH$ should be abandoned in favor of 2,2'-oxydiethanol and 2-ethoxyethanol, respectively; likewise, names such as 1-O-methylresorcinol are not recommended, although this type of name is used in specialized nomenclature of carbohydrates (*see* Chapter 30).

A few trivial names for ethers are recognized in the 1965 IUPAC Rules (*see* Examples 10,11,13,14,15 and 16) but these are falling into disuse and should be abandoned.

### Table 26.2. Examples Of Acceptable Usage

1.  $CH_3OCH_3$

    Methoxymethane
    Dimethyl ether
    *not* Methyl ether

2.  $CH_3CH_2OCH_2CH_2Br$

    1-Bromo-2-ethoxyethane

3.  $ClCH_2CH_2OCH_2CH_2Cl$

    1-Chloro-2-(2-chloroethoxy)ethane
    Bis(2-chloroethyl) ether

4.  $CH_3OCH_2CH_2CH_2OCH_3$

    1,3-Dimethoxypropane

5.  $CH_3CH(OCH_2CH_2CH_2CH_2CH_3)_2$

    1,1-Bis(pentyloxy)ethane

6.  $CH_3OCH_2OCH_2CH_2OCH_2OCH_3$

    2,4,7,9-Tetraoxadecane

7.

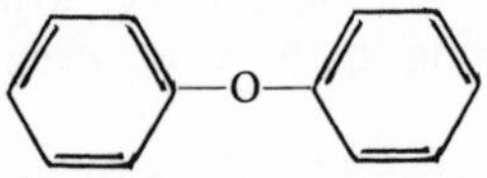

    Phenoxybenzene
    Diphenyl ether
    *not* Phenyl ether

8.

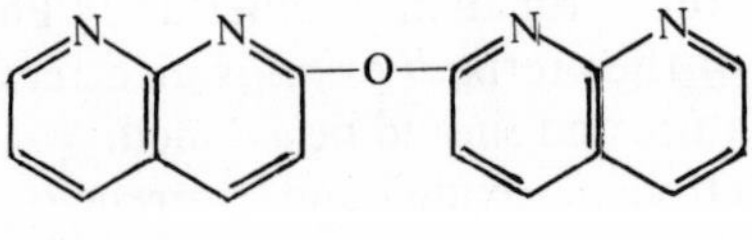

    2-(1,8-Diazanaphth-2-yloxy)-1,8-diazanaphthalene
    Di-1,8-naphthyridin-2-yl ether

9.

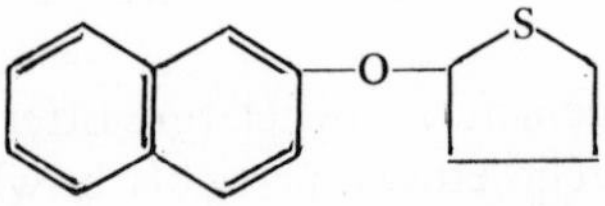

    2-(2-Naphthyloxy)thiacyclopentane
    Tetrahydro-2-(2-naphthyloxy)thiophene

10.

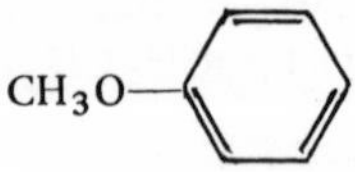

    $CH_3O-$

    Methoxybenzene
    *not* Anisole

11.

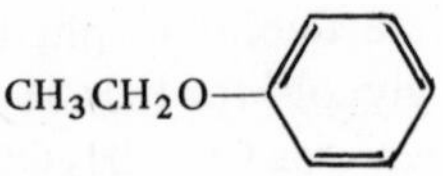

    $CH_3CH_2O-$

    Ethoxybenzene
    *not* Phenetole

12.

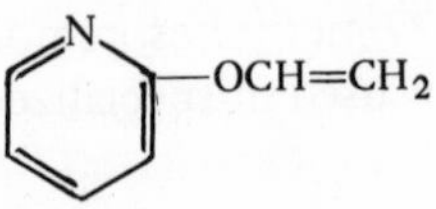

    $-OCH=CH_2$

    2-(Ethenyloxy)azabenzene
    2-(Vinyloxy)pyridine

13.

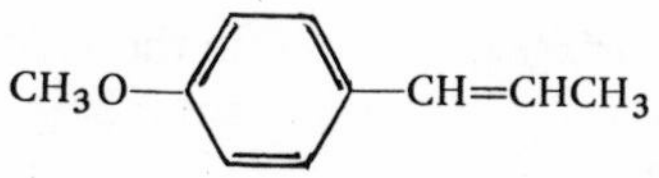

$CH_3O$—⬡—$CH{=}CHCH_3$

1-Methoxy-4-(1-propenyl)benzene
*not* Anethole

14.

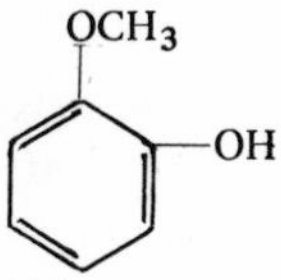

$OCH_3$
⬡—OH

2-Methoxybenzenol
2-Methoxyphenol
*not* Guaiacol

15.

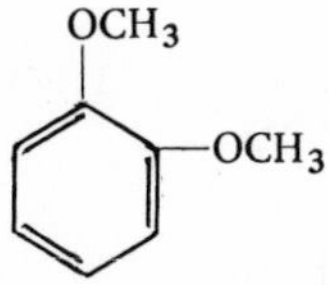

$OCH_3$
⬡—$OCH_3$

1,2-Dimethoxybenzene
*not* Veratrole

16.

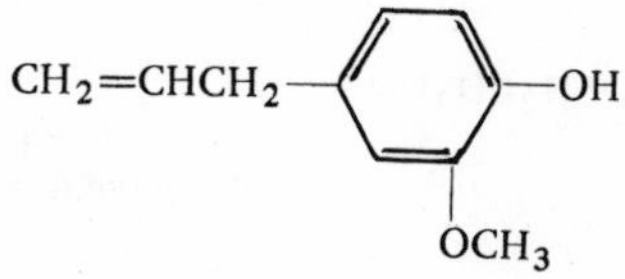

$CH_2{=}CHCH_2$—⬡—OH
$OCH_3$

2-Methoxy-4-(2-propenyl)benzenol
4-Allyl-2-methoxyphenol
*not* Eugenol

17.

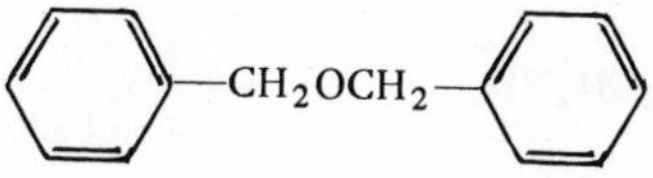

⬡—$CH_2OCH_2$—⬡

(Phenylmethoxymethyl)benzene
Dibenzyl ether
*not* Benzyl ether

18.

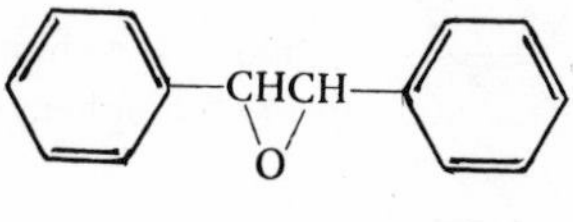

⬡—CHCH—⬡
O

2,3-Diphenyloxirane
1,2-Epoxy-1,2-diphenylethane
*not* Stilbene oxide

19.

$CH_2CHCH_2Cl$
O

(Chloromethyl)oxirane
1-Chloro-2,3-epoxypropane
*not* Epichlorohydrin

20.

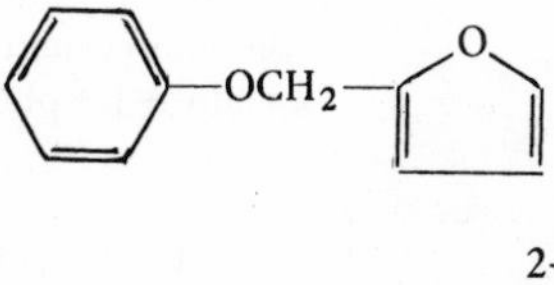

⬡—$OCH_2$—furan

2-(Phenoxymethyl)oxacyclopenta-2,4-diene
2-(Phenoxymethyl)furan

21.

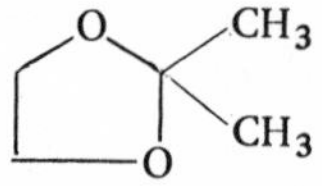

2,2-Dimethyl-1,3-dioxacyclopentane
2,2-Dimethyl-1,3-dioxolane

22.

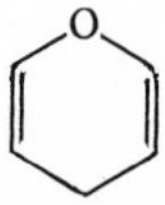

Oxacyclohexa-2,5-diene
4H-Pyran

23.

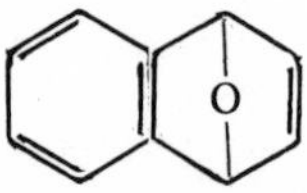

1,4-Dihydro-1,4-epoxynaphthalene

24.    $HOCH_2CH_2OCH_2CH_2OH$    2,2'-Oxydiethanol
                                  *not* Diethylene glycol

25.    $HOCH_2CH_2OCH_2CH_2OCH_2CH_2OH$

                                  2,2'-(Ethylenedioxy)diethanol
                                  *not* Triethylene glycol

26.    $CH_3CH_2OCH_2CH_2OH$     2-Ethoxyethanol
                                  *not* Ethylene glycol monoethyl ether

27.    $H_2NCH_2CH_2OCH_2CH_2NH_2$    2,2'-Oxydiethanamine
                                      2,2'-Oxybis(ethylamine)

28.

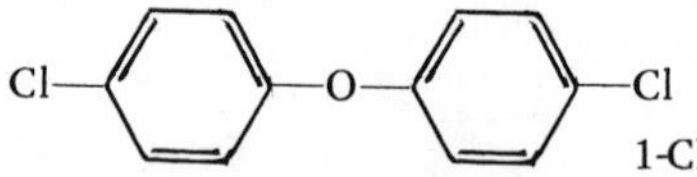

1-Chloro-4-(4-chlorophenoxy)benzene
Bis(4-chlorophenyl) ether

29.

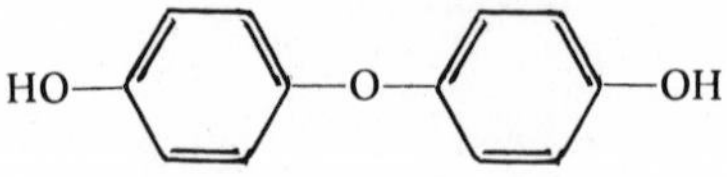

4,4'-Oxydibenzenol
4,4'-Oxydiphenol

30.

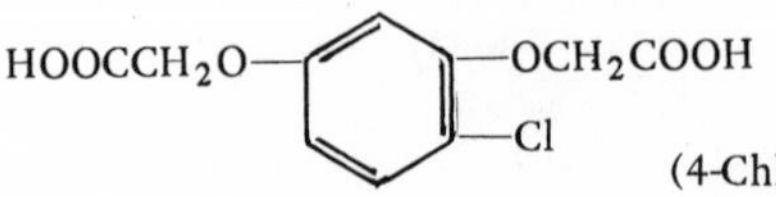

(4-Chloro-1,3-phenylenedioxy)-
diethanoic acid
(4-Chloro-1,3-phenylenedioxy)diacetic
acid

31.    $CH_3C(OCH_2CH_3)_3$     1,1,1-Triethoxyethane
                                *not* Triethyl orthoacetate

# Halogen, Nitro, Nitroso and Azido Compounds

I n substitutive nomenclature certain functional groups have not been assigned suffix names and must therefore always be designated by prefixes (*see* Chapter 8). The most frequently occurring of these groups are halogen atoms and the groups $-NO_2$, $-NO$, and $-N_3$. The naming of compounds containing these substituents is discussed in this chapter, except for acid halides and acid azides (*see* Chapter 17) and ionic compounds of the quaternary ammonium type (*see* Chapter 24).

## Recommended Nomenclature Practice

Compounds containing one or more F, Cl, Br, I, $NO_2$, NO, or $N_3$ groups each singly bound to a non-acyl carbon atom are named substitutively by combining the appropriate prefixes (*see* Table 8.1) with the name of the parent chain or ring system.

$CH_3NO_2$                    Nitromethane

$$\begin{array}{c} CH_3 \\ | \\ CH_3C-N_3 \\ | \\ CH_3 \end{array}$$          2-Azido-2-methylpropane

$C_6H_5CH_2CH_2Cl$          (2-Chloroethyl)benzene

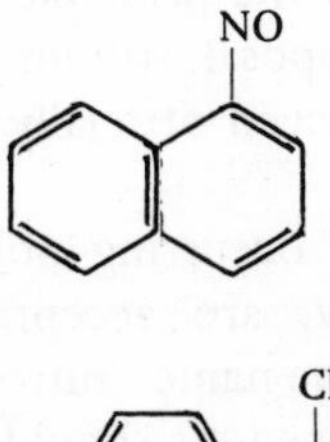

1-Nitrosonaphthalene

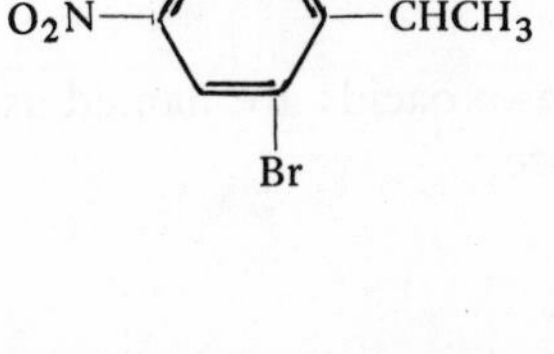

2-Bromo-1-isopropyl-4-nitrobenzene

As illustrated by the last example, prefix names of functional and nonfunctional substituting groups are considered together for arrangement in alphabetical order. The presence of multiplying prefixes does not affect this order of citation (*see* Chapter 8).

Names of several other groups containing halogen or nitrogen that are always cited as prefixes are listed in Table 8.1.

### *Discussion*

Simple halogen and azido compounds have often been given radico-functional names (*see* Chapter 9)—*e.g.*, ethyl chloride and benzyl azide—but modern systematic nomenclature calls for the substitutive method, which is of much broader applicability.

For two of the groups listed in the first section of Table 8.1 *Chemical Abstracts* has used prefix names other than those given in the 1965 IUPAC Rules: *iodoso* (in place of *iodosyl* for —IO and *iodoxy* (in place of *iodyl*) for —IO$_2$. The IUPAC names are to be preferred, however, for general use and were adopted by *Chemical Abstracts* in 1972.

When all or nearly all of the hydrogen atoms in an organic compound have been replaced by halogen, citation of a locant for each halogen atom may often be avoided by use of the special multiplying prefix *per* (*1*). This prefix denotes replacement of all hydrogen atoms in a parent compound or substituting group by halogen atoms of a single kind, with the exception of hydrogen whose replacement would alter the identity of a functional group, as in perfluoro-1-butanol for $F_3CCF_2CF_2CF_2OH$ and perchloroethyl for $Cl_3CCCl_2$ —. Although elimination of lengthy series of locants in this manner results in worthwhile simplification of names, usage of *per* has unfortunately been loose and inconsistent to date. In recent years, accelerated research on highly fluorinated organic compounds has increased the need for new rules, and the Nomenclature Committee of the Division of Fluorine Chemistry of the American Chemical Society has now prepared a proposal whereby use of the prefix *perfluoro* would be abandoned in favor of the letter prefix *F* (italicized). The meaning of *F* would be similar to that given above for *per,* and the principle could, of course, be extended to the other halogens. The proposal, still under review by other nomenclature committees, has not yet been officially accepted by the Society.

A few trivial names for simple, commonly occurring polyhalogen compounds, as shown in the examples below, are acceptable, but systematic names are to be preferred. However, the name "nitroglycerin" for 1,2,3-propanetriyl trinitrate should be abandoned, as should "chloropicrin" for trichloronitromethane.

Esters of halogen oxoacids and nitrogen oxoacids are named as such —*e.g.*, *tert*-butyl hypochlorite and ethyl nitrate.

## Table 27.1. Examples of Acceptable Usage

1. $CH_3I$ — Iodomethane
Methyl iodide

2. $CH_2{=}CHCl$ — Chloroethene
Vinyl chloride

3. $HCCl_3$ — Trichloromethane
Chloroform

4. $Cl_3CNO_2$ — Trichloronitromethane
*not* Chloropicrin

5. $Cl_3CCHO$ — Trichloroethanal
Trichloroacetaldehyde
Chloral

6.

$$\begin{array}{c} CH_3 \\ | \\ CH_3C{-}Br \\ | \\ CH_3 \end{array}$$

2-Bromo-2-methylpropane
*tert*-Butyl bromide

7. $F_3CCF_2CF_2CF_2CF_3$ — Dodecafluoropentane
Perfluoropentane

8. $CH_2{=}CHCH_2CH_2NO_2$ — 4-Nitro-1-butene

9.

Cl—⟨ ⟩—Cl

1,4-Dichlorobenzene
*p*-Dichlorobenzene

10.

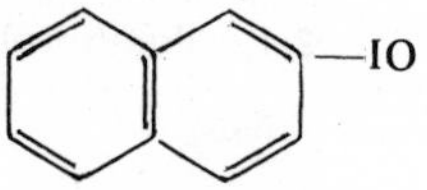

—IO

2-Iodosylnaphthalene
2-Iodosonaphthalene

11.

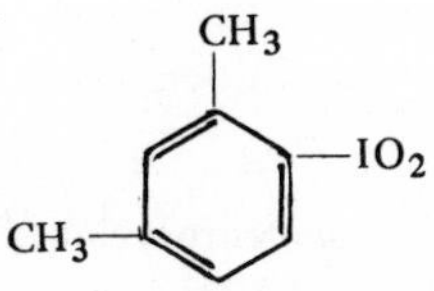

$CH_3$

—$IO_2$

$CH_3$—

1-Iodyl-2,4-dimethylbenzene
1-Iodoxy-2,4-dimethylbenzene
4-Iodoxy-*m*-xylene

12.

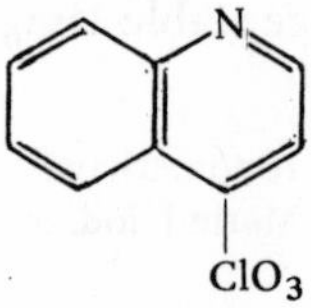

4-Perchloryl-1-azanaphthalene
4-Perchlorylquinoline

13.

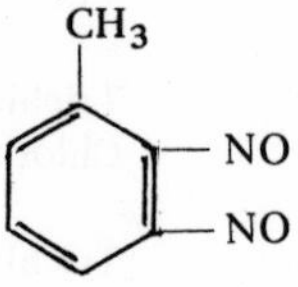

1-Methyl-2,3-dinitrosobenzene
2,3-Dinitrosotoluene

14.

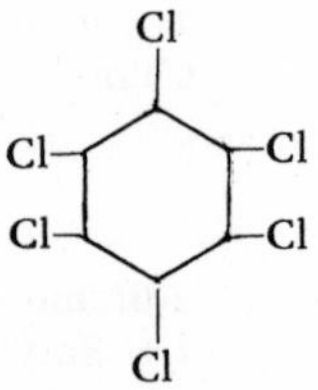

1,2,3,4,5,6-Hexachlorocyclohexane

15.   $C_6H_5CH_2Cl$

(Chloromethyl)benzene
$\alpha$–Chlorotoluene
Benzyl chloride

16.   $C_6H_5CHCl_2$

(Dichloromethyl)benzene
$\alpha, \alpha$–Dichlorotoluene
Benzylidene chloride
*not*  Benzal chloride

17.

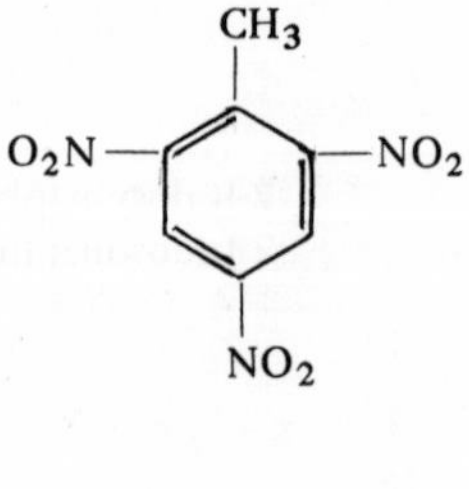

2-Methyl-1,3,5-trinitrobenzene
2,4,6-Trinitrotoluene

18.

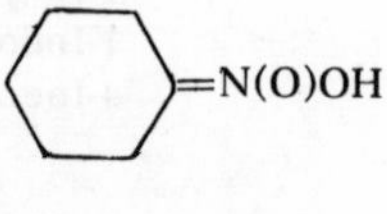

=N(O)OH   *aci*-Nitrocyclohexane

19.

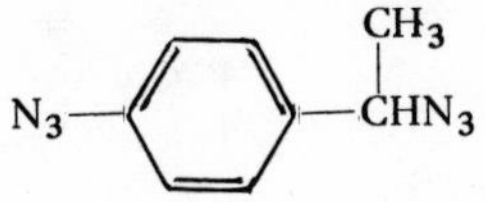

$$N_3 - \underset{\underset{CHN_3}{\overset{CH_3}{|}}}{\bigcirc} $$

1-Azido-4-(1-azidoethyl)benzene

## Literature Cited

1. "Report of ACS Nomenclature, Spelling and Pronunciation Committee for First Half of 1952, Section C", *Chem. Eng. News* (1952) **30**, 4514.

# 28

# Ions, Free Radicals, and Radical-Ions

The naming of organic chemical structures that contain one or more unpaired electrons or that carry a formal positive or negative charge is discussed in this chapter. Included among such structures are many that may be called "reactive intermediates" or, indeed, are purely hypothetical; however, this does not eliminate the need for unambiguous systematic names. Because nomenclature in this field has been inconsistent and poorly defined in some areas, continuing study and development of naming methods to accommodate both known and new structural classes have taken place in recent years.

The recommendations presented here are the result of careful consideration of available official rules, published proposals, and past usage; they deviate to some extent from practices recognized by the 1965 IUPAC Rules, and in some cases they introduce names not previously approved or widely used. These "new" names, however, are formed systematically by extending established nomenclature principles; they are not simply invented.

Neutral organic molecular structures containing one or more atoms carrying unpaired electrons are called **free radicals**. An unpaired electron is indicated in a structural formula by a dot alongside the appropriate atom and in a molecular formula by a superscript dot following the formula. The terms **radical-cation** and **radical-anion** are used to denote free radicals that carry a net positive or negative charge; for such structures a superscript plus or minus sign follows the dot in the molecular formula. Organic cations and anions are depicted in the usual manner. The class names **carbocations** (*1*) and **carbanions** denote structures in which the charge is on carbon.

Neutral molecular structures in which a dicovalent carbon atom has only a sextet of electrons in its bonding shell are called **carbenes**; structures in which a monocovalent nitrogen is similarly electron-deficient are preferably called **nitrenes** (*see* Discussion section).

The nomenclature of **cations** formed by the addition of a proton or some other cationic group to a **hetero atom** in an organic molecule is discussed in Chapter 24; naming of **anions** formed by the loss of a proton from an organic **hetero atom** is covered in Chapters 17, 18, 19, and 35.

## Recommended Nomenclature Practice

**Uncharged Species.** The name of an organic free radical is preferably formed by citing the prefix name of the corresponding substituting group and adding the word *radical;* if the prefix name ends in *y* or *o*, this terminal letter is changed to *yl.*

$CH_3CH_2^{\bullet\bullet}$                                   Ethyl radical

$CH_3\overset{\bullet}{C}HCH_2CH_2^{\bullet\bullet}$        1-Methyltrimethylene diradical

$CH_3O^{\bullet}$                                           Methoxyl radical

$CH_3CH_2CONH^{\bullet}$                                    Propanamidyl radical

$(CH_3)_2N^{\bullet}$                                       Dimethylaminyl radical

In naming **acyclic carbenes** and **nitrenes**, groups attached to the parent structures $H_2C$: and $HN$: are designated by the usual prefixes. To distinguish the singlet from the triplet state in the name, the word *singlet* or *triplet* may be added.

$Cl_2C^{\bullet}_{\bullet}$                                 Dichlorocarbene

$(CH_3)_2\overset{\bullet}{C}{\bullet}$                     Dimethylcarbene triplet

$C_6H_5\overset{\bullet\bullet}{N}{}^{\bullet}_{\bullet}$   Phenylnitrene

When a carbene group is part of a cyclic system, it is best designated by the nonseparable replacement prefix *carbena.*

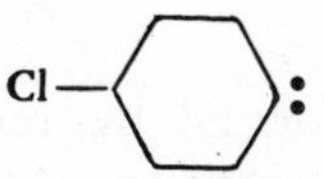

Cl—⟨ ⟩:                                                    4-Chlorocarbenacyclohexane

$(C_6H_9Cl)$

**Charged Species.** An organic **cation** conceptually formed from a free radical by the loss of its unpaired electron is named by changing the terminal word *radical* to *cation*.

$CH_3CH_2{}^+$                              Ethyl cation

$CH_3CH_2\overset{+}{C}HCH_2CH_3$          1-Ethylpropyl cation

${}^+CH_2\,CH_2{}^+$                        Ethylene dication

$CH_3NH^+$                                  Methylaminyl cation

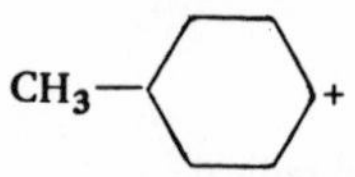

$CH_3\!-\!\langle\;\rangle\!-\!{}^+$        4-Methylcyclohexyl cation

Similarly, an organic **anion** conceptually formed by addition of an electron to a free radical so as to produce an unshared electron pair is named by changing the terminal word *radical* to *anion*.

$CH_3CH_2{}^-$                              Ethyl anion

$CH_3\,\bar{C}HCH{=}CH^-$                   3-Methylpropenylene dianion

$C_6H_5CH^{2-}$                             Phenylmethylene dianion

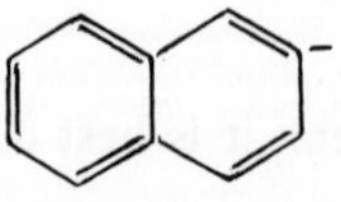

2-Naphthyl anion

An organic **radical-cation** conceptually formed by removal of an electron from a nonradical molecule is best named by adding the word *radical-cation* to the name of the reference molecule.

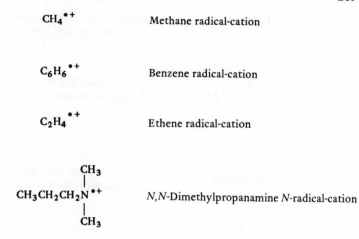

$CH_4^{\bullet +}$      Methane radical-cation

$C_6H_6^{\bullet +}$      Benzene radical-cation

$C_2H_4^{\bullet +}$      Ethene radical-cation

$$\begin{array}{c} CH_3 \\ | \\ CH_3CH_2CH_2N^{\bullet +} \\ | \\ CH_3 \end{array}$$      *N,N*-Dimethylpropanamine *N*-radical-cation

A **radical-anion** resulting from conceptual addition of an electron to a nonradical molecule is best named similarly.

$CH_4^{\bullet -}$      Methane radical-anion

$C_6H_6^{\bullet -}$      Benzene radical-anion

A **cation** whose structure is not derivable by conceptual loss of an electron from a free radical or molecule, but may be viewed as resulting from protonation of a carbon atom in a saturated molecule, is best named by using the adjective *protonated*.

$CH_5^+$      Protonated methane

$C_6H_{13}^+$      Protonated cyclohexane

**Components of Ionic Molecules.** In naming charged organic structures as components of ionic molecules, one-word names containing the suffix *ium* (denoting addition of $H^+$) or *ide* (denoting loss of $H^+$) are preferred. For radical-ions the compound suffix *iumyl* or *idyl* is best used. A protonated carbene group that is part of a ring is best designated by the prefix *carbenia*.

$CH_3CH_2CH_2{}^+$

Propyl cation (as single species)
Ethylcarbenium (as ionic component)

$CH_3-\langle\;\rangle+$

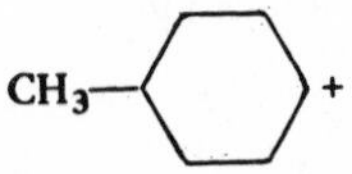

4-Methylcyclohexyl cation (as single species)
4-Methylcarbeniacyclohexane (as ionic component)

$CH_3\bar{C}HCH_3$

Isopropyl anion (as single species)
2-Propanide (as ionic component)

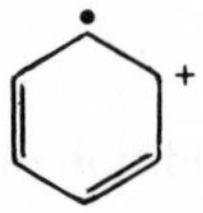

Benzene 1,2-radical-cation (as single species)
Phen-2-ium-1-yl (as ionic component)

Benzene 1,2-radical-anion (as single species)
3,5-Cyclohexadien-2-id-1-yl ( (as ionic component)

$CH_5{}^+$

Protonated methane (as single species)
Methanium (as ionic component)

## Discussion

In extending the principles of systematic organic nomenclature to the naming of ions and free radicals, certain special needs are encountered. Since charges and unpaired electrons are often considered to be delocalized, some flexibility in nomenclatural specificity is desirable so that an appropriate degree of structural ambiguity can be implied. However, because organic chemists use classical structures as a basis for depicting ions and free radicals, unambiguous names for such reference structures are required. If the inadequacy of one canonical form to represent a structure is to be pointed out, the word *hybrid* may be appended. Another means of introducing ambiguity is the deliberate omission of locants denoting charges or unpaired electrons.

It is also desirable in naming organic ions and free radicals to provide systematically formed names for these species as independent entities, not associated with or attached to any other atom, ion, or group. Thus, the free radical $CH_3^\cdot$ is best called methyl radical, whereas the group $CH_3$ as a substituent replacing H in a parent compound is simply designated methyl. Similarly, $(CH_3)_3C^+$ considered alone is preferably named $t$-butyl cation, but in the name of an ion pair it is designated trimethylcarbenium (*see* Example 9, Table 28.1); analogous treatment of $(C_6H_5)_3C^-$ results in the names triphenylmethyl anion and triphenylmethanide (*see* Example 16). The principles applied in forming these names are given below.

Gaps in the present recommendations remain; for example, provision is not made for many cationic fragments, such as $C^+$ and $ClC^+$ observed in mass spectrometry. Another area needing further study is that of cations considered as having two-electron three-center bonds. Much development work remains to be done.

As used in forming the systematic names recommended in this chapter, the terminal word *radical* simply serves to confirm and emphasize the presence of an unpaired electron in a group of atoms having a free valence; when accompanied by the appropriate locant, it can also serve to specify the position of the unpaired electron in a radical-ion (*see* Example 19). Names of cations and anions are derived from those of the corresponding free radicals by replacing the word *radical* with *cation* or *anion* to signify conceptual loss or gain of an electron. This approach affords systematic names for organic free radicals and ions of conventional stoichiometric composition suitable for designating these species an independent entities.

When an organic group carrying a positive or negative charge is to be named as a component of an ionic molecule, the aforementioned type of name is not suitable; for this purpose one-word names wherein the charge is denoted by a suffix are preferred. Such names are readily derived by viewing the ionic group as having been formed by conceptual gain or loss of a proton by a neutral molecule and adding the suffix *ium* or *ide* to the name of the latter. Thus $CH_3^+$ is named *carbenium* (addition of $H^+$ to $H_2C$:), and $CH_3^-$ is named *methanide* (loss of $H^+$ from $CH_4$); both names are recognized in the 1965 IUPAC Rules.

While there has been much usage of the term "carbonium" in naming organic structures containing a positively charged carbon atom, this practice has been inconsistent and often confusing (2). It is recommended that "carbonium" no longer be used, either as a class name—*e.g.*, "carbonium ions"—or in names of individual cations—*e.g.*, "trimethylcarbonium"; the class name carbocations and the systematic name trimethylcarbenium are preferred.

The naming of organic radical-ions recommended here deviates in part from that presented in the 1965 IUPAC Rules (*see* Rules C-83.3 and C-84.4). The name benzene radical-cation is more thoroughly descriptive than benzene cation for $C_6H_6^{\cdot+}$, and the compound suffix *idyl* is preferred to *ylide* for sake of consistency with its counterpart *iumyl*, recommended by IUPAC. Furthermore, since free radicals stand at the top of the list in the *Chemical Abstracts* order of precedence of functions, it is logical that the two-part suffixes used in naming both radical-cations and radical-anions should end in *yl—i.e., iumyl* and *idyl*. The seniority of free radicals relative to ionic groups also suggests that the former should receive first consideration for assignment of lowest-numbered locants when there is a choice, as in phen-2-ium-1-yl for $C_6H_6^{.+}$. It should be noted that 1965 IUPAC Rule C-83.1 authorizes use of the suffix *ylium* for naming nonradical cations—*e.g.*, "ethylium" for $CH_3CH_2^+$; this practice should be discontinued. The alternative IUPAC-recognized name "ethenium" (Rule C-82.4) is even less desirable.

The terms carbene and nitrene form the basis for a specialized nomenclature applicable to the two important classes of electron-deficient species represented by the parent structures $H_2C$: and $H\ddot{N}$: . In the triplet state these species are diradicals and may be so named—*e.g.*, dichlorocarbene triplet or dichloromethylene diradical; the term carbene does not itself distinguish between the singlet and triplet states. The name "azene" has been proposed as an alternative to nitrene, but the latter is preferred on the basis of usage.

The trivial radical-anion names ketyl, as in benzophenone ketyl for $(C_6H_5)_2\dot{C}$-$O^-$, and *p*-benzosemiquinone anion for $^.OC_6H_4O^-$ are well established by usage, but the systematic names diphenylmethanone *C*-radical-*O*-anion and 2,5-cyclohexadiene-1,4-dione *O*-radical-*O'*-anion are preferable.

### Table 28.1. Examples Of Acceptable Usage

1.        $C_6H_5^.$             Phenyl radical

2.        $CH_3\dot{C}HCH_2^.$        Methylethylene diradical

3.

                                2-Naphthyloxyl radical

4.        $^.HNCH_2CH_2CH_2NH^.$      Trimethylenediaminyl diradical

5.            $CH_3NHNH^.$        2-Methyldiazanyl radical <br> 2-Methylhydrazinyl radical

6.        $CH_3CH_2CH_2CH_2\overset{\displaystyle O}{\overset{\|}{C}}{}^.$        Pentanoyl radical

7. $CH_3CH_2CH_2\overset{\bullet}{C}H\bullet$      Propylcarbene triplet
Propylmethylene diradical

8. $Cl_3C\overset{\bullet\bullet}{N}\colon$      (Trichloromethyl)nitrene

9. $(CH_3)_3C^+$      *t*-Butyl cation
Trimethylcarbenium (as ionic component)

10. $C_6H_5NH^+$      Phenylaminyl cation

11. 2.5-Cyclohexadien-1-yl cation
Carbeniacyclohexa-2,5-diene (as ionic component)

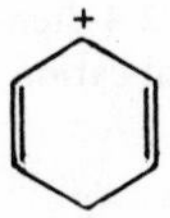

12. $CH_2^{+2}$      Methylene dication

13. $CH_3CH_2S^+$      Ethylthiyl cation

14. Methoxycarbonyl cation
Methoxyoxocarbenium (as ionic component)

15. $CH_3CH_2CH_4^+$      1-Protonated propane
1-Propanium (as ionic component)

16. $CH_3CH_2\overset{\overset{\displaystyle CH_3}{\displaystyle |}}{C}HCH_2^-$      2-Methylbutyl anion
2-Methylbutanide (as ionic component)

17. 1,4-Cyclohexylene dianion
1,4-Cyclohexanediide (as ionic component)

18.

Cyclohexane radical-cation
Cyclohexaniumyl (as ionic component)

19.

Naphthalene 1,2-radical-anion
1,2-Dihydronaphthalen-2-id-1-yl (as
   ionic component)

20.

Azacyclopenta-2,4-diene 1-radical-cation
Pyrrole 1-radical-cation

## *Literature Cited*

1. G. A. Olah, *J. Amer. Chem. Soc.* (1972) **94**, 808-820.
2. C. D. Hurd, *J. Chem. Educ.* (1971) **48**, 490.

# Ketones

As already mentioned in Chapter 20, which deals with aldehydes, compounds containing the carbonyl group ($>C=O$, commonly written CO) attached to two carbon atoms are called **ketones**. In that chapter the close structural relationship between aldehydes and ketones is pointed out, and the need in systematic nomenclature to distinguish between these two types of carbonyl compounds is questioned. Nevertheless, since modern usage continues to reflect the desire of organic chemists to treat aldehydes and ketones as two separate functional classes and since no official approval has yet been given to proposals for a unified nomenclature, the traditional approach is retained in this book.

Also included in this chapter are **thioketones**, which contain the thiocarbonyl group ($>C=S$), and **quinones**, which are a sub-class made up of cyclic unsaturated ketones having two carbonyl groups directly attached to a six-membered ring containing two double bonds. Structures in which an ester, amide, or anhydride group is part of a ring and which, as a consequence, are named as heterocyclic ketones are discussed in Chapters 17 and 21.

*Recommended Nomenclature Practice*

**Monoketones** in which the carbon of the CO group is a member of a chain or ring system are preferably named substitutively by adding the suffix *one* to the name of the corresponding hydrocarbon, with elision of the final *e* of the hydrocarbon name.

$CH_3CH_2COCH_2CH_3$         3-Pentanone

$CH_2=CHCOCH_2CH_2CH_3$      1-Hexen-3-one

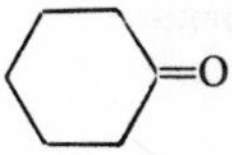

Cyclohexanone

When the ring system is named as an aromatic hydrocarbon, it may be necessary to show the position of added hydrogen.

225

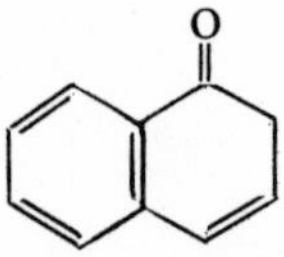

1(2*H*)-Naphthalenone

**Polyketones** are named substitutively by adding the appropriate suffix *dione, trione, tetrone,* etc. to the name of the corresponding hydrocarbon.

$CH_3COCH_2CH_2CH_2COCH_3$       2,6-Heptanedione

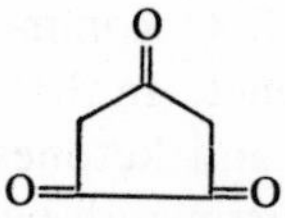

1,2,4-Cyclopentanetrione

**Heterocyclic ketones** are named similarly. Where necessary to avoid ambiguity the location of added hydrogen is shown.

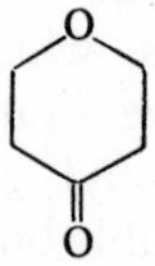

Oxacyclohexan-4-one

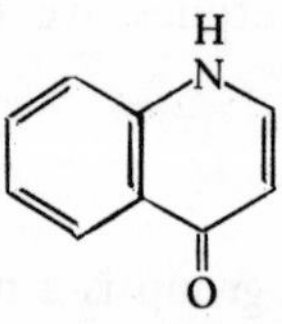

1-Azanaphthalen-4(1*H*)-one

**Ketones** in which one or more **cyclic systems** are directly connected to the carbonyl group are preferably named substitutively using the suffixes *one, dione, trione,* etc. However, in current usage such structures are commonly given radicofunctional names ending in the words *ketone, diketone, triketone,* etc.

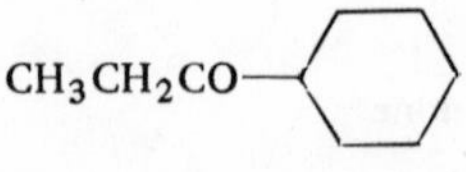

$CH_3CH_2CO-$⟨ ⟩       1-Cyclohexyl-1-propanone
                                    Cyclohexyl ethyl ketone

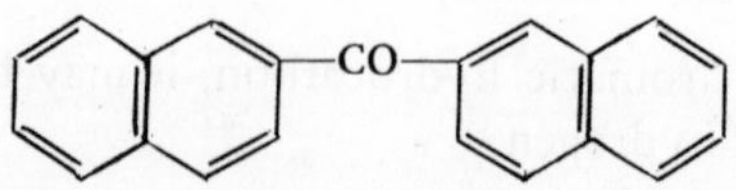

$-CO-$       Di-2-naphthylmethanone
                              Di-2-naphthyl ketone

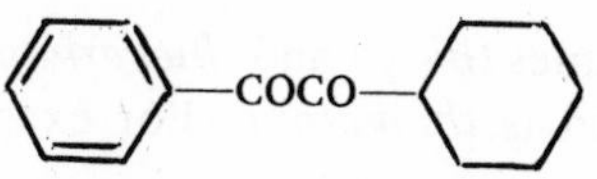

[structure: phenyl—COCO—cyclohexyl]  1-Cyclohexyl-2-phenylethanedione
Cyclohexyl phenyl diketone

When the ketone group must be named as a substituent, the prefix name *oxo* (denoting O= ) is generally used. However, when attachment is through the CO group, acyl names (see Chapter 17) are commonly employed.

$CH_3COCH_2CH_2CH_2CN$    5-Oxohexanenitrile

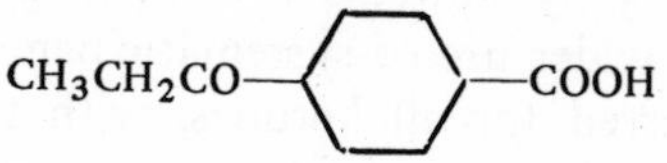

$CH_3CH_2CO$—[cyclohexane ring]—COOH    4-(1-Oxopropyl)cyclohexanecarboxylic acid
4-Propanoylcyclohexanecarboxylic acid

A number of **trivial names** for the more frequently encountered ketones are well established by usage. Among these are (1) names ending in *quinone* for aromatic unsaturated cyclic diketones or tetraketones characterized by complete conjugation, and (2) names ending in *phenone* or *naphthone* for structures containing the $C_6H_5CO$ group or the $C_{10}H_7CO$ group.

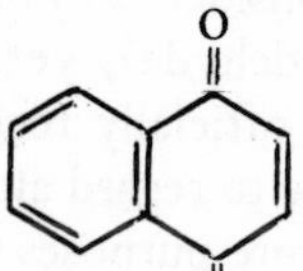

[structure: naphthalene ring with two O groups at 1,4 positions]

1,4-Naphthalenedione
1,4-Naphthoquinone

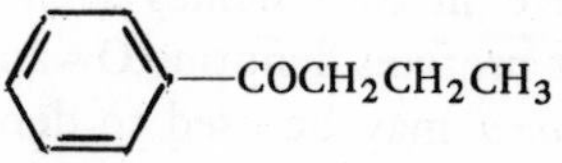

[structure: phenyl—COCH$_2$CH$_2$CH$_3$]    1-Phenyl-1-butanone
Butyrophenone

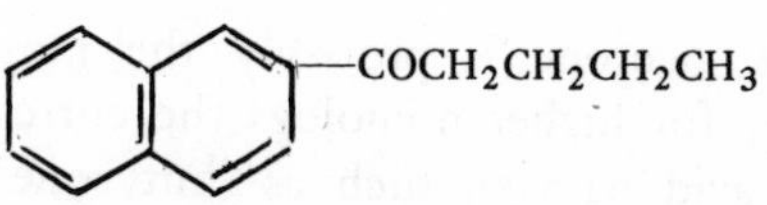

[structure: naphthalene—COCH$_2$CH$_2$CH$_2$CH$_3$]    1-(2-Naphthyl)-1-pentanone
2'-Valeronaphthone

Although trivial names of these two types are acceptable, systematic names (as shown) are to be preferred. Most other trivial names for ketones are no longer considered acceptable.

**Thioketones** are named systematically in a manner analogous to that described above for ketones by using the substitutive suffixes *thione,* *dithione, trithione,* etc. and the prefix names *thioxo* and *thiocarbonyl.* In radicofunctional names the terminal word is *thioketone.* For examples, *see* Table 29.1.

## Discussion

Although substitutive nomenclature is now generally accepted as the preferred method of forming systematic names for organic compounds, radicofunctional nomenclature is still widely used in naming ketones, particularly those structures in which the CO group is directly attached to one or to two ring systems. Indeed, for the latter type of compound, substitutive nomenclature has as yet found only limited use. Nevertheless, in keeping with the current trend toward wider use of systematic names, substitutive nomenclature is to be preferred for all ketones, with the radicofunctional approach as an acceptable second choice in many instances.

For acyclic ketones, as well as for even the simplest acyclic-cyclic ketones, *Chemical Abstracts* now uses substitutive names in preference to radicofunctional names—*e.g.,* 2-butanone rather than ethyl methyl ketone, and 1-cyclopropyl-1-butanone rather than cyclopropyl propyl ketone. It must be admitted, however, that the 1965 IUPAC Rules did not see fit to recommend substitutive names for cyclic-cyclic ketones; this is undoubtedly because such names are based on the parent structure $H_2CO$, for which the ketone name, methanone, could be considered to be in conflict with the aldehyde name, methanal (or formaldehyde); yet, for higher homologs, parent names such as 1-butanone are officially recognized and used. A simple solution to the problem would be to regard aldehydes and ketones as a single functional class for nomenclature purposes (*cf.* Chapter 20), but this proposal has never progressed beyond the discussion stage.

The term "keto" should no longer be used either as a prefix in naming individual compounds or as an adjective in class names such as "keto acids." *Oxo* and *carbonyl* are the correct prefixes denoting O= and >C=O, respectively. As stated in Chapter 20, *oxo* may be used to denote both aldehyde and ketone groups when these occur in the same chain system (*see* Example 31).

Among trivial names for ketones, acetone is probably the most common, and is acceptable. However, for higher homologs the corresponding names (formed from trivial acid names), such as "butyrone" and "valerone," should not be used. Also, contracted names for heterocyclic ketones such as 4-piperidone for 4-piperidinone, although recognized in the 1965 IUPAC Rules, are not recommended.

Although a few trivial names containing acid-name stems as prefixes— *e.g.,* acetophenone for $CH_3COC_6H_5$ and benzophenone for $C_6H_5COC_6$-$H_5$ —are acceptable, the corresponding systematic names are to be preferred. In any event, trivial names of this kind should be used only when the ring system cited in the main portion of the names is benzene or naphthalene. It should also be noted that in this type of name the contracted prefix *propio* is used rather than *propiono* (*see* Example 22).

The 1965 IUPAC Rules recognize the trivial names benzil for $C_6H_5COCOC_6H_5$, benzoin for $C_6H_5CH(OH)COC_6H_5$, deoxybenzoin for $C_6H_5CH_2COC_6H_5$, and chalcone for $C_6H_5CH=CHCOC_6H_5$, but these names should be abandoned (*see* Examples 24-27).

The word "quinone" has often been used as a specific name for the compound whose properly formed common name is 1,4-benzoquinone or *p*-benzoquinone; this practice should be discontinued. "Quinone" is used correctly either as a combining form in trivial names of individual compounds or as a generic term denoting the class of unsaturated cyclic diketones or polyketones characterized by a completely conjugated all-carbon system of one or more six-membered rings to which are attached an even number of O= groups in a manner such that no more than two of these are connected to any one ring. The word hydroquinone is recognized as a well-established trivial name for the compound 1,4-benzenediol (*see* Chapter 35).

Although acyl names (*e.g.,* hexanoyl for $CH_3CH_2CH_2CH_2CH_2CO-$) are commonly used for substituting groups carrying an O= at the point of attachment, the systematic method of naming such groups is to treat them in the same manner as substituting groups in which the O= is not at the point of attachment—*e.g.,* 1-oxobutyl for $CH_3CH_2CH_2CO-$, 3-oxobutyl for $CH_3COCH_2CH_2-$, 1,3-dioxobutyl for $CH_3COCH_2CO-$. The trivial names acetonyl for $CH_3COCH_2-$, acetonylidene for $CH_3COCH<$, phenacyl for $C_6H_5COCH_2-$, and phenacylidene for $C_6H_5COCH<$ should be abandoned.

Although a few trivial names of thioketones in which the prefix *thio* is attached to the name of the analogous oxygen compound, *e.g.,* thioacetone and thiobenzophenone, are recognized in the 1965 IUPAC Rules, such names are not recommended for use.

The terms "ketal" and "hemiketal" denoting an acetal or hemiacetal derived from a ketone rather than from an aldehyde are no longer officially recognized (1965 IUPAC Rules) and should be abandoned both as class names and in names of individual compounds. Recommendations for naming acetals and hemiacetals as provided in Chapter 20 may be applied to these types of ketone derivative (*see* Examples 35-38).

### Table 29.1.   Examples of Acceptable Usage

| | | |
|---|---|---|
| 1. | $CH_3COCH_3$ | 2-Propanone<br>Acetone<br>Dimethyl ketone |
| 2. | $CH_3COCH_2CH_3$ | 2-Butanone<br>Ethyl methyl ketone |

3.  $CH_3COCH_2CHCH_3$ with $CH_3$ substituent
    4-Methyl-2-pentanone
    Isobutyl methyl ketone

4.  $CH_3COCH_2Cl$
    1-Chloro-2-propanone
    Chloroacetone

5.  $CH_3COCH{=}CH_2$
    3-Buten-2-one
    Methyl vinyl ketone

6.  $CH_3COCCH_3$ with two $CH_3$ substituents
    3,3-Dimethyl-2-butanone
    *not* Pinacolone

7.  $CH_3COCOCH_3$
    2,3-Butanedione
    *not* Biacetyl

8.  $CH_3COCH_2COCH_3$
    2,4-Pentanedione
    *not* Acetylacetone

9.  $CH_3COCH_2CH_2COCH_3$
    2,5-Hexanedione
    *not* Acetonylacetone

10. $CH_2{=}C{=}O$
    Ethenone
    Ketene

11.
    Cyclopentanone

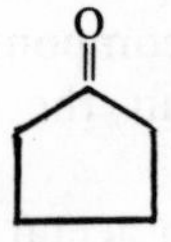

12. 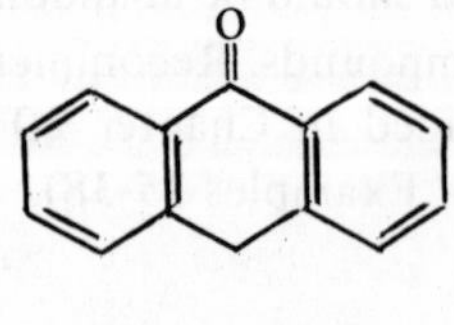 9(10H)-Anthracenone
    *not* Anthrone

13. 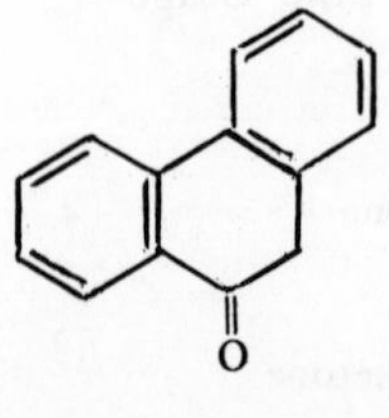 9(10H)-Phenanthrenone
    *not* Phenanthrone

14.

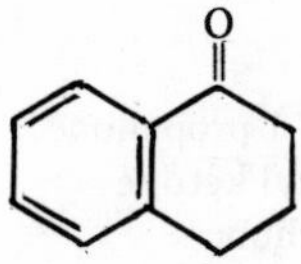

3,4-Dihydro-1(2H)-naphthalenone
*not* α-Tetralone

15.

O=⬡=O

1,4-Cyclohexanedione

16.

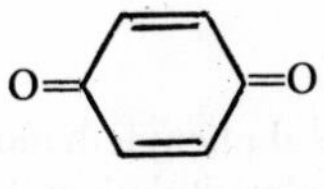

O=⬡=O

2,5-Cyclohexadiene-1,4-dione
1,4-Benzoquinone

17.

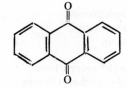

9,10-Anthracenedione
Anthraquinone

18.

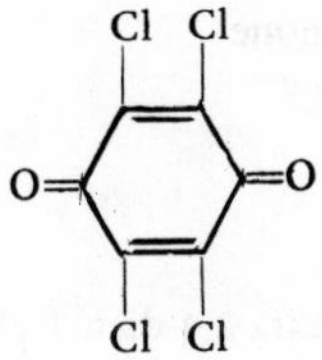

Cl  Cl
O=⬡=O
Cl  Cl

Tetrachloro-2,5-cyclohexadiene-1,4-dione
Tetrachloro-1,4-benzoquinone
*not* Chloranil

19.

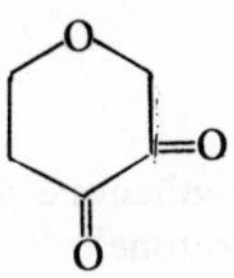

Oxacyclohexane-3,4-dione
Dihydro-2H-pyran-3,4-dione

20.

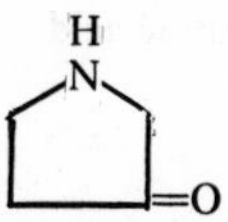

Azacyclopentan-3-one
3-Pyrrolidinone
*not* 3-Pyrrolidone

21. $C_6H_5COCH_3$

1-Phenylethanone
Methyl phenyl ketone
Acetophenone

22.

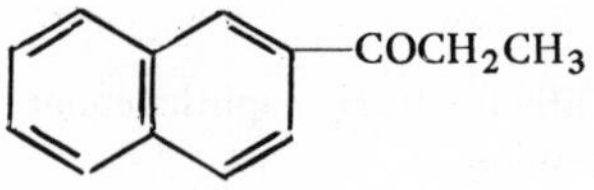

—COCH$_2$CH$_3$

1-(2-Naphthyl)-1-propanone
Ethyl 2-naphthyl ketone
2'-Propionaphthone

23.    C$_6$H$_5$COC$_6$H$_5$

Diphenylmethanone
Diphenyl ketone
Benzophenone

24.    C$_6$H$_5$COCOC$_6$H$_5$

Diphenylethanedione
Diphenyl diketone
*not* Benzil

25.    C$_6$H$_5$CHCOC$_6$H$_5$
              |
             OH

2-Hydroxy-1,2-diphenylethanone
Hydroxyphenylmethyl phenyl ketone
*not* Benzoin

26.    C$_6$H$_5$CH$_2$COC$_6$H$_5$

1,2-Diphenylethanone
Benzyl phenyl ketone
*not* Deoxybenzoin

27.    C$_6$H$_5$CH=CHCOC$_6$H$_5$

1,3-Diphenyl-2-propen-1-one
Phenyl 2-phenylethenyl ketone
*not* Chalcone

28.    (C$_6$H$_5$)$_2$C=C=O

Diphenylethenone
Diphenylketene

29.

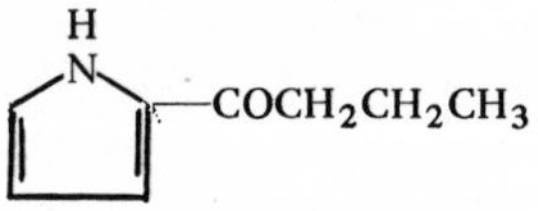

—COCH$_2$CH$_2$CH$_3$

1-(Azacyclopenta-2,4-dien-2-yl)-1-butanone
Propyl 2-pyrrolyl ketone

30.

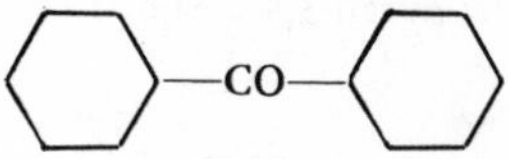

—CO—

Dicyclohexylmethanone
Dicyclohexyl ketone

31.    OHCCH$_2$CH$_2$COCH$_2$COOH

3,6-Dioxohexanoic acid

32    CH$_3$CO— —COOH
                                        —COOH

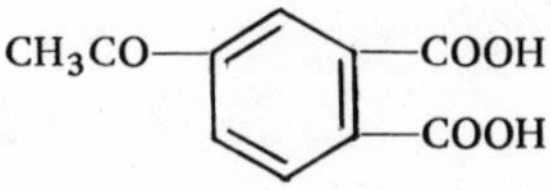

4-(1-Oxoethyl)-1,2-benzenedicarboxylic acid
4-Acetylphthalic acid

33.

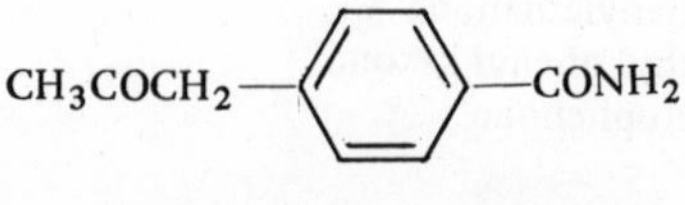

CH$_3$COCH$_2$— —CONH$_2$

4-(2-Oxopropyl)benzenecarboxamide
4-(2-Oxopropyl)benzamide

34.   CH₃CO—⟨benzene⟩—COCH₃     1,1'-(1,4-Phenylene)diethanone

35.
$$CH_3CH_2\overset{\displaystyle OCH_3}{\underset{\displaystyle OH}{C}}CH_2CH_2CH_3$$
3-Methoxy-3-hexanol

36.
$$CH_3\overset{\displaystyle OCH_3}{\underset{\displaystyle OCH_3}{C}}CH_2CH_2CH_3$$
2,2-Dimethoxypentane

37.
$$C_6H_5\overset{\displaystyle OCH_3}{\underset{\displaystyle OCH_3}{C}}CH_2CH_2CH_3$$
(1,1-Dimethoxybutyl)benzene

38.
$$C_6H_5\overset{\displaystyle OCH_3}{\underset{\displaystyle OCH_3}{C}}C_6H_5$$
Dimethoxydiphenylmethane

39. 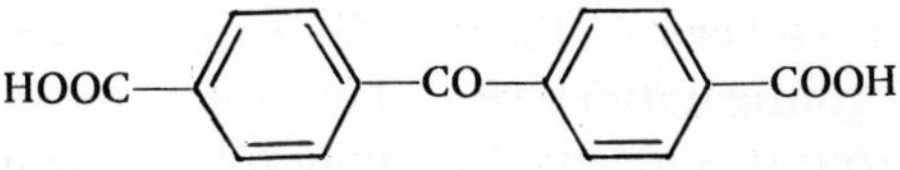  HOOC—⟨benzene⟩—CO—⟨benzene⟩—COOH

4,4'-Carbonyldibenzenecarboxylic acid
4,4'-Carbonyldibenzoic acid

40.  CH₃CSCH₃

2-Propanethione
Dimethyl thioketone
*not*  Thioacetone

41.  C₆H₅CSC₆H₅

Diphenylmethanethione
Diphenyl thioketone
*not*  Thiobenzophenone

42. 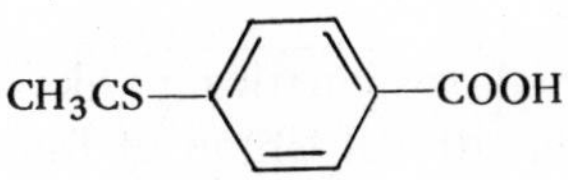  CH₃CS—⟨benzene⟩—COOH

4-(1-Thioxoethyl)benzenecarboxylic acid
4-Ethanethioylbenzoic acid
*not*  4-Thioacetylbenzoic acid

43. 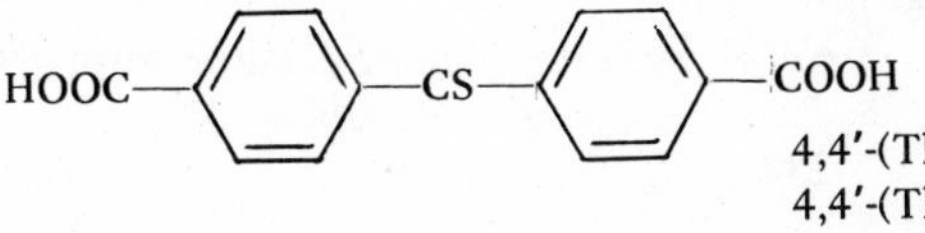  HOOC—⟨benzene⟩—CS—⟨benzene⟩—COOH

4,4'-(Thiocarbonyl)dibenzenecarboxylic acid
4,4'-(Thiocarbonyl)dibenzoic acid

# 30

# Natural Products

**B** roadly speaking, an organic natural product is a compound or mixture of compounds extracted from plant, animal, or (rarely) mineral substances. Usually the term refers only to substances of fairly complex structure. Although the isolation procedure may cause some chemical alteration, particularly hydrolysis, natural products are regarded in principle as pre-existent in the material from which they were extracted.

For reasons that will appear later, natural products are usually designated by short nonsystematic names that do not describe their structures fully. Rules for assigning such names are far from comprehensive; they usually legalize trivial names and often deal with structural families rather than individual compounds.

Because of the wide diversity of natural products and the specialized nature of their chemistry, no attempt is made in this chapter to give detailed rules and principles. In lieu of this, reference is made to selected publications providing up-to-date guidance for naming these materials, including official rules, tentative rules, and unofficial proposals. It is especially important to remember that in nearly all cases the short nonsystematic names that have been established for natural products are specific with respect to configuration—*i.e.*, they refer to individual stereoisomers and are therefore not usually applicable to compounds synthesized in the laboratory.

*Recommended Nomenclature Practice*

Whenever practicable, a natural product should be designated by the systematic name for its structure; needless creation of new trivial names is undesirable.

In the absence of other rules, the trivial name of a new natural product should be based on a material source or a property of the product and should bear an ending representing some aspect of its chemical structure, preferably the principal functional group. If this is unknown, the name should end in *in (1)*.

**Lipids.** Glycerides are preferably named systematically like other esters. The IUPAC-IUB (IUB stands for International Union of Biochemistry) Commission on Biochemical Nomenclature has recently promulgated definitive rules for naming lipids *(2)*.

**Carbohydrates.** The *Rules of Carbohydrate Nomenclature* approved by the American Chemical Society (*3*) are detailed, as are the more recent tentative rules published by the IUPAC Commission on the Nomenclature of Organic Chemistry and the IUPAC-IUB Commission (*4*) and the IUPAC-IUB tentative rules on cyclitols (*5*). There have been proposals to systematize the naming of dextrans (*6*), pectins (*7*), and mucopolysaccharides (*8*).

**Amino Acids, Polypeptides, and Proteins.** Definitive rules for the nomenclature of natural amino acids and related substances have been published by the IUPAC Commission (*9, 10*). Polypeptides are named by affixing the prefix names of the amino acid units in the sequence in which they occur to the name of the chain-terminating *N*-acylated amino acid.

$$HOOCCHNH_2CH_2CH_2CONHCH(CH_2SH)CONHCH_2COOH$$

$\gamma$-Glutamylcysteinylglycine

Tentative rules for naming synthetic modifications of natural peptides have been established recently (*11*).

Fully systematic names can also be applied to polypeptides, but this is much more cumbersome than the "consecutive unit" method. Protein structures are so complex that they are invariably designated by trivial names.

**Porphyrins.** Porphyrins are derived from the structure shown, and may be named systematically as substituted porphines (*12*).

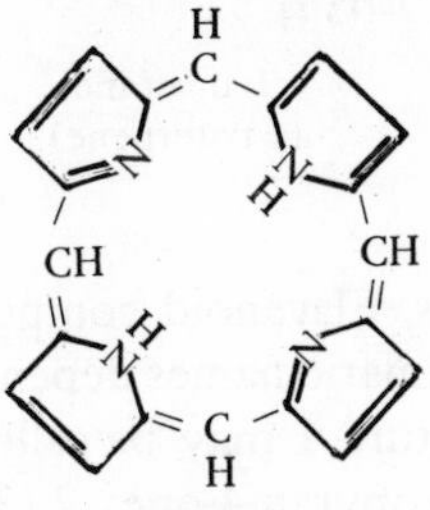

Porphine

In common practice, derivatives of porphine are called porphyrins—*e.g.*, hematoporphyrin and the etioporphyrins. Nomenclature for the related corrinoids has been prescribed (*13*).

**Carotenoids.** Tentative rules for nomenclature of carotenoids are available (*9*).

**Terpenoids.** The naming of many monoterpene hydrocarbons is described by IUPAC Rules A-71 to A-75. These rules establish the names menthane, thujane, carane, pinane, bornane, norcarane, norpinane, and norbornane for parent compounds, and authorize the use of a few trivial names. They are derived from the much more detailed presentation in "System of Nomenclature for Terpene Hydrocarbons" (14).

Similar lists of parent hydrocarbon names have been published for sesquiterpenes (15), diterpenes (16), and triterpenes (17, 18), and derivatives of these may be named accordingly.

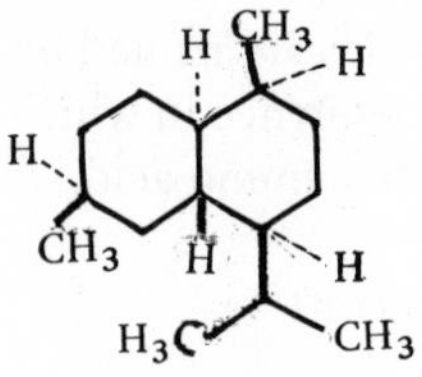

Cadinane
(a sesquiterpene)

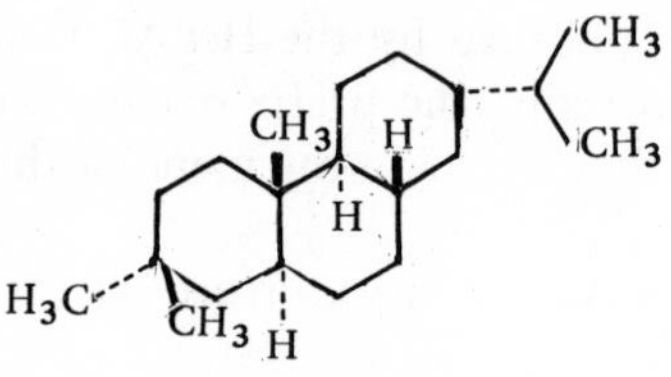

Abietane
(a diterpene)

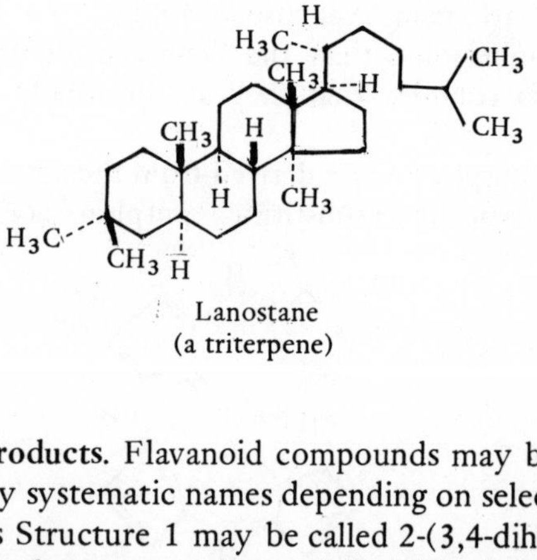

Lanostane
(a triterpene)

**Phenolic Plant Products.** Flavanoid compounds may be given various systematic or partially systematic names depending on selection of the parent compound. Thus Structure 1 may be called 2-(3,4-dihydroxyphenyl)-3,5,7-trihydroxy-4H-1-benzopyran-4-one, 2-(3,4-dihydroxyphenyl)-3,5,7-trihydroxychromone, 3,3',4',5,7-pentahydroxyflavone, or quercetin. The proposal of Freudenberg and Weinges (19) to establish Structure 2, called flavan, as a parent compound produces the name 3,3',4',5,7-pentahydroxyflav-2-en-4-one for 1.

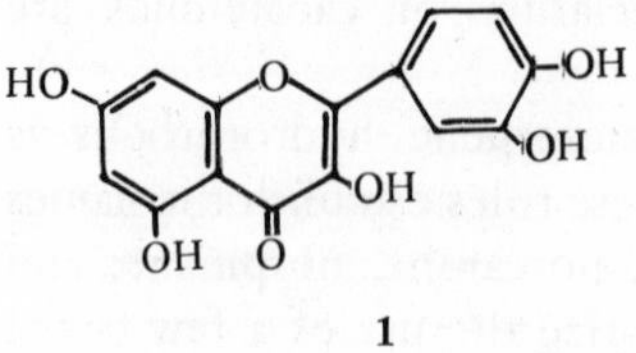

1

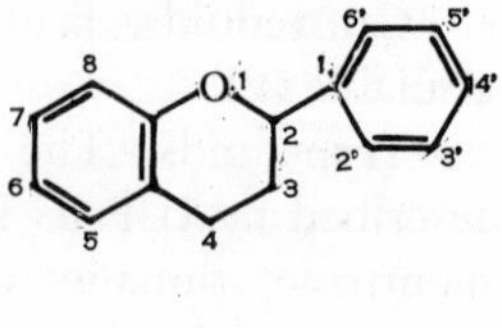

2

The **lignans** present a similar situation. The systematic name for (–) galbacin, Structure 3, is tetrahydro-3,4-dimethyl-2,5-bis(3,4-methylene-dioxyphenyl)furan, but this name does not define the stereochemistry.

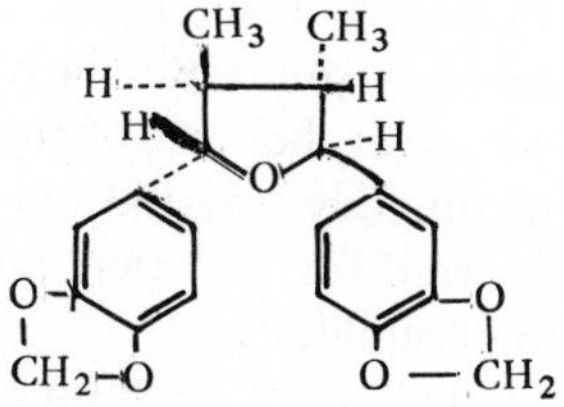

3

If 2,3-dimethyl-1,4-diphenylbutane is defined as lignan, as proposed by Freudenberg and Weinges (*20*), and their stereochemical symbols are used, the name becomes (3,4) (3′,4′)-bis(methylenedioxy)-7,7′-epoxy-α7,β8,β7′-α8′-lignan.

The nomenclature of **tannins** presents no special features.

**Nucleosides and Related Compounds.** Names of these compounds may be based on those of the parent purines and pyrimidines: adenine, cytosine, guanine, thymine, and uracil.

**Steroids.** These may be named by the recent "Definitive Rules for Nomenclature of Steroids" (*18*).

**Alkaloids.** In spite of the magnitude and importance of this family, there is no system of nomenclature for these compounds as such. The recent IUPAC-IUB rules on steroids make some suggestions for naming alkaloids.

**Enzymes** are named according to rules established by the International Union of Biochemistry (*21*).

**Vitamins** are named either by systematic organic nomenclature or by rules established by the IUPAC–IUB (*13*).

## Discussion

Because organic compounds were isolated and studied for many decades before an adequate structural theory and notation system were devised, the early nomenclature of natural products was completely nonsystematic—*i.e.*, not descriptive of structure. The vast majority of natural products were named, and continue to be named, for their sources (*22, 23*). This may involve either the genus name, the species name, or the common name of a plant or animal, or some other name of a naturally occurring material. Examples could be given by the hundreds: eugenol and caryophyllene from *Eugenia caryophyllata* (cloves), echinochrome A from the echinoderm *Arbacia pustulosa*, butyric acid from butter (Latin *butyrum*), uric acid from urine, and many others.

Another type of trivial name for natural products is descriptive of some property of a designated compound. Thus sterols are named from Greek *stereos*, solid, and durene from Latin *durus*, hard, because these materials were early isolated as solids. Similarly, glycine, glycerol, etc. are named from Greek *glykys*, sweet, because of their taste.

Very rarely, and fortunately so, natural products are named for persons or places: pelletierine for Pelletier, mexogenin for Mexico, and truxillic acid for Truxillo (Trujillo, Peru).

Closely related substances have sometimes been named merely by adding a serial symbol to a family name: α-carotene, β-carotene; vitamin A, vitamin B; protoporphyrin IX. Usually, as has happened for the vitamins, these names become displaced by better ones.

There are at least two good reasons for continuing to assign trivial names to new natural products and to use such names for old and new ones. The isolation of a new compound from a natural source must precede determination of structure and sometimes does so by many years; during that time some means of designating the compound is required. Even when the structure has been established, the systematic name describing it is usually so complex (partly for stereochemical reasons) that chemists continue to use the trivial name. (The same convenience of brevity often leads to common names, often merely jargon, for synthetic compounds as well; compare drug names.)

As systematic organic nomenclature has been developed, some of the old trivial names have been retained as official. For example, the Geneva conference adopted the methane, ethane, propane, butane sequence over the systematic monane, diane, triane, tetrane names. This principle of retention of well-established names has been adopted by the IUPAC, but it has been found necessary to make explicit the list of names thus retained; there is otherwise no definition of what is "well established."

The expression of function in the names of organic natural products began early, first for acids and bases (alkaloids). The development of suffixes to designate function, which went on gradually in the nineteenth century (24), made this easier, and in 1924 the IUC (1) declared: "In the event of the constitution of a natural substance being too complex or not well known, the name with which it is to be designated should, at any rate, feature a termination in agreement with the principal chemical group."

This dictum has been followed moderately well, but for one reason or another new natural products are still often assigned names ending in the noncommittal suffix *in*. Formerly this was confused with the ending *ine*, but the same IUC committee pointed out the need for reserving the *ine* suffix for basic substances.

A recurring question is the desirability of elevating one or another coined name to approved status, or more frequently that of assigning a parent name from which a family of natural products can be named systematically by the usual addition of affixes. For example, in the systematic naming of steroids (*9*), the hydrocarbon shown as Structure 4 is called 5β-estrane.

4

Proposals for such systematization usually arise for families of natural products in which development has been rapid and recent. An IUPAC statement is apropos: "If a new trivial name is created, any syllables in it that are used in systematic nomenclature must bear their systematic meaning and be used by the systematic methods." Unfortunately, however, the common practice of adding prefixes and suffixes to trivial names often is not definitive of structure. For example, even if the formula of gibberic acid is known, the name ketonorallogibberic acid does not describe fully the structure of the derivative. Thus, care should be exercised to make sure that when trivial names are used as parent names, the principles of systematic nomenclature are applied properly in constructing names of derivatives.

### Literature Cited

1. J. E. Courtois, ADVAN. CHEM. SER., **1953**, 8, 86.
2. "The Nomenclature of Lipids," *European J. Biochem.* **1967**, 2, 127; *Biochem.* **1967**, 6, 3287; *J. Biol. Chem.* **1967**, 242, 2485.
3. "Rules of Carbohydrate Nomenclature," *J. Org. Chem.* (1963) 28, 281.
4. "Tentative Rules for Carbohydrate Nomenclature" (CNOC-CBN), *Biochemistry* **1971**, 10, 3983; *Biochem. J.* **1972**, 27, 741; and other places.
5. "The Nomenclature of Cyclitols," *European J. Biochem.* **1968**, 5, 1; *Biochem. Biophys. Acta* **1968**, 165, 1; *Arch. Biochem. Biophys.* **1968**, 128, 269.
6. H. J. Koepsell *et al.*, *J. Biol. Chem.* **1953**, 200, 793.
7. F. A. Henglein, *Makromol. Chem.* **1953**, 10, 89.
8. R. W. Jeanloz, *Arthritis Rheumat*, **1960**, 3, 233; *Chem. Abstr.* **1960**, 54, 21211.
9. "Tentative Rules for the Nomenclature of Carotenoids (CNOC-CBN), *Biochem. J.* **1972**, 127, 741; and other places.

10. "Addendum to Definitive Rules for the Nomenclature of Natural Amino Acids and Related Substances," *J. Org. Chem.* **1963**, 28, 291.

11. "Rules for Naming Synthetic Modifications of Natural Peptides. Tentative Rules" *J. Biol Chem.* **1967**, 242, 555; *Biochemistry* **1967**, 6, 362; *Arch. Biochem. Biophys.* **1967**, 121, 6.

12. C. Rimington, *Endeavour* **1955**, 14, 126.

13. "Nomenclature of Vitamins, Coenzymes, and Related Compounds," *J. Biol. Chem.* **1966**, 241, 2987; *Biochem. J.* **1967**, 102, 15; *European J. Biochem.* **1967**, 2, 1.

14. "System of Nomenclature for Terpene Hydrocarbons," ADVAN. CHEM. SER. **1955**, 14, 4.

15. F. Sôrm, *Progr. Chem. Org. Nat. Prod.* **1961**, 19, 1; *Pure Appl. Chem.* **1961**, 2, 533.

16. R. MCrindle, K. H. Overton, *Advan. Org. Chem.* **1965**, 5, 47.

17. S. Allard, G. Ourisson, *Tetrahedron* **1957**, 1, 277.

18. "Definitive Rules for Steroid Nomenclature," *Pure Appl. Chem.* **1972**, 31, 283; and other places

19. K. Freudenberg, K. Weinges, *Tetrahedron* **1960**, 8, 336.

20. K. Freudenberg, K. Weinges, *Tetrahedron* **1961**, 15, 115.

21. "Enzyme Nomenclature Recommendations of the International Union of Biochemistry," Elsevier, New York, **1965**.

22. M. P. Crosland, "Historical Studies in the Language of Chemistry," Harvard University Press, Cambridge, **1962**, p. 287.

23. G. Bertrand, *Bull. Soc. Chim. Biol.* **1923**, 5, 94.

24. M. P. Crosland, *Bull. Soc. Chim. Biol.* **1923**, 5, 299.

# Nitriles

A s a class, **nitriles** are characterized by the presence of the CN ($-C\equiv N$) group. However, the term nitrile, when used as a suffix in names of individual compounds, denotes only $\equiv N$—*i.e.*, nitrogen triply bound to a single carbon atom. Nitriles may be viewed formally as being derived either by simultaneous replacement of all three hydrogen atoms of a $-CH_3$ group by nitrogen or by like replacement of both the $=O$ and $-OH$ portions of a $-COOH$ group. From a chemical viewpoint, nitriles are commonly regarded as derivatives of carboxylic acids, and this relationship has led to the formation of many trivial names based on those of the corresponding acids.

Functional isomers of nitriles represented by the generic formula $R-NC$ (often called "isocyanides") are considered in Chapter 32.

*Recommended Nomenclature Practice*

**Acyclic mononitriles** and **dinitriles** are preferably named substitutively by adding the suffixes *nitrile* and *dinitrile*, respectively, to the name of the corresponding hydrocarbon.

$$CH_3CH_2CH_2CH_2CH_2CN$$  Hexanenitrile

$$\begin{array}{c} CH_3 \\ | \\ NCCH_2CHCH_2CH_2CH_2CN \end{array}$$  3-Methylheptanedinitrile

$$CH_3CH=CHCH_2CN$$  3-Pentenenitrile

**Cyclic nitriles** are preferably named substitutively by adding the suffixes *carbonitrile, dicarbonitrile, tricarbonitrile,* etc. to the name of the ring system to which the CN group(s) are attached.

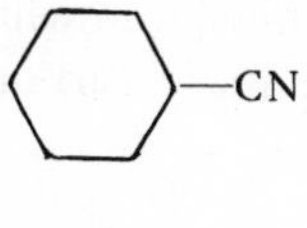

Cyclohexanecarbonitrile

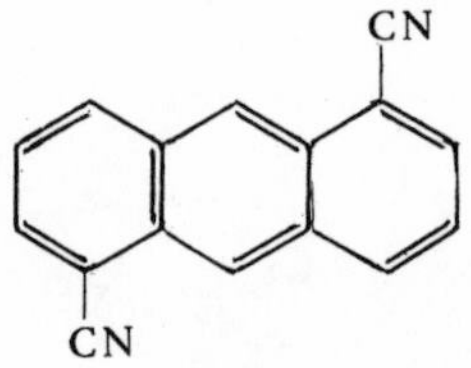

1,5-Anthracenedicarbonitrile

This type of name is also used for **acyclic polynitriles** in which more than two CN groups are attached to the same straight chain.

$$NCCH_2\overset{\overset{\textstyle CN}{|}}{C}HCH_2CH_2CN \qquad \text{1,2,4-Butanetricarbonitrile}$$

When the nitrile function must be treated as a substituting group, the prefix *cyano*, denoting —CN, is used.

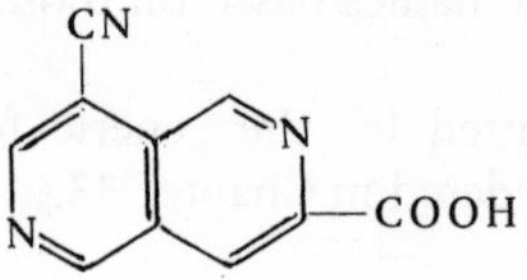

8-Cyano-2,6-diazanaphthalene-3-carboxylic acid

$$NCCH_2\overset{\overset{\textstyle CH_2CN}{|}}{\underset{\underset{\textstyle CH_2CN}{|}}{C}}CH_2CH_2CH_2CN \qquad \text{3,3-Bis(cyanomethyl)heptanedinitrile}$$

As alternatives, **trivial names** may be used for nitriles in instances where the corresponding acid has an acceptable trivial name (*see* Chapter 16). Such names are formed by changing the ending *ic acid* or *oic acid* to *onitrile*,— *e.g.*, acetonitrile for $CH_3CN$ and succinonitrile for $NCCH_2CH_3CN$. Names such as succinonitrile imply that all COOH groups have been changed to CN groups.

**Acyl cyanides** corresponding to the generic formula RCOCN are often named analogously to acyl halides (*see* Chapter 17),—*e.g.*, hexanoyl cyanide for $CH_3CH_2CH_2CH_2CH_2COCN$—but substitutive names—*e.g.*, 2-oxoheptanenitrile—are to be preferred.

*Discussion*

As with other functional classes, conjunctive names (*see* Chapter 9) are used by *Chemical Abstracts* for acyclic nitriles in which CN groups are separated from a ring system by one or more carbon atoms—*e.g.*,

4-morpholineacetonitrile—but substitutive names are generally to be preferred. Also, reflecting widespread usage, *Chemical Abstracts* has used the short form propionitrile for $CH_3CH_2CN$ rather than propiononitrile.

Radicofunctional names for nitriles (*see* Chapter 9), such as methyl cyanide for $CH_3CN$, have appeared frequently in the literature, but these should no longer be employed. The long-known simplest nitrile, HCN, is commonly called hydrogen cyanide. The name "hydrocyanic acid" for HCN, although also widely used, is not recommended. In deference to usage, the trivial name cyanogen for NCCN is acceptable; this compound has seldom been named as a nitrile.

Trivial names for amino-substituted nitriles should not be formed by combining the suffix *nitrile* with the trivial name of the corresponding amino acid, as in "alaninenitrile" or "alaninonitrile." Although *Chemical Abstracts* formerly made a single exception in the case of "glycinonitrile" (for $H_2NCH_2CN$), this name is now cross-referred to aminoacetonitrile.

Hydroxy-substituted nitriles, often named as "cyanohydrins", should be given systematic names; cyanohydrin names should be abandoned (*see* Examples 14 and 15 in Table 31.1).

The prefix *nitrilo,* which would logically denote the substituent $\equiv N$ (by analogy with *amino* for $-NH_2$ and *imino* for $=NH$), has not been much used with this meaning; usage has been heavily in favor of *cyano* (for $-C\equiv N$), as recommended above. However, nitrilo is officially recognized and used as the name of the multivalent substituting group in which nitrogen is singly bound to each of three discrete groups (*see* Chapter 11).

### Table 31.1. Examples of Acceptable Usage

| | | |
|---|---|---|
| 1. | HCN | Methanenitrile |
| | | Formonitrile |
| | | Hydrogen cyanide |
| 2. | $CH_3CH_2CN$ | Propanenitrile |
| | | Propiononitrile[a] |
| | | Propionitrile (*Chemical Abstracts*) |
| 3. | $CH_3$<br>$\vert$<br>$CH_3CHCH_2CN$ | 3-Methylbutanenitrile |
| | | Isovaleronitrile |
| 4. | $CH_2{=\!=}CHCN$ | Propenenitrile |
| | | Acrylonitrile |

5.  NCCN

Ethanedinitrile
Oxalonitrile
Cyanogen

6.  $NCCH_2CH_2CH_2CH_2CN$

Hexanedinitrile
Adiponitrile

7.  $C_6H_5CN$

Benzenecarbonitrile
Benzonitrile

8.  $C_6H_5CH_2CN$

Phenylethanenitrile
Phenylacetonitrile

9.

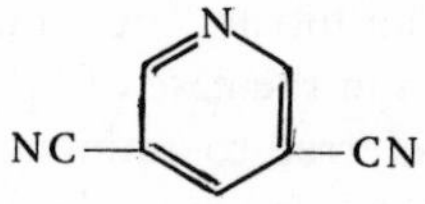

Azabenzene-3,5-dicarbonitrile
3,5-Pyridinedicarbonitrile

10.  $NCCH_2CH_2COOH$

3-Cyanopropanoic acid
3-Cyanopropionic acid

11.  $NCCH_2CHCH_2CN$ (CN)

1,2,3-Propanetricarbonitrile

12.

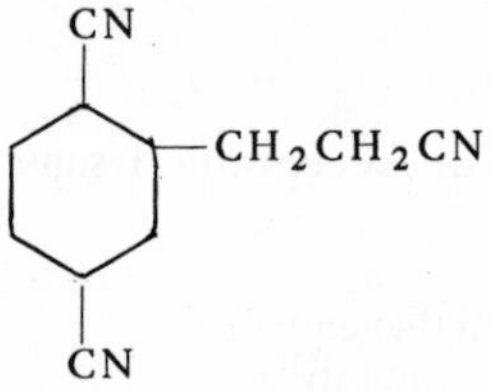

2-(2-Cyanoethyl)-1,4-cyclo-
hexanedicarbonitrile

13.  $NCCH_2CHCN$ ($NH_2$)

2-Aminobutanedinitrile
2-Aminosuccinonitrile
*not*  Aspartonitrile

14.  $HOCH_2CH_2CN$

3-Hydroxypropanenitrile
3-Hydroxypropiononitrile[a]
*not*  Ethylene cyanohydrin

15.

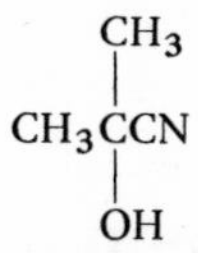

$$CH_3$$
$$|$$
$$CH_3CCN$$
$$|$$
$$OH$$

2-Hydroxy-2-methylpropanenitrile
2-Hydroxy-2-methylpropiononitrile[a]
*not* Acetone cyanohydrin

16. $C_6H_5COCN$

Oxophenylethanenitrile
Benzoyl cyanide

[a]Recommended in 1965 IUPAC Rules. *See* Discussion Section.

# 32

# Nitrogen Compounds (Miscellaneous)

This chapter deals with classes of organic nitrogen compounds not treated elsewhere in this book. Consideration is given first to structures in which nitrogen is linked to carbon, then to those having nitrogen attached to oxygen or sulfur. Finally, compounds containing chains of two or more nitrogen atoms are discussed.

*Recommended Nomenclature Practice*

The inorganic names **cyanic acid** for HOC≡N, **fulminic acid** for HONC, and **thiocyanic acid** for HSC≡N are the basis for naming esters and salts of these acids.

| | |
|---|---|
| $C_6H_5OCN$ | Phenyl cyanate |
| $CH_3CH_2CH_2ONC$ | Propyl fulminate |
| NaSCN | Sodium thiocyanate |

Compounds belonging to the classes R-N=C=O, R-N=C=S, and R-NC are preferably named substitutively by using the prefix names *carbonylamino*, *thiocarbonylamino*, and *carbylamino*. As noted in Table 8.1, suffix names have not been assigned to these functional groups.

| | |
|---|---|
| $CH_3CH_2CH_2NCO$ | 1-(Carbonylamino)propane |

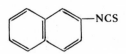—NCS    2-(Thiocarbonylamino)naphthalene

| | |
|---|---|
| $CH_3NC$ | (Carbylamino)methane |

**Acyl derivatives** of the acids HOCN and HSCN are best named as mixed anhydrides; on the other hand, when an acyl group replaces the H of HNCO, HNCS or HNC, the compound is preferably named like an acid halide (*see* Chapter 17).

| | |
|---|---|
| $C_6H_{11}COOCN$ | Cyanic cyclohexanecarboxylic anhydride |
| $CH_3CH_2COSCN$ | Cyanic propanoic thioanhydride |
| $CH_3CH_2CONCS$ | Propanoyl isothiocyanate |

The tautomeric parent structures $H_2NCN$ and $HN{=}C{=}NH$ are preferably given the systematic names **cyanic amide** and **methanediimine** when they carry organic substituents.

$(CH_3CH_2CH_2CH_2)_2NCN$ — Cyanic dibutylamide

$C_6H_{11}N{=}C{=}NC_6H_{11}$ — Dicyclohexylmethanediimine

The compound $H_2NOH$, traditionally called hydroxylamine, is better assigned the parent name *azanol* in naming its organic derivatives; the corresponding name for $H_2NSH$ is *azanethiol* (*see* Discussion).

$CH_3CH_2NHOH$ — Ethylazanol

$C_6H_5CH_2CH_2NHSH$ — (2-Phenylethyl)azanethiol

$CH_3CH_2CH_2NHOCH_3$ — Methoxypropylazane

*N*-Acylated derivatives are best named as substituted amides and *O*-acylated derivatives as esters.

$CH_3CH_2CH_2CH_2CH_2CONHOH$ — *N*-Hydroxyhexanamide

$H_2NOCOC_6H_{11}$ — Azanyl cyclohexanecarboxylate

Compounds of the type $RCH{=}NOH$ and $R_2C{=}NOH$, commonly named as oximes of aldehydes and ketones, are preferably treated systematically.

$CH_3CH_2CH{=}NOH$ — Propylideneazanol

$$\overset{\displaystyle CH_3}{\underset{\textstyle C_6H_5C{=}NOH}{|}}$$ — (1-Phenylethylidene)azanol

$ClCH_2CH{=}NOCH_3$ — (2-Chloroethylidene)methoxyazane

Examples of several less common types of structure containing the group $={=}NOH$, previously given specialized class names, are included in Table 32.3.

**Salts** of organic derivatives of $H_2NOH$ and $H_2NSH$ are preferably named systematically like salts of amines, alcohols, and thioalcohols (*see* Chapters 24 and 19).

$C_6H_5CH_2CH_2\overset{+}{N}H_2OH\ Cl^-$ — Hydroxy(2-phenylethyl)azanium chloride

$(CH_3CH_2)_2NSNa$ — Sodium diethylazanethiolate

Compounds in which a chain of two or more nitrogen atoms carries organic substituents are best treated substitutively as derivatives of the nitrogen chain. The names diazane for $H_2NNH_2$, triazane for $H_2NNHNH_2$, tetrazane for $H_2NNHNHNH_2$, etc. are used. Names for unsaturated chains correspond to those of unsaturated carbon chains—*i.e.,* diazene for $HN=NH$, triazene for $H_2NN=NH$, etc. Substituents may be functional or nonfunctional.

| | |
|---|---|
| $C_6H_5NHNH_2$ | Phenyldiazane |
| $CH_3NHNHCOOH$ | 2-Methyldiazanecarboxylic acid |
| $CH_3CH_2N=NCH_3$ | Ethylmethyldiazene |
| $C_6H_5N=NOH$ | Phenyldiazenol |
| $C_6H_5N=NN(CH_3)_2$ | 3,3-Dimethyl-1-phenyltriazene |
| $CH_3CH=NNHC_6H_5$ | 1-Ethylidene-2-phenyldiazane |
| $C_6H_5CH=NN=CHC_6H_5$ | Bis(phenylmethylene)diazane |
| $C_6H_5N=N(O)C_6H_5$ | Diphenyldiazene oxide |

Acylated derivatives not containing higher-ranking functional groups are preferably treated systematically as substituted ketones, thioketones, or imines. Additional examples of structures containing nitrogen chains for which various class names have been used are included in Table 32.3.

| | |
|---|---|
| $CH_3CH_2CONHNH_2$ | 1-Diazanyl-1-propanone |
| $C_6H_5CSNHNHCH_3$ | (2-Methyldiazanyl)phenylmethanethione |
| $CH_3C(=NH)N=NNH_2$ | 1-(1-Triazen-1-yl)-1-ethanimine |

Salts in which a member of a nitrogen chain carries a positive charge are best named by replacing the terminal *e* of diazane, triazane, etc. with *ium* and citing the anionic group as a separate word (*see* Chapter 24).

$$CH_3CH_2N\overset{+}{N}H_2CH_2CH_3 \ Br^-  \qquad \text{1,2-Diethyldiazanium bromide}$$

When it becomes necessary to treat the nitrogen-containing groups considered in this chapter as substituents, **prefix names** as shown in Tables 32.1 and 32.2 are used. The recommended names shown are to be preferred in forming systematic names; however, some of those reflecting current usage are acceptable alternatives. Many names in these Tables are placed in parentheses according to principles given in Chapter 13.

*Discussion*

Unfortunately, little has been done officially to date to systematize and simplify the nomenclature of the several types of organic nitrogen compounds covered in this chapter. By usage a large variety of class names, most of them semisystematic, has evolved that severely taxes the chemist's memory. In the absence of better alternatives these names have been used by *Chemical Abstracts* and officially recognized by the IUPAC Commission. The recommendations made here constitute an attempt to improve matters by applying modern principles of systematic organic nomenclature so as to eliminate the need for many specialized names designating nitrogen-containing groups. In most cases one is urged to abandon such specialized names, as will be seen from Table 32.3.

Among the functional classes considered here, only those derived from the parent structures HOCN, HSCN, and HONC are properly named as acid derivatives. The officially approved inorganic names cyanic acid, isocyanic acid, and fulminic acid are retained as the basis for naming esters and organic anhydrides derived from these structures. Thus, although the 1970 IUPAC Inorganic Rules recognize isocyanic acid as the name for HNCO, the treatment as esters of structures in which the "esterifying" organic group is attached to *nitrogen* deviates from established principles and usage. Therefore, names such as "phenyl isocyanate" and "methyl isothiocyanate" are no longer recommended.

### Table 32.1 Some Monovalent Nitrogen-Containing Substituting Groups (*No Longer Recommended)

| Formula | Prefix Name |
|---|---|
| —OCN | Cyanato |
| —SCN | Thiocyanato |
| —ONC | Fulminato |
| —NCO | (Carbonylamino) |
| | *Isocyanato |
| —NCS | [(Thiocarbonyl)amino] |
| | *Isothiocyanato |
| —NC | (Carbylamino) |
| | *Isocyano |
| —NHCN | (Cyanoamino) |
| —N=C=NH | [(Iminomethylene)amino] |
| —NHOH | (Hydroxyazyl) |
| | (Hydroxyamino) |
| —NHSH | (Mercaptoazyl) |
| | (Mercaptoamino) |
| —ONH$_2$ | (Azyloxy) |
| | (Aminooxy) |
| —SNH$_2$ | (Azylthio) |
| | (Aminothio) |
| —NHNH$_2$ | Diazanyl |
| | Hydrazino |
| —NHNHNH$_2$ | 1-Triazanyl |
| | *Triazano |
| —N=NH | Diazenyl |
| | *Diazeno |
| —NHN=NH | 2-Triazen-1-yl |

Members of the class RNC, properly called isocyanides, should be named substitutively. Radicofunctional names such as "phenyl isocyanide" are not recommended nor is the class name "isonitriles." Older names based on "carbylamine" have largely disappeared and should not be revived.

### Table 32.2 Some Multivalent Nitrogen-Containing Substituting Groups (*No Longer Recommended)

| Formula | Prefix Name |
|---|---|
| $\overset{\mid}{-}$NCN or $=$NCN | (Cyanoazylene) (Cyanoimino) |
| $\overset{\mid}{-}$NOH or $=$NOH | (Hydroxyazylene) (Hydroxyimino) |
| $\overset{\mid}{-}$NSH or $=$NSH | (Mercaptoazylene) (Mercaptoimino) |
| $-$NHNH$-$ | 1,2-Diazanediyl *Hydrazo |
| $=$NNH$_2$ | 1,1-Diazanediyl *Hydrazono |
| $-$NH$\overset{\mid}{N}$NH$-$ | 1,2,3-Triazanetriyl |
| $-$N$=$N$-$ | Diazenediyl *Azo |
| $-$NHN$=$N$-$ | 1-Triazene-1,3-diyl *Diazoamino |
| $-$N$=$N(O)$-$ | (Oxidodiazenediyl) *Azoxy |
| $=$NN$=$ | Diazanetetrayl *Azino |

Since compounds with the structure ROCN and RSCN are correctly named as esters, those of the classes RCOOCN, RCSOCN, and RC(=NH)-OCN are properly treated as mixed anhydrides; those of the type RCOSCN, RCSSCN, and RC(=NH)SCN are mixed thioanhydrides (*see* Chapter 17). Therefore, the previous practice of naming these classes as acyl esters, as exemplified by "benzoyl cyanate" for $C_6H_5$COOCN, although officially recognized, should now be abandoned. The corresponding isocyanato and isothiocyanato compounds, such as $C_6H_5$CONCS, do not qualify for treatment as anhydrides and are best named like acid halides—*i.e.*, benzoyl isothiocyanate.

The name cyanamide for $H_2$NCN is recognized as an acceptable shortened form of cyanic amide; although the latter might be preferred by inference from the 1970 IUPAC Inorganic Rules, it has been used little. Similarly, for the hypothetical tautomer, HN=C=NH, the name carbodiimide, though well established, should now be replaced by the systematic form, methanediimine.

The inorganic name hydroxylamine for $H_2$NOH is officially recognized and used in naming organic derivatives; for the sulfur analog, $H_2$NSH, *Chemical Abstracts* has preferred hydrosulfamine in recent years, but thiohydroxylamine is a better choice for consistency. Similarly, the name hydrazine for $H_2$NNH$_2$ is well established as a parent name in organic nomenclature. However, systematic parent names for these inorganic compounds formed by extension of officially approved principles. are to be preferred—*i.e.,* azanol for $H_2NOH$, azanethiol for $H_2NSH$, diazane for $H_2NNH_2$, and diazene for $HN=NH$. By using these parent names and by applying the fundamental principle that chains of identical atoms are held intact, the naming of many of the organic nitrogen classes covered in this chapter is both unified and enormously simplified.

The present recommendation involves extending the systematic treatment already officially approved for $H_2NNHNH_2$, $H_2NNHNHNH_2$, etc. to the first two members of the homologous series and, when appropriate, using the resulting names (azane for $NH_3$ and diazane for $H_2NNH_2$) as the stems of regularly formed substitutive names. Naming of parent structures in which a single nitrogen atom is attached only to carbon and hydrogen—*i.e.,* amides, imides, amidines, amines, and imines— is not affected (*see* Chapters 21-23), and the usual seniority ranking of functional groups (Table 8.1) is preserved.

In treating substituting groups having a single nitrogen atom attached to oxygen or sulfur, systematic names corresponding to those of the parent compounds are preferred—*i.e., (azyloxy)* for $H_2NO-$, *(hydroxyazylene)* for $HON<$, etc., as shown in Tables 32.1 and 32.2 By this treatment $H_2N-$ is the counterpart of $CH_3-$, just as $NH_3$ (as a parent structure) is the counterpart of $CH_4$, and the prefix name *amino* is reserved for substituting groups where nitrogen is attached only to carbon and hydrogen. Also, for consistency with names of higher homologs, the prefix names *diazanyl* and *diazenyl* are preferable to *hydrazino* and *diazeno.*

No attempt is made here to discuss individually the plethora of special semisystematic names developed to distinguish among the various structural types produced by formal replacement of $-OH$ and $=O$ in $RCOOH$ with $-NHOH$, $=NOH$, $-NHNH_2$, and $=NNH_2$. Although these specialized names have received official recognition by IUPAC and *Chemical Abstracts,* they are difficult to remember and do not offer any real advantages. Examples representing most of the possible varieties are included in Table 32.3.

The traditional and still popular practice of naming compounds of the structure $RCH=NOH$ and $R_2C=NOH$ as oximes of aldehydes and ketones should be discontinued. The same applies to the hydrazone type of name for $RCH=NNH_2$ and $R_2C=NNH_2$. Names such as "acetaldehyde oxime" and "benzophenone phenylhydrazone" that use a terminal modifying word to denote conversion of one kind of functional group to another by replacement have no basis in principle in systematic organic nomenclature. Therefore, although the terms oxime, aldoxime, ketoxime, hydrazone, azine, and osazone are acceptable as class names, they should no longer be used to name individual compounds.

Organic derivatives of the parent structures HN=NH and HN=N(O)H have traditionally been named as azo and azoxy compounds, but this treatment should be discontinued in favor of systematic names based on diazene and diazene oxide. Names such as "azobenzene" for $C_6H_5N=NC_6H_5$ are not fully descriptive ("azodibenzene" would be better), and the use of both "azo" and "diazo" to denote the presence of two nitrogen atoms, as in —N=N— and =N$_2$, respectively, is an obvious inconsistency. To make matters worse, the group HN=N— has been assigned the name "diazeno" when unsubstituted and "azo" when substituted, as in "phenylazo" for $C_6H_5N=N-$. The recommendation here is to reserve *diazo* for the group =N$_2$ (*see* Chapter 27) and to use *diazenyl* for HN=N—, whether or not the latter is substituted. The prefix names (*1-oxidodiazenyl*) for HN=N(O)—, (*2-oxidodiazenyl*) for H(O)-N=N—, *diazenediyl* for —N=N—, and (*oxidodiazenediyl*) for —N=N(O)— follow logically, and the terms "azo," "diazeno," and "azoxy" are no longer needed. For the parent structure HN=NH, the name "diimide," though previously used by *Chemical Abstracts,* should also be abandoned, as should the functional suffix "diazoic acid" denoting —N=NOH.

Other types of names that have been accorded official recognition but are not recommended for continued use are exemplified by "methoxyamine" for H$_2$NOCH$_3$, "acetonitrolic acid" for CH$_3$C(=NOH)NO$_2$, and "benzonitrosolic acid" for $C_6H_5C$ (= NOH)NO.

The prefix names "oximido" and "isonitroso" for HON=, "hydroxylamino" for HONH—, and "aminoxy" for H$_2$NO—, although sometimes encountered, are not acceptable alternatives to their properly formed counterparts; other prefixes, officially recognized, but not recommended for continued use are "hydrazo" for —NHNH—, "hydrazono" for H$_2$NN=, and "diazoamino" for —N=NNH— (*see* Table 32.2).

## Table 32.3. Examples of Acceptable Usage

1.  HOCH$_2$CH$_2$CH$_2$OCN                3-Hydroxypropyl cyanate

2.  OCN—⟨benzene ring⟩—NCO

    1,4-Bis(carbonylamino)benzene
*not* 1,4-Phenylene diisocyanate

3.  CH$_3$O—⟨benzene ring⟩—NC

    1-(Carbylamino)-4-methoxybenzene

4.  CH$_3$CH$_2$CH$_2$COOCN              Butanoic cyanic anhydride
*not* Butanoyl cyanate

5.  (CH$_3$CH$_2$)$_2$NCN                 Diethylcyanamide

6.

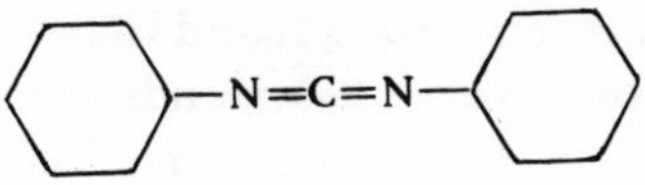

Dicyclohexylmethanediimine
Dicyclohexylcarbodiimide

7. $C_6H_5CH_2NHOH$

(Phenylmethyl)azanol
N-Benzylhydroxylamine

8.

$$\overset{\overset{\displaystyle CH_3}{|}}{CH_3CH_2NOCH_3}$$

Ethylmethoxymethylazane
N-Ethyl-N,O-dimethylhydroxylamine

9.

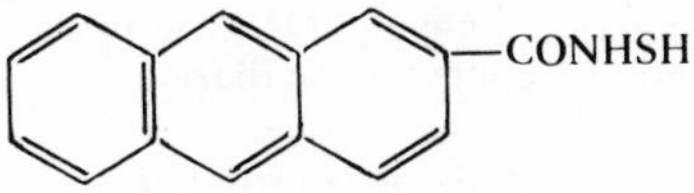

—CONHSH

N-Mercapto-2-anthracenecarboxamide

10. $CH_3CH_2CH_2CONHOH$

N-Hydroxybutanamide
*not* Butyrohydroxamic acid

11. $C_6H_5CH{=}NOH$

(Phenylmethylene)azanol
Benzylidenehydroxylamine
*not* Benzaldehyde oxime

12.

$$\overset{\overset{\displaystyle CH_3}{|}}{CH_3CH_2CH_2C{=}NOH}$$

(1-Methylbutylidene)azanol
(1-Methylbutylidene)hydroxylamine
*not* 2-Pentanone oxime

13.

$$\underset{\underset{\displaystyle OH}{|}}{C_6H_5C{=}NOH}$$

N-Hydroxybenzenecarboximidic acid
N-Hydroxybenzimidic acid
*not* Benzohydroximic acid

14.

$$\underset{\underset{\displaystyle NH_2}{|}}{CH_3C{=}NOH}$$

$N^2$-Hydroxyethanamidine
$N^2$-Hydroxyacetamidine
*not* Acetamidoxime

15.

$$\underset{\underset{\displaystyle NOH}{\|}}{CH_3CCH_2COOCH_2CH_3}$$

Ethyl 3-(hydroxyazylene)butanoate
Ethyl 3-(hydroxyimino)butyrate

16. $CH_3\overset{+}{N}H_2OCH_2CH_3$ $Br^-$

Ethoxymethylazanium bromide

17.   ⬡—COONHC$_6$H$_5$

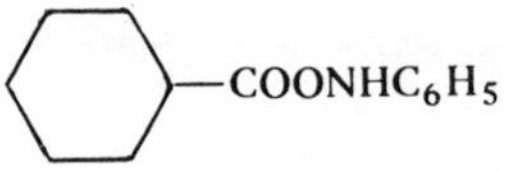

Phenylazyl cyclohexanecarboxylate
*O*-Cyclohexanecarbonyl-*N*-phenylhydroxyl-
amine


18.
$$CH_3CH_2\overset{\overset{\displaystyle CH_3}{|}}{N}NHCH_3$$

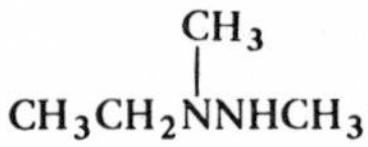

1-Ethyl-1,2-dimethyldiazane
1-Ethyl-1,2-dimethylhydrazine


19.   CH$_3$NHNHCOOH

2-Methyldiazanecarboxylic acid
2-Methylhydrazinecarboxylic acid
*not* 3-Methylcarbazic acid


20.   C$_6$H$_5$NHNHCONHCH$_3$

*N*-Methyl-2-phenyldiazanecarboxamide
*N*-Methyl-2-phenylhydrazinecarboxamide
*not* 4-Methyl-1-phenylsemicarbazide


21.   CH$_3$OOCNHNHCOOCH$_3$

Dimethyl 1,2-diazanedicarboxylate
Dimethyl 1,2-hydrazinedicarboxylate
*not* Dimethyl bicarbamate


22.   C$_6$H$_5$NHCONHNHCONHC$_6$H$_5$

*N,N'*-Diphenyl-1,2-diazanedicarboxamide
*N,N'*-Diphenyl-1,2-hydrazinedicarboxamide
*not* 1,6-Diphenylbiurea


23.

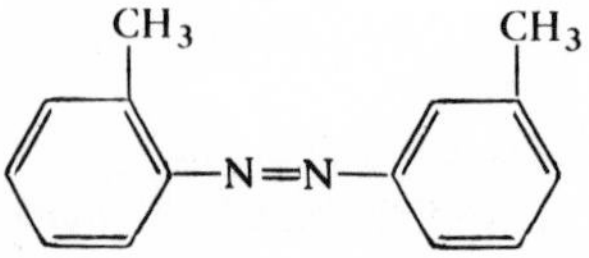

(2-Methylphenyl)(3-methylphenyl)diazene
*not* *m,o'*-Azotoluene


24.   C$_6$H$_5$N=N(O)CH$_2$CH$_3$

Ethylphenyldiazene 1-oxide
*not* (Ethyl-*ONN*-azoxy)benzene


25.   CH$_3$CH$_2$OCON=NCOOCH$_3$

Ethyl methyl diazenedicarboxylate
*not* Ethyl methyl azodiformate


26.   O$_2$N—⬡—N=NOH

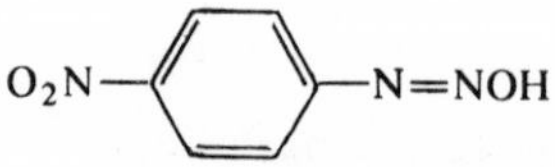

(4-Nitrophenyl)diazenol
*not* 4-Nitrobenzenediazoic acid


27.   Cl—⬡—CH=NNH$_2$

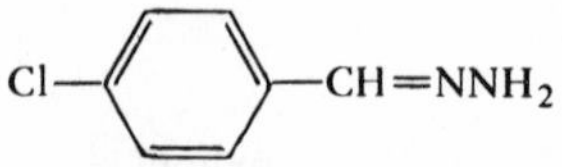

[(4-Chlorophenyl)methylene]diazane
(4-Chlorobenzylidene)hydrazine
*not* 4-Chlorobenzaldehyde hydrazone

28. $(CH_3CH_2)_2C=NNHCH_3$

1-(1-Ethylpropylidene)-2-methyldiazane
1-(1-Ethylpropylidene)-2-methylhydrazine
*not* 3-Pentanone methylhydrazone

29. $C_6H_5CH=NN=CHC_6H_5$

Bis(phenylmethylene)diazane
Dibenzylidenehydrazine
*not* Benzaldehyde azine

30. $C_6H_5CONHNHCH_3$

(2-Methyldiazanyl)phenylmethanone
$N'$-Methylbenzohydrazide

31. 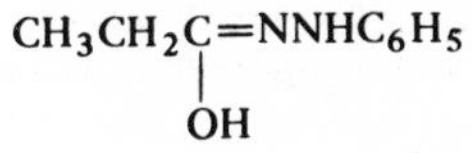 $CH_3CH_2\underset{\underset{\displaystyle OH}{|}}{C}=NNHC_6H_5$

1-(2-Phenyl-1,1-diazanediyl)-1-propanol
*not* $N'$-Phenylpropanehydrazonic acid
*not* Propionic acid phenylhydrazone

32. 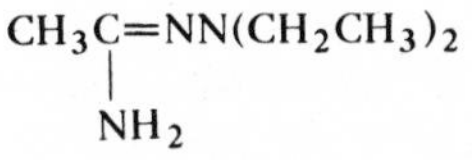 $CH_3\underset{\underset{\displaystyle NH_2}{|}}{C}=NN(CH_2CH_3)_2$

1-(2,2-Diethyl-1,1-diazanediyl)ethanamine
*not* Acetamide diethylhydrazone
*not* $N',N'$-Diethylacetamidrazone

33. 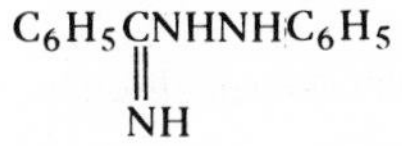 $C_6H_5\underset{\underset{\displaystyle NH}{\|}}{C}NHNHC_6H_5$

Phenyl(2-phenyldiazanyl)methanimine
$N'$-Phenylbenzimidohydrazide

34. 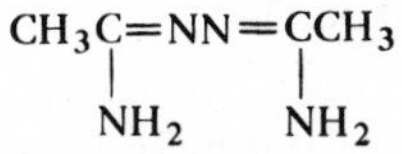 $CH_3\underset{\underset{\displaystyle NH_2}{|}}{C}=NN=\underset{\underset{\displaystyle NH_2}{|}}{C}CH_3$

1,1,1',1'-Diazanetetraylbis(ethanamine)
*not* Acetamide azine

35. 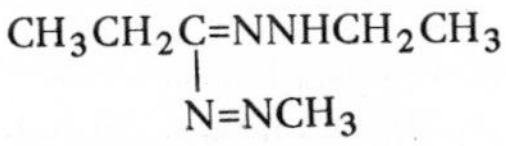 $CH_3CH_2\underset{\underset{\displaystyle N=NCH_3}{|}}{C}=NNHCH_2CH_3$

[1-(Ethyl-1,1-diazanediyl)propyl] methyl-
diazene
*not* 3,5-Diethyl-1-methylformazan

36. 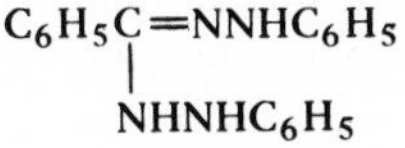 $C_6H_5\underset{\underset{\displaystyle NHNHC_6H_5}{|}}{C}=NNHC_6H_5$

1-Phenyl-2-[phenyl(2-phenyldiazanyl)-
methylene] diazane
*not* $N'$-Phenylbenzohydrazide phenylhydrazone

37. 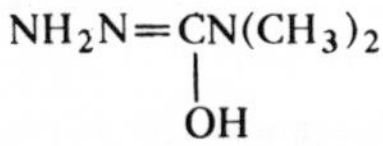 $NH_2N=\underset{\underset{\displaystyle OH}{|}}{C}N(CH_3)_2$

1,1-Diazanediyl(dimethylamino)methanol
*not* 4,4-Dimethylisosemicarbazide

38. $CH_3NHNHCONHNHCH_2CH_3$

(2-Ethyldiazanyl)(2-methyldiazanyl)-
methanone
1-Ethyl-5-methylcarbohydrazide
*not* 1-Ethyl-5-methylcarbazide

39. $C_6H_5N=NCON=NC_6H_5$

Bis(phenyldiazenyl)methanone
*not* 1,5-Diphenylcarbodiazone

40.  $(CH_3)_2C{=}NNHCONHC_6H_5$

2-(1-Methylethylidene)-$N$-phenyldiazane-
carboxamide
*not* Acetone 4-phenylsemicarbazone

41.  $CH_3N{=}NCONHN(CH_3)_2$

(2,2-Dimethyldiazanyl)(methyldiazenyl)-
methanone
*not* 1,1,5-Trimethylcarbazone

42.

$$CH_3$$
$$CH_3CONHNCOCH_3$$

1,1'-(Methyl-1,2-diazanediyl)diethanone
2'-Acetyl-2'-methylacetohydrazide
*not* 1,2-Diacetyl-1-methylhydrazine

43.  $(CH_3CS)_2NNHCH_3$

1,1'-(Methyl-1,1-diazanediyl)diethanethione
$N$-Ethanethioyl-$N'$-methylethanethiohydra-
zide
*not* 1,1-Diethanethioyl-2-methylhydrazine

44.

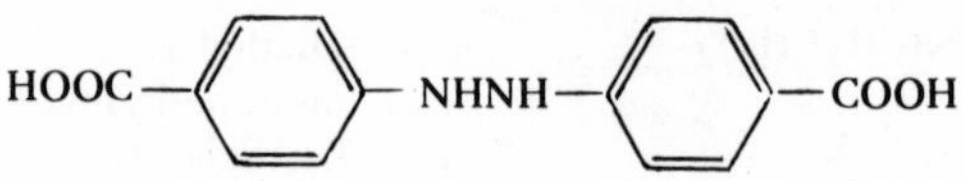

HOOC—⟨benzene⟩— NHNH—⟨benzene⟩—COOH

4,4'-(1,2-Diazanediyl)dibenzenecarboxylic
acid
*not* 4,4'-Hydrazodibenzenecarboxylic acid

45.  $H_2NCH_2CH_2N{=}NCH_2CH_2NH_2$   2,2'-Diazenediylbis(etnanamine)
*not* 2,2'-Azodiethylamine

46.

$$CH_3CCH_2CH_2CONH_2$$
$$\underset{NNHCH_3}{\overset{\|}{}}$$

4-(Methyl-1,1-diazanediyl)pentanamide
*not* 4-(Methylhydrazono)pentanamide

47.  $C_6H_5N\overset{+}{H}NH_2C_6H_5$  $Cl^-$

1,2-Diphenyldiazanium chloride
1,2-Diphenylhydrazinium chloride
*not* 1,2-Diphenylhydrazonium chloride

48.  $CH_3CH_2\overset{+}{N}H_2\overset{+}{N}H_2CH_3$  $SO_4{}^{2-}$

1-Ethyl-2-methyldiazanediium sulfate
1-Ethyl-2-methylhydrazinium(2+) sulfate

49.

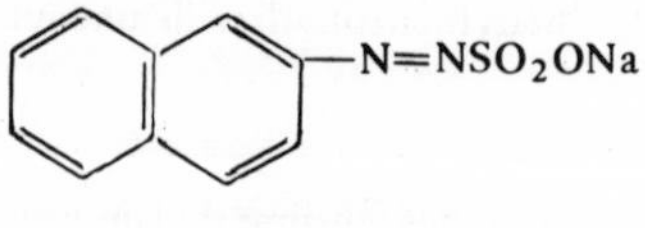

⟨naphthalene⟩—N=NSO$_2$ONa

Sodium 2-naphthyldiazenesulfonate

50.  $C_6H_5NHN{=}NNHNHC_6H_5$      1,5-Diphenyl-2-pentazene

51.

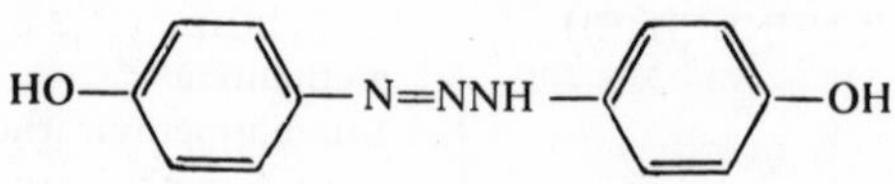

HO—⟨benzene⟩—N=NNH—⟨benzene⟩—OH

4,4'-(1-Triazene-1,3-diyl)dibenzenol
*not* 4,4-(Diazoamino)dibenzenol

# Organometallic Compounds

C ompounds discussed in this chapter contain at least one carbon-to-metal or carbon-to-metalloid bond. For present purposes, elements considered to be metallic or metalloid are all those in the Periodic Table *except* hydrogen, the noble gases, boron, carbon, silicon, nitrogen, phosphorus, oxygen, sulfur, and the halogens. Throughout the chapter the word "metals" is used as a collective term denoting both metals and metalloids.

In this area of nomenclature, which deals with structures containing both organic and inorganic portions, attention must be paid to officially approved rules and principles developed by both organic and inorganic chemists. In a good many instances an **organometallic compound**, as defined above, can be assigned at least two properly formed systematic names—one organic and the other inorganic in derivation. Since the subject of this book is organic nomenclature, the emphasis here is on the application of organic rules and principles and on names representing usage by organic chemists. However, in Table 33.1, names conforming to official inorganic nomenclature practice are shown as alternatives when they differ from the recommended organic names. Also, for structures wherein a **central metal atom** is attached to a larger number of atoms or groups (called **ligands**) than can be accounted for by its classical or stoichiometric valency, organic nomenclature is not applicable; such structures can be named only as **coordination compounds**.

Only a few fundamental principles used in naming coordination compounds are presented in this chapter. For more detailed and comprehensive information the 1970 IUPAC Inorganic Rules (*1*) should be consulted.

*Recommended Nomenclature Practice*

Organometallic compounds of As(III), Sb(III), Bi(III), Ge(IV), Sn(IV), and Pb(IV) are acceptably named as substitution products of their corresponding hydrides: arsine, stibine, bismuthine, germane, stannane, and plumbane.

$(C_6H_5)_3Sb$ Triphenylstibine

$(C_2H_5)_3SnSn(C_2H_5)_3$ Hexaethyldistannane

(CH$_3$)$_2$AsCl                                  Chlorodimethylarsine

Organometallic compounds of arsenic and germanium of the appropriate types may be named as substituted derivatives of the hypothetical parent acids corresponding to those of phosphorus (*see* Chapter 36).

(C$_6$H$_5$)$_2$As(O)OH                            Diphenylarsinic acid

$$\overset{\displaystyle O}{\overset{\displaystyle \|}{CH_3GeOH}}$$                                    Methylgermanonic acid

C$_6$H$_5$As(O)(OH)OK                             Potassium hydrogen phenylarsonate

Organometallic derivatives of selenium and tellurium may also be named in the same way as the corresponding sulfur compounds (*see* Chapters 18, 19, 39, and 40).

C$_6$H$_5$TeH                                     Benzenetellurol

(C$_2$H$_5$)$_2$Se                                 Diethyl selenide

C$_6$H$_5$SeO$_2$(OH)                              Benzeneselenonic acid

Compounds in which metals other than those mentioned above are joined to one or more organic groups through carbon are named by prefixing the names of the organic groups, arranged alphabetically, to the name of the metal.

C$_6$H$_5$ZnCH$_3$                                 Methylphenylzinc

If other atoms or groups are also present, these are designated by separate words following the rest of the name.

CH$_3$MgI                                         Methylmagnesium iodide

(C$_6$H$_5$)$_2$Sb(O)OH                            Diphenylantimony hydroxide oxide

Compounds in which organic molecules such as olefinic and aromatic compounds, carbon monoxide, and isocyanides are bound to a central metallic atom are best named as coordination compounds according to the 1970 IUPAC Inorganic Rules. Thus a name consists of the names of ligands in alphabetical order, the name of the metal, the suffix *ate* (for anions only), and designation of oxidation number of the metal (Stock system) or the charge on the ion (Ewens-Bassett system).

| | |
|---|---|
| $[Cr(C_6H_6)_2]$ | Bis(benzene)chromium(0) |
| $K[PtCl_3(C_2H_4)]$ | Potassium trichloro(ethylene)-<br>platinate(II)<br>Potassium trichloro(ethylene)-<br>platinate(1−) |
| $[In(CH_3)_2][Co(CO)_4]$ | Dimethylindium (1+)<br>tetracarbonylcobaltate(1−) |

Here, as in all coordination nomenclature, anionic ligands have names ending in *o* and thus differing sometimes from the corresponding organic group names.

| | | | |
|---|---|---|---|
| $Cl^-$ | chloro | $CH_3S^-$ | methylthio |
| $H^-$ | hydrido | $C_6H_5O^-$ | phenolato |
| $OH^-$ | hydroxo | $CH_3COO^-$ | acetato |
| $CN^-$ | cyano | $(CH_3)_2N^-$ | dimethylamido |
| $CH_3O^-$ | methoxo | | |

Linkage of *some or all* atoms of a chain or ring jointly to the central atom is designated by the prefix $\eta$, whereas such linkage through a *single* atom is similarly designated by $\sigma$.

| | |
|---|---|
| $Fe(C_5H_5)_2$ | Di-$\eta$-cyclopentadienyliron(II) |

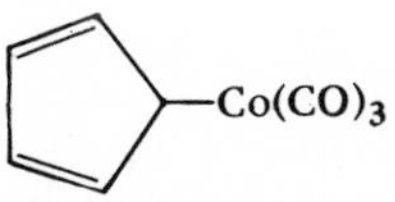

Tricarbonyl($\sigma$-cyclopenta-2,4-dien-1-yl)-cobalt

Except for As, Sb, Bi, Ge, Sn, and Pb, names of monovalent groups having free bonds on metal atoms end in *io*, this suffix replacing the final *ium, um,* or *y* of the element name. For Cu, Ag, and Au, the Latin forms of the element names are modified to give cuprio, argentio, and aurio. For tungsten or wolfram, zinc, cobalt, and nickel the prefix names are formed by adding the suffix *io* to the element names. Manganio is the corresponding term derived from manganese.

| | |
|---|---|
| $(OC)_3MnC_5H_4COOH$ | (Tricarbonylmanganio)-<br>cyclopentadienecarboxylic acid |
| $NaCH(COOC_2H_5)_2$ | Diethyl sodiomalonate |

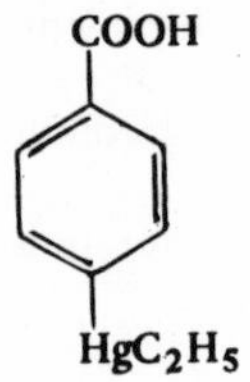

COOH

4-(Ethylmercurio)benzoic acid

$HgC_2H_5$

For As(III), Sb(III), Bi (III), Ge(IV), Sn(IV), and Pb(IV) the names of monovalent groups are derived from those of the hydrides, and organometallic compounds are named accordingly: $AsH_2$, arsino; $SbH_2$, stibino; $BiH_2$, bismuthino; $GeH_3$, germyl; $SnH_3$, stannyl; $PbH_3$, plumbyl.

$(CH_3)_2AsCH_2CH_2OH$            2-(Dimethylarsino)ethanol

Divalent and trivalent groups based on these elements are named by adding the suffixes *diyl* and *triyl* to the hydride names.

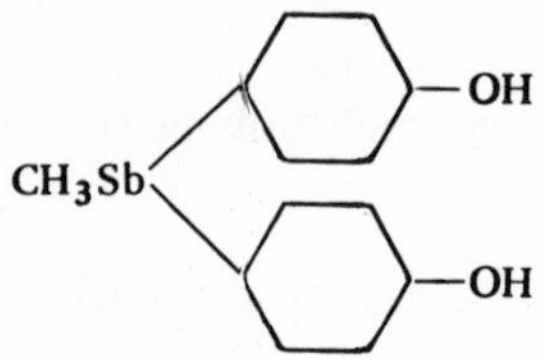

$CH_3Sb$ —OH  —OH            4,4'-(Methylstibinediyl)dicyclohexanol

## Discussion

*All* organometallic compounds can be named by applying rules of coordination nomenclature, and such names are usually preferred by inorganic chemists. Thus, while preference is expressed here for names based on organic usage, many coordination names are shown as acceptable alternatives in Table 33.1.

If an organometallic compound is ionic and this characteristic is to be emphasized, a two-word name such as disodium acetylide ($Na_2C_2$) and sodium methanide ($NaCH_3$) is appropriate (*see* Chapter 28). Cations consisting of four organic groups bonded to one arsenic, antimony, or bismuth atom, or three such groups bonded to selenium or tellurium, are named as onium ions, as specified in Chapter 24.

The recommended names for organometallic acids of arsenic correspond to those of acids of phosphorus (*see* Chapter 36). Similar names are sometimes still used for acids of antimony (stibinous, stibinic, stibonous, stibonic), but such compounds are better named otherwise; the same is true for RSn(O)OH and RPb(O)OH, where names ending in "stannonic acid" and "plumbonic acid" are not to be preferred. Organometallic acids are sometimes named as derivatives of hydrocarbons—*e.g.*, $C_6H_5As(O)(OH)_2$, benzenearsonic acid — but use of the inorganic acids as parents in naming leads to better consistency with nomenclature of phosphorus compounds.

Di-$\eta$-cyclopentadienyliron is acceptably called ferrocene, and substituted ferrocenes may be so named, but fused-ring derivatives of ferrocene should be named systematically.

Some currently used prefix names for metals and metal-containing groups do not conform to the rules given for their formation. A number of such names appear in the *Chemical Abstracts* list of names of radicals.

## Table 33.1. Examples of Acceptable Usage

| | | |
|---|---|---|
| 1. | $CH_3Li$ | Methyllithium |
| 2. | $C_6H_5CHNaCH_2CH_2CHNaC_6H_5$ | (1,4-Diphenyltetramethylene)disodium<br>Disodium 1,4-diphenyl-1,4-butanediide |
| 3. | $NaC{\equiv}CNa$ | Ethynylenedisodium<br>Disodium ethynediide<br>Disodium acetylide |
| 4. | $[CH_3CH_2CH_2CH_2CH(C_2H_5)CH_2]_2Zn$ | Bis(2-ethylhexyl)zinc |
| 5. | $CH_3CH_2CH_2(C_2H_5)(CH_3)Al$ | Ethylmethylpropylaluminum |
| 6. | $(C_6H_5CH_2)_3Bi$ | Tribenzylbismuthine<br>Tribenzylbismuth |
| 7. | $(C_2H_5)_4Pb$ | Tetraethylplumbane<br>Tetraethyllead |
| 8. | $(C_6H_5)_2Te$ | Diphenyl telluride<br>Diphenyltellurium |
| 9. | $CH_3CH_2CH_2CH_2MgI$ | Butylmagnesium iodide<br>Butyliodomagnesium |
| 10. | $(CH_3)_2CHCH_2CH{:}CHHgOOCCH_3$ | (4-Methyl-1-pentenyl)mercuric acetate<br>Acetoxy(4-methyl-1-pentenyl)mercury |
| 11. | $CH(HgI)_3$ | Methylidynetris(mercuric iodide)<br>Methylidynetris(iodomercury) |
| 12. | $(C_6H_5)_3SbCl_2$ | Dichlorotriphenylantimony |

13.  $C_6H_5Sb(SCH_3)SC_2H_5$                    (Ethylthio)(methylthio)phenylstibine
                                                  (Ethylthio)(methylthio)phenylantimony

14.  $[(CH_3)_3Sn]_2SO_4$                         Bis(trimethylstannyl) sulfate
                                                  Sulfatobis(trimethyltin)
                                                  Sulfatobis(trimethylstannane)

15.  $[(CH_3)_2CH]_2AlH$                          Diisopropylaluminum hydride
                                                  Hydridodiisopropylaluminum

16.

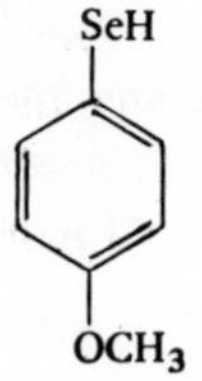

4-Methoxybenzeneselenol
Hydrido(4-methoxyphenyl)selenium

17.

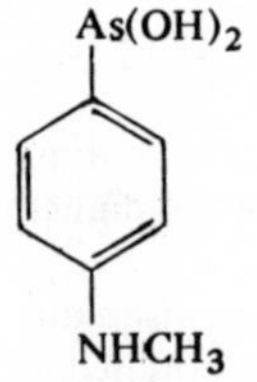

4-(Methylamino)phenylarsonous acid
Dihydroxo[4-(methylamino)phenyl]arsenic

18.

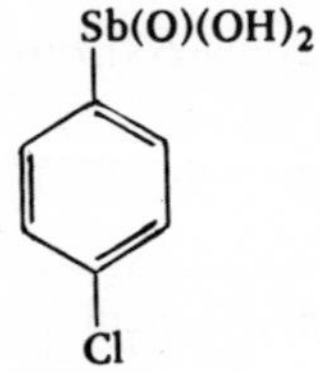

(4-Chlorophenyl)dihydroxystibine oxide
(4-Chlorophenyl)dihydroxooxoantimony

19.  $C_2H_5GeN$                                  Ethylgermanium nitride
                                                  Ethylnitridogermanium

20.  $(C_5H_5)_2Ni$                               Dicyclopentadienylnickel

21.  $C_5H_6Mn(CO)_3$                             Tricarbonyl($\eta$-cyclopentadiene)manganese

22.  $C_6H_6Cr(CO)_3$                             (Benzene)tricarbonylchromium

23.  $[(C_2H_5O)_3P]_4Ni$                         Tetrakis(triethyl phosphite)nickel

24.  $[(C_2H_5)_3Bi]_2Ni(CO)_2$                   Dicarbonylbis(triethylbismuthine)nickel
                                                  Dicarbonylbis(triethylbismuth)nickel

25. HCOOPbCo(CO)$_4$

Methanoatolead(II) tetracarbonyl-
cobaltate(I)

26.

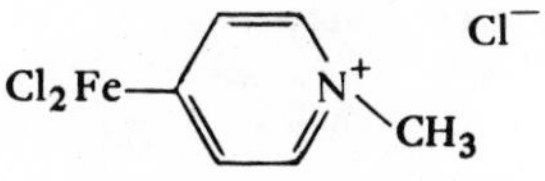

Cl$_2$Fe—⟨ ⟩—N$^+$—CH$_3$  Cl$^-$

4-(Dichloroferrio)-1-methylpyridinium
chloride
Dichloro-4-pyridyliron methochloride

27. [(C$_6$H$_5$)$_3$Pb]$_2$Fe(CO)$_4$

Bis(triphenyllead) tetracarbonylferrate (2−)
Tetracarbonylbis(triphenylplumbyl)iron
(if the compound is nonionic)

## Literature Cited

1. "Nomenclature of Inorganic Chemistry. Definitive Rules," IUPAC, 1970, Butterworths, London, 1971; *Pure Appl. Chem.* (1971) **28**, 1-110.

# 34

# Peroxy Compounds

**O**rganic compounds containing the group —OO— are broadly classified as peroxides; more specifically, those containing the group —OOH are termed **hydroperoxides**, and those represented by the generic formula RCO-OOH are called **peroxy acids**; compounds of the type RCO-OO-COR are known as **diacyl peroxides**.

As explained further in the Discussion section, the application of substitutive nomenclature to peroxy compounds, although lacking precedent, has not been officially proscribed; thus, for reasons of consistency with treatment of other functional classes, substitutive names are recommended here in preference to the traditional radicofunctional ones. The preferred nomenclature of this chapter closely parallels that of ethers (Chapter 26), to which disubstituted nonacyl peroxides are structurally similar.

## Recommended Nomenclature Practice

**Disubstituted peroxides** having the structure ROOR', where R and R' are the same or different acyclic or cyclic groups, are preferably named substitutively as *dioxy* derivatives of the senior chain or ring system to which the —OO— group is attached. Principles to be followed in selecting the parent chain or ring are given in Chapter 26. Symmetrical structures containing only one —OO— group may also be named radicofunctionally by using the terminal word *peroxide.*

$CH_3CH_2CH_2OOCH_2CH_2CH_3$    1-(Propyldioxy)propane
Dipropyl peroxide

$\overset{\displaystyle CH_3}{\underset{\displaystyle |}{\phantom{x}}}$
$CH_3CH_2OOCH_2CHCH_3$    1-(Ethyldioxy)-2-methylpropane

$CH_3OOCH_2CH_2OOCH_3$    1,2-Bis(methyldioxy)ethane

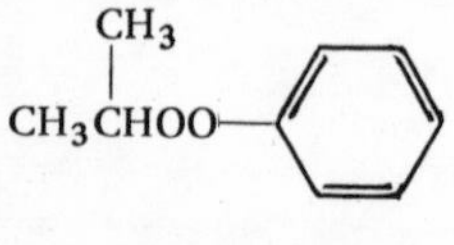

$\overset{\displaystyle CH_3}{\underset{\displaystyle |}{\phantom{x}}}$
$CH_3CHOO$—⟨benzene ring⟩    (Isopropyldioxy)benzene

264

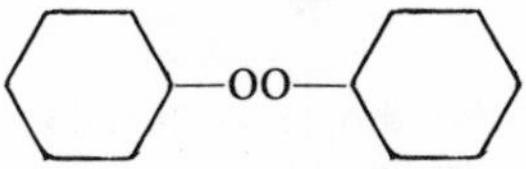

(Cyclohexyldioxy)cyclohexane
Dicyclohexyl peroxide

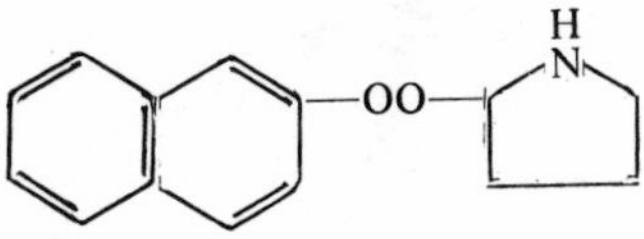

2-(2-Naphthyldioxy)azacyclopentane

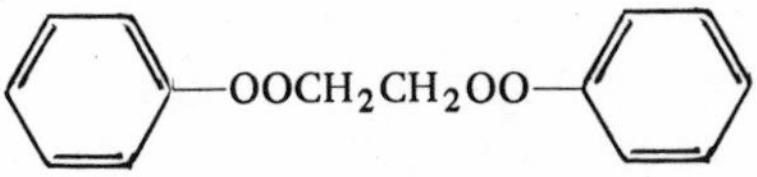

1,2-Bis(phenyldioxy)ethane

Structures in which one or more —OO— groups occur in a ring system are preferably named as heterocyclic compounds (*see* Chapter 6).

1,2-Dioxacyclohexane

An —OO— group forming a bridge between two carbon atoms of a polycyclic aromatic ring system is denoted by the nondetachable prefix *epidioxy.*

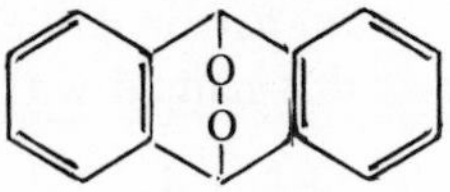

9,10-Dihydro-9,10-epidioxyanthracene

Structures in which the —OO— group is attached to two acyl groups are named radicofunctionally by using the terminal word *peroxide.*

$CH_3CH_2CO\text{-}OO\text{-}COCH_2CH_3$          Dipropanoyl peroxide

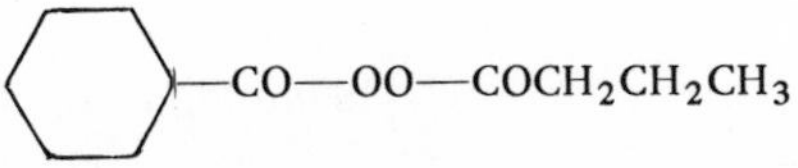

—CO—OO—$COCH_2CH_2CH_3$

Butanoyl cyclohexanecarbonyl peroxide

Hydroperoxides are preferably named substitutively as *hydroperoxy* derivatives of the chain or ring system to which the —OOH group is attached. Simple monohydroperoxides may also be named radicofunctionally by using the terminal word *hydroperoxide.*

$$CH_3\overset{\displaystyle CH_3}{\underset{\displaystyle CH_3}{\overset{|}{\underset{|}{C}}}}OOH$$

2-Hydroperoxy-2-methylpropane
*tert*-Butyl hydroperoxide

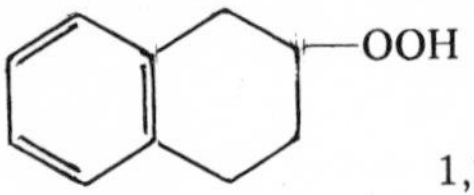

—OOH

1,2,3,4-Tetrahydro-2-hydroperoxynaphthalene

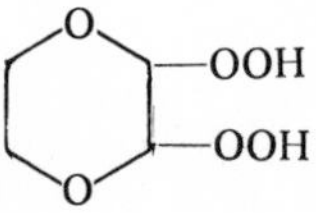

—OOH
—OOH

2,3-Dihydroperoxy-1,4-dioxacyclohexane

Salts of hydroperoxides are given regular radicofunctional names ending in *peroxide*.

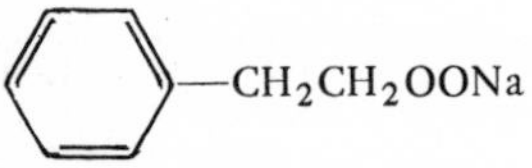

—CH$_2$CH$_2$OONa          Sodium 2-phenylethyl peroxide

Monoacyl hydroperoxides, which have the generic formula RCO-OOH, are named as **peroxy acids** by prefixing the term *peroxy* either to the full name of the corresponding acid, when the acid is named with *oic* suffix, or to the ending *carboxylic acid.*

$CH_3CH_2CH_2CH_2CH_2CO-OOH$          Peroxyhexanoic acid

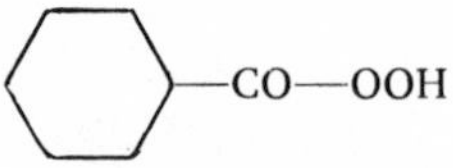

—CO—OOH          Cyclohexaneperoxycarboxylic acid

When more than one acid group is present, the prefixes *mono, di, tri,* etc. are required to avoid ambiguity.

$$\overset{3}{C}H_2\overset{2}{C}H_2\overset{1}{C}O-OOH$$
$$\underset{4}{C}H_2\underset{5}{C}H_2\underset{6}{C}OOH$$

Monoperoxyhexanedioic acid

CH$_2$CH$_2$CO—OOH
|                                    Diperoxyhexanedioic acid
CH$_2$CH$_2$CO—OOH

1,2-Cyclohexanemonoperoxy-
dicarboxylic acid

**Esters** of **peroxy acids** are named in the usual way (*see* Chapter 17); when it is necessary to designate an esterifying group at the peroxy acid function, the prefix *OO* is used.

CH$_2$CH$_2$CO-OOCH$_2$CH$_3$
|
CH$_2$CH$_2$COOCH$_3$        *OO*-Ethyl *O*-methyl monoperoxyhexanedioate

Examples of acceptable **trivial names** for peroxy compounds are included in Table 34.1. In contrast to its meaning in systematic names, the prefix *peroxy,* when used with the trivial name of a polycarboxylic acid, denotes conversion of all COOH groups to the peroxy structure—*e.g.,* peroxysuccinic acid for HOO-COCH$_2$CH$_2$CO-OOH.

*Discussion*

The use of substitutive nomenclature for peroxy compounds is not mentioned in the 1965 IUPAC Rules, perhaps because members of this class are encountered less frequently than other structurally similar classes. To date, *Chemical Abstracts* has held to the traditional radicofunctional names ending in *peroxide.* However, the modern trend in systematic organic nomenclature is toward substitutive treatment for all functional classes, and there is no official basis for deviating from this principle for peroxy compounds, except for diacyl peroxides, which bear a structural resemblance to anhydrides of acids; anhydrides have never been given substitutive names (*see* Chapter 17).

Usage has been inconsistent in its inclusion or omission of *di* or *bis* in forming radicofunctional names for symmetrical peroxides. Although *Chemical Abstracts* has indexed simple structures of this type under names such as "Ethyl peroxide" and "Benzoyl peroxide", such omission of the multiplying prefix is incorrect by modern principles. On the other hand, in *trivial* names of polyperoxycarboxylic acids containing no COOH groups the prefix *peroxy* is sufficient, as in peroxyphthalic acid; here the determining principle is the same as that used in forming the names phthalaldehyde and phthalamide (*see* Chapters 20 and 21). For acids in which not all of the COOH groups have been converted to the peroxy structure, systematic names should be used. Names such as "phthalic monoperoxyacid" are not recommended.

Prior to its 1947 Indexes, *Chemical Abstracts* used the prefix *"per"* rather than *peroxy;* although the names "performic acid", peracetic acid", and "perbenzoic acid" are recognized in the 1965 IUPAC Rules, this kind of name should not be used.

In the older literature hydroperoxides are sometimes named as derivatives of hydrogen peroxide—*e.g.,* "ethyl hydrogen peroxide" for $CH_3CH_2OOH$. This type of name should not be revived.

### Table 34.1. Examples of Acceptable Usage

1.  $CH_3OOCH_3$

    (Methyldioxy)methane
    Dimethyl peroxide

2.  $ClCH_2CH_2OOCH_2CH_3$

    1-Chloro-2-(ethyldioxy)ethane
    (2-Chloroethyldioxy)ethane

3.

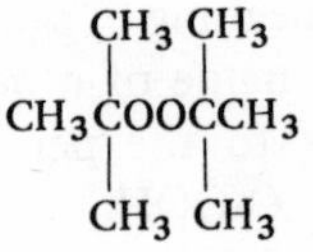

    2-(1,1-Dimethylethyldioxy)-2-methyl-
      propane
    Di-*tert*-butyl peroxide

4.

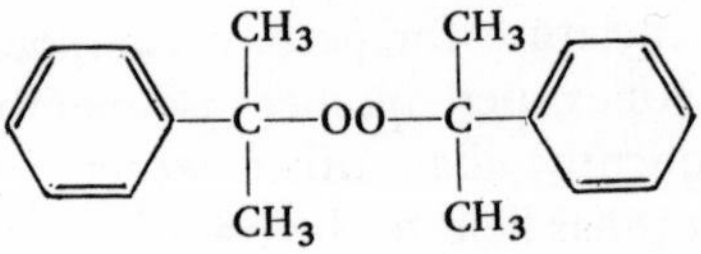

    [1-Methyl-1-(1-methyl-1-phenyl-
      ethyldioxy)ethyl] benzene
    Bis(1-methyl-1-phenylethyl) peroxide
    *not* Cumene peroxide
    *not* Dicumyl peroxide

5.

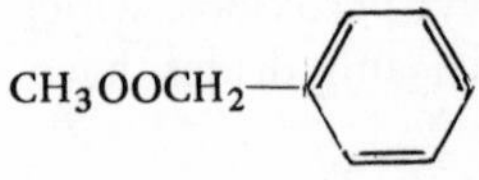

    $CH_3OOCH_2$—

    (Methyldioxymethyl)benzene

6.

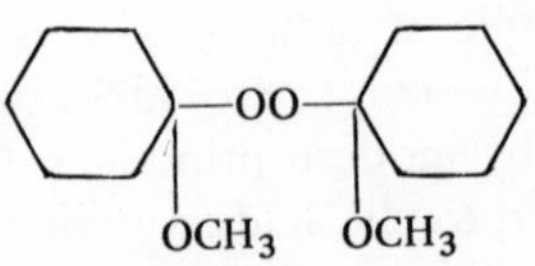

    1-Methoxy-1-(1-methoxycyclohexyl-
      dioxy)cyclohexane
    Bis(1-methoxycyclohexyl) peroxide

7.

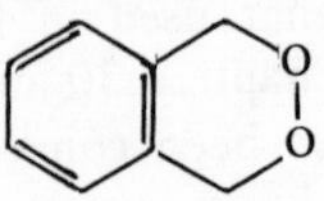

    1,4-Dihydro-2,3-dioxanaphthalene

8.

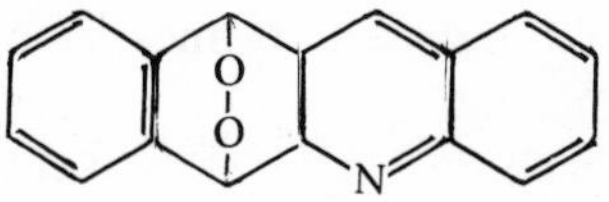

6,11-Dihydro-6,11-epidioxy-5-aza-
naphthacene
6,11-Dihydro-6,11-epidioxybenz[*b*] acridine

9.     $CH_3CO\text{-}OO\text{-}COCH_3$     Diethanoyl peroxide
Diacetyl peroxide
*not*  Acetyl peroxide

10.

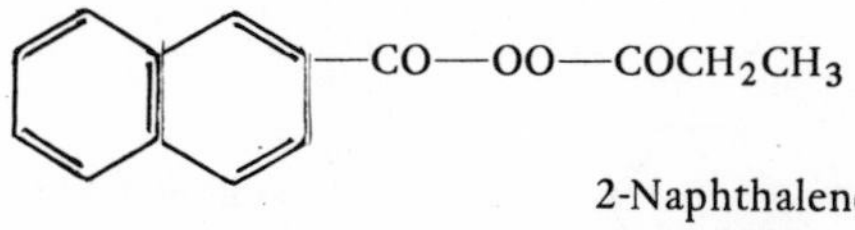

$-CO\text{—}OO\text{—}COCH_2CH_3$

2-Naphthalenecarbonyl propanoyl
peroxide
2-Naphthoyl propionyl peroxide

11.

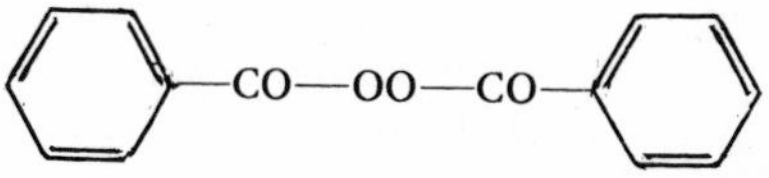

$-CO\text{—}OO\text{—}CO-$

Bis(benzenecarbonyl) peroxide
Dibenzoyl peroxide
*not*  Benzoyl peroxide

12.

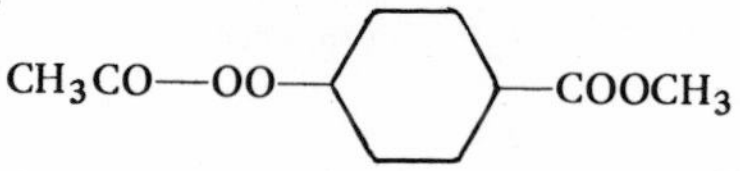

$CH_3CO\text{—}OO-$     $-COOCH_3$

Methyl 4-(ethanoyldioxy)cyclohexane-
carboxylate
Methyl 4-(acetyldioxy)cyclohexane-
carboxylate

13.

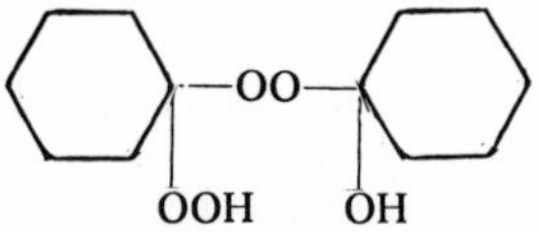

$-OO-$

$OOH$     $OH$

1-(1-Hydroperoxycyclohexyldioxy)-
cyclohexanol

14.

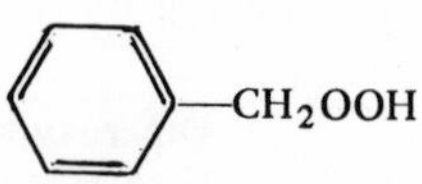

$-CH_2OOH$

(Hydroperoxymethyl)benzene
Phenylmethyl hydroperoxide
Benzyl hydroperoxide
*not*  Benzyl hydrogen peroxide

15.

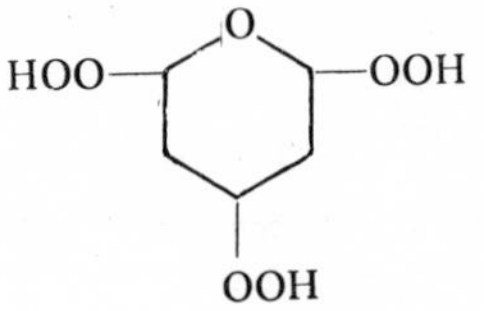

2,4,6-Trihydroperoxyoxacyclohexane
Tetrahydro-2,4,6-trihydroperoxy-2H-pyran

16.

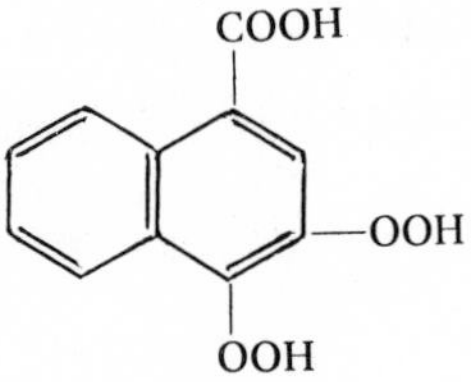

3,4-Dihydroperoxy-1-naphthalene-carboxylic acid
3,4-Dihydroperoxy-1-naphthoic acid

17.

Disodium 1,4-phenylene diperoxide

18.        HCO—OOH

Peroxymethanoic acid
Peroxyformic acid
*not* Performic acid

19.        $CH_3CO$—OOH

Peroxyethanoic acid
Peroxyacetic acid
*not* Peracetic acid

20.

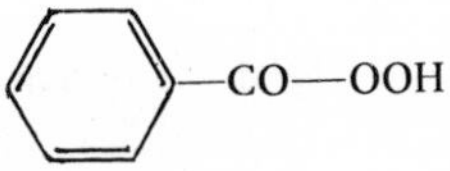

Benzeneperoxycarboxylic acid
Peroxybenzoic acid
*not* Perbenzoic acid

21.        HO-CO-OOH

Monoperoxycarbonic acid

22.        HOO-CO-OOH

Diperoxycarbonic acid

23.        HOOC-OO-COOH

Dioxydimethanoic acid
Dioxydiformic acid

24.
$\overset{5}{H}OO\overset{4}{C}CH_2\overset{3}{C}H_2\overset{2}{C}H_2\overset{1}{C}O$—OOH

Monoperoxypentanedioic acid

25.
HOO—$COCH_2CH_2CH_2CO$—OOH

Diperoxypentanedioic acid
Peroxyglutaric acid

26.

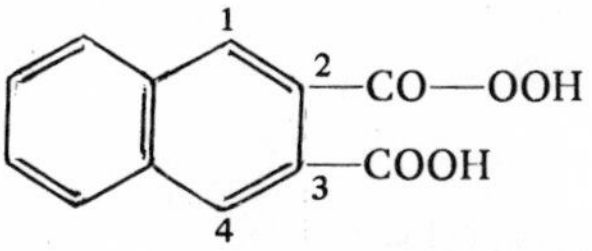

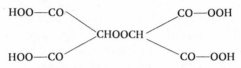

$-CO-OOH$

$-COOH$

2,3-Naphthalenemonoperoxy-
dicarboxylic acid

27.

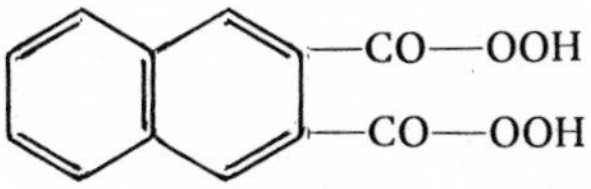

$-CO-OOH$

$-CO-OOH$

2,3-Naphthalenediperoxydicar-
boxylic acid

28.

$HOO-CO$  $CO-OOH$

$CHOOCH$

$HOO-CO$  $CO-OOH$

Dioxybis(diperoxypropanedioic acid)
Dioxybis(peroxymalonic acid)

29.

$CH_2CO-OONa$

$CH_2CO-OOH$

Sodium hydrogen diperoxybutanedioate
Sodium hydrogen peroxysuccinate

30.

$CH_2CO-OOCH_3$

$CH_2COOCH_2CH_3$

$O$-Ethyl $OO$-methyl monoperoxybutane-
dioate

# 35

# Phenols and Thiophenols

Compounds containing one or more OH or SH groups attached to carbon of an aromatic ring are called phenols and thiophenols, respectively. Although phenol and thiophenol are the common names of the compounds $C_6H_5OH$ and $C_6H_5SH$, these terms have also become well established as generic names for aromatic hydroxy and mercapto compounds. As noted in Chapter 19, the class name thiols denotes the sulfur analogs of both phenols and alcohols and replaces the older term "mercaptans," now officially abandoned.

### Recommended Nomenclature Practice

**Phenols.** Phenols are named substitutively by adding the suffix *ol* to the name of the corresponding aromatic hydrocarbon or heterocyclic system, any final *e* in the hydrocarbon or heterocyclic compound name being elided.

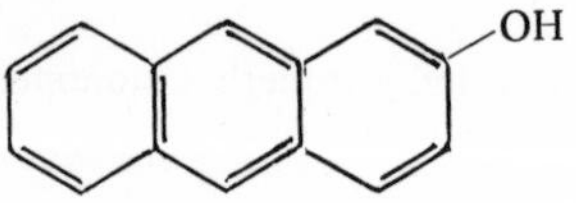

2-Anthracenol

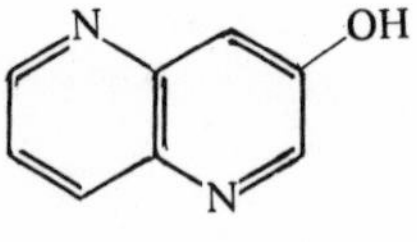

1,5-Diazanaphthalen-3-ol

Multiplying prefixes are used as necessary in the usual way.

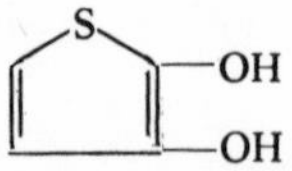

Thiacyclopenta-2,4-diene-2,3-diol

When one or more hydroxy groups must be designated as substituents, the prefix names *hydroxy, dihydroxy,* etc. are used.

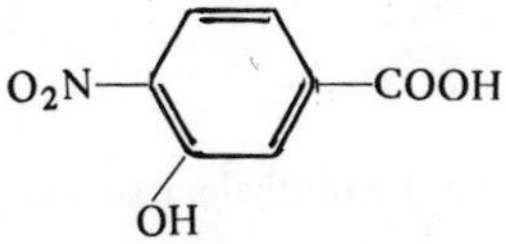

$O_2N$—⬡—COOH    3-Hydroxy-4-nitrobenzenecarboxylic acid 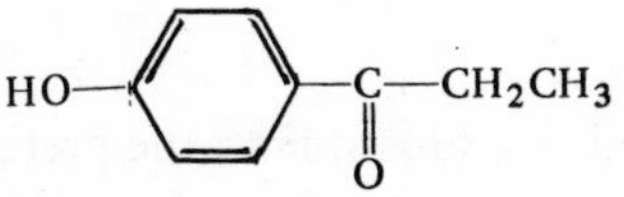
|
OH

HO—⬡—C—$CH_2CH_3$    1-(4-Hydroxyphenyl)-1-propanone
‖
O

The contracted name naphthol for $C_{10}H_7OH$ and the trivial name phenol for $C_6H_5OH$ are still acceptable. Some other trivial names, whose use should be discontinued, appear in Table 35.1.

**Thiophenols.** Thiophenols are named substitutively according to the principles just given for phenols by using the suffixes *thiol, dithiol,* etc.

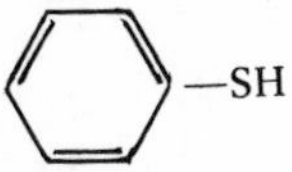

⬡—SH    Benzenethiol
*not* Phenyl mercaptan

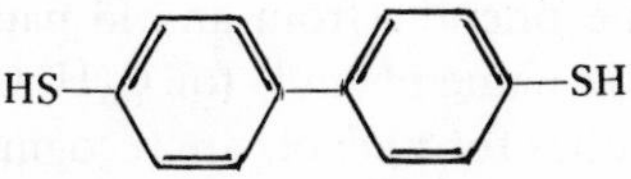

HS—⬡—⬡—SH    4,4'-Biphenyldithiol

The prefix name for the SH group is *mercapto.*

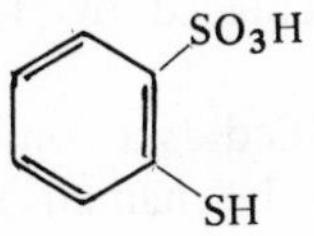

⬡ with $SO_3H$ and SH    2-Mercaptobenzenesulfonic acid

**Salts.** Ionic compounds in which a metal replaces the H of a phenolic hydroxy or mercapto group are named as **salts** of the corresponding phenol or thiophenol. The names of anions are formed by changing the parent suffixes *ol* and *thiol* to *olate* and *thiolate,* respectively. Some abbreviated names ending in *oxide* are also recognized in the 1965 IUPAC Rules (*see* Example 13).

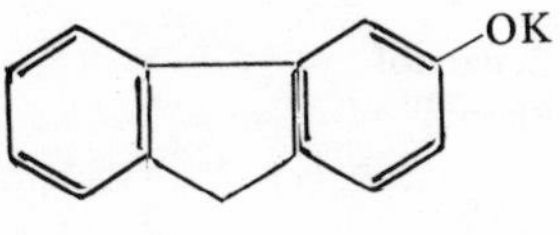

⬡—⬡(CH₂)—OK    Potassium 3-fluorenolate

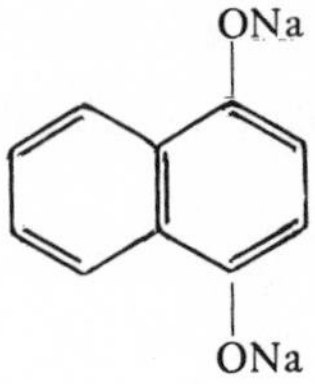

Disodium 1,4-naphthalenediolate

When the anionic group must be named as a substituent, the prefix names *oxido* and *sulfido* are used.

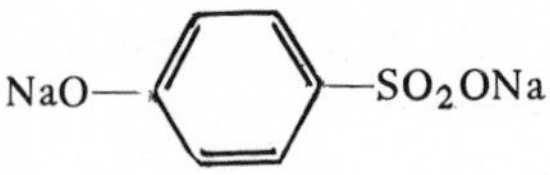

Disodium 4-oxidobenzenesulfonate

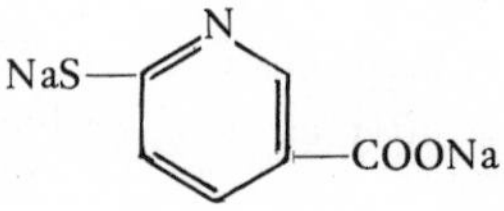

Disodium 5-sulfidoazabenzene-3-carboxylate

## Discussion

Substitutive names for phenols are generally applicable and widely used. The most conspicuous exception to fully systematic nomenclature for this class of compounds is the name phenol (from an old name for benzene, "phene," whence also the prefix name phenyl) for $C_6H_5OH$. A few other more or less familiar trivial names for phenols are recognized in the 1965 IUPAC Rules, but these should no longer be used. Names such as toluenol and xylenol, based on trivial names of substituted aromatic hydrocarbons, should also be abandoned. The same applies to " $\alpha$-naphthol" and "$\beta$-naphthol", which are now replaced by 1-naphthol and 2-naphthol.

The word "thiophenol" is not recommended as a common name for $C_6H_5SH$; all individual thiophenols should be named systematically. However, in appending the suffix *ol* to the heterocycle name thiophene, the final *e* is retained to avoid ambiguity—*i.e.*, thiopheneol. Radicofunctional names terminating in the class name "mercaptan" should not be used.

### Table 35.1. Examples of Acceptable Usage

1.

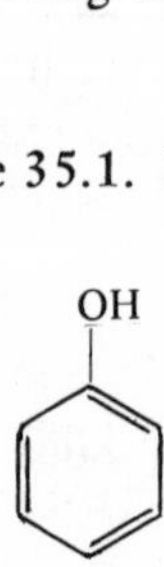

Benzenol
Phenol

2.

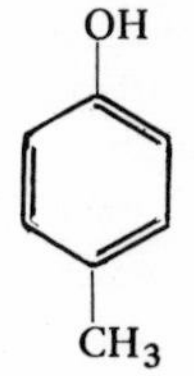

4-Methylbenzenol
4-Methylphenol
*not* p-Cresol

3.

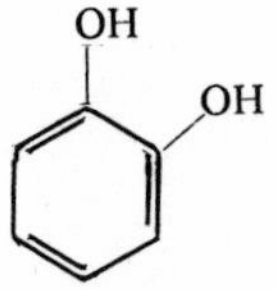

1,2-Benzenediol
*not* Pyrocatechol

4.

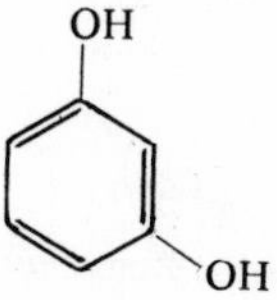

1,3-Benzenediol
*not* Resorcinol

5.

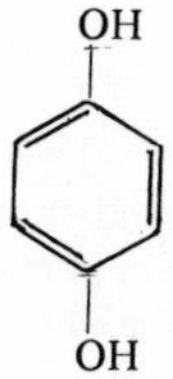

1,4-Benzenediol
*not* Hydroquinone

6.

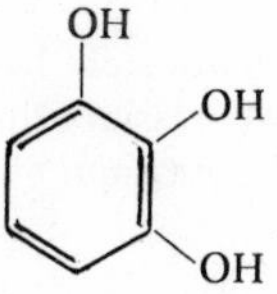

1,2,3-Benzenetriol
*not* Pyrogallol

7.

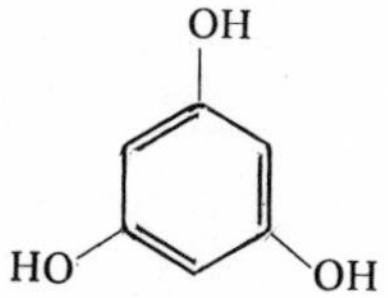

1,3,5-Benzenetriol
*not* Phloroglucinol

8.

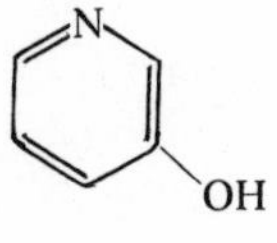

Azabenzen-3-ol
3-Pyridinol

9.

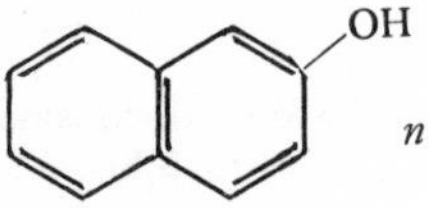

2-Naphthalenol
2-Naphthol
*not*  β-Naphthol

10.

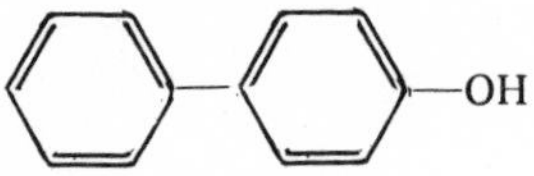

—OH  4-Biphenylol

11.

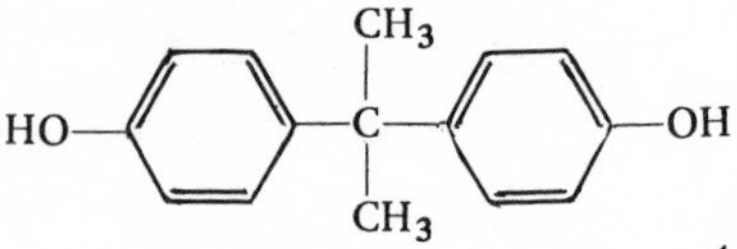

4,4'-Isopropylidenebisbenzenol
4,4'-Isopropylidenebisphenol

12.

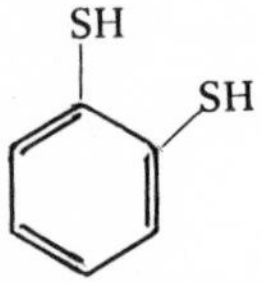

1,2-Benzenedithiol

13.

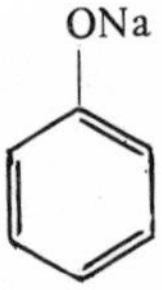

Sodium benzenolate
Sodium phenolate
Sodium phenoxide

14.

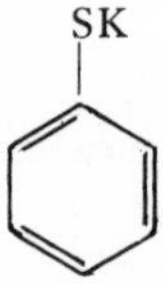

Potassium benzenethiolate
*not*  Potassium thiophenolate
*not*  Potassium phenyl mercaptide

15.

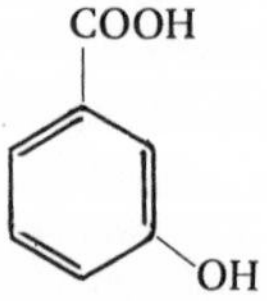

3-Hydroxybenzenecarboxylic acid
3-Hydroxybenzoic acid

16.

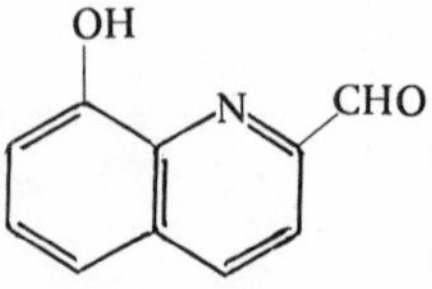

8-Hydroxy-1-azanaphthalene-2-
  carboxaldehyde
8-Hydroxy-2-quinolinecarboxaldehyde

17.

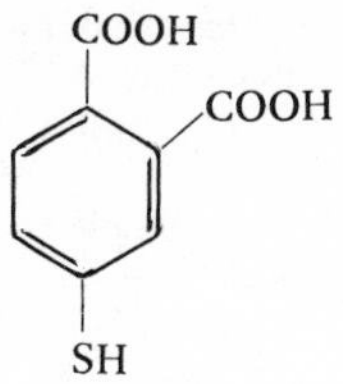

COOH

COOH

SH

4-Mercapto-1,2-benzenedicarboxylic acid

4-Mercaptophthalic acid

18.

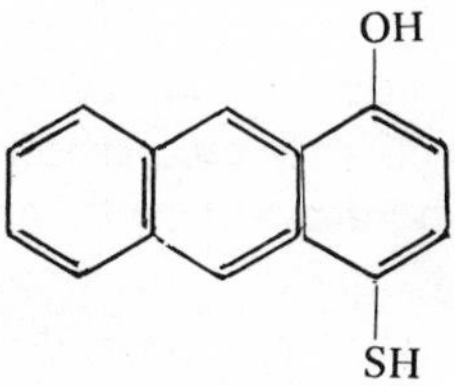

OH

SH

4-Mercapto-1-anthracenol

19.

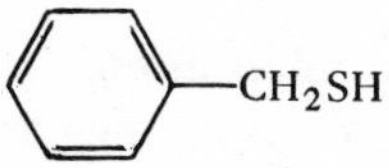

—CH$_2$SH

Phenylmethanethiol

*not* Benzyl mercaptan

# 36

# Phosphorus Compounds

An organic phosphorus compound contains at least one phosphorus atom connected directly or indirectly to the carbon skeleton of the molecule. The phosphorus atom may be a member of an organic chain or ring system, or it may be the central atom in a functional group that is in turn attached to a chain or ring. Although such functional groups commonly have one or more oxygen atoms or OH groups surrounding the phosphorus atom, analogous structures wherein sulfur, selenium, tellurium, or nitrogen (as :NH or :N) may be regarded as having replaced oxygen are also considered in this chapter, as are structures wherein halogen or nitrogen (as $NH_2$) has replaced OH.

The methods of naming organic phosphorus compounds recommended here may be extended to the corresponding **arsenic** compounds, but they apply only in part to organic derivatives of **antimony** and **bismuth**.

### Recommended Nomenclature Practice

**Monophosphorus Compounds.** PARENT STRUCTURES. **Acyclic phosphorus compounds** containing only one phosphorus atom, as well as compounds in which each of several phosphorus-containing functional groups contains only a single phosphorus atom, are preferably named as derivatives of the sometimes hypothetical parent structures listed in Table 36.1. When hydrogen attached to phosphorus is replaced by a hydrocarbon group, the derivative is named substitutively.

| | |
|---|---|
| $C_6H_5PH_2$ | Phenylphosphine |
| $(CH_3)_2PH_3$ | Dimethylphosphorane |
| $CH_3CH_2CH_2CH_2P{=}O$ | Butyloxophosphine |
| $H_2(O)PCH_2CH_2CH_2P(O)H_2$ | Trimethylenebis(phosphine oxide) |

When hydrogen of an OH group is replaced, the derivative is named radicofunctionally, similarly to esters of carboxylic acids.

| | |
|---|---|
| $(CH_3CH_2O)_3PO$ | Triethyl phosphate |
| $(C_6H_5O)_2POCH_3$ | Methyl diphenyl phosphite |
| $(C_6H_5)_2POCH_3$ | Methyl diphenylphosphinite |
| $(CH_3CH_2CH_2O)_2P(O)OH$ | Dipropyl hydrogen phosphate |
| $(CH_3O)_2POCH_2CH_2OP(OCH_3)_2$ | Ethylene tetramethyl diphosphite |

Ionic compounds are named as salts of phosphorus-containing acids or analogously to ammonium compounds (*see* Chapter 24).

## Table 36.1. Phosphorus-Containing Parent Structures

*Hydrides*

| | | | |
|---|---|---|---|
| $PH_3$ | Phosphine | $H_3PO$ | Phosphine oxide |
| $PH_5$ | Phosphorane | $H_3PS$ | Phosphine sulfide |
| | | $H_3PNH$ | Phosphine imide |

*Acids*

| | | | |
|---|---|---|---|
| $P(OH)_3$ | Phosphorous acid | $P(O)(OH)_3$ | Phosphoric acid |
| $HP(OH)_2$ | Phosphonous acid | $HP(O)(OH)_2$ | Phosphonic acid |
| $H_2POH$ | Phosphinous acid | $H_2P(O)OH$ | Phosphinic acid |
| $O=POH$ | Phosphenous acid | $O=P(O)OH$ | Phosphenic acid |
| | $H_4POH$ | Phosphoranoic acid[a] | |
| | $H_3P(OH)_2$ | Phosphoranedioic acid[a] | |
| | $H_2P(OH)_3$ | Phosphoranetrioic acid[a] | |
| | $HP(OH)_4$ | Phosphoranetetroic acid[a] | |
| | $P(OH)_5$ | Phosphoranepentoic acid[a] | |

*Bases and Salts*

| | |
|---|---|
| $PH_4^+\ OH^-$ | Phosphonium hydroxide |
| $PH_4^+\ Cl^-$ | Phosphonium chloride |
| (and others containing other anions) | |

[a]These names, though here recommended and prescribed by the 1952 Rules (see Discussion Section), are not recognized by IUPAC and are no longer used by *Chemical Abstracts*.

$(CH_3CH_2CH_2O)_2P(O)O^-$  $Na^+$      Sodium dipropyl phosphate

$(CH_3CH_2)_4P^+$ $Cl^-$              Tetraethylphosphonium chloride

Acyclic structures containing more than two phosphorus atoms as non-terminal members of an organic chain system are preferably named by replacement nomenclature (see Chapter 7). The prefix *phospha* denotes a trivalent phosphorus atom.

$$CH_3CH_2PHCH_2PCH_2CH_2NHCH_2CH_2PCH_2PHCH_2CH_3$$

              |                    |

              $CH_3$                 $CH_3$

5,11-Dimethyl-8-aza-3,5,11,13-tetraphosphapentadecane

Cyclic phosphorus compounds in which one or more phosphorus atoms are members of an organic ring system are preferably named as heterocyclic structures (see Chapter 6).

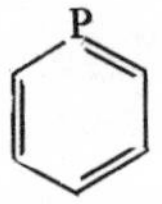

Phosphabenzene

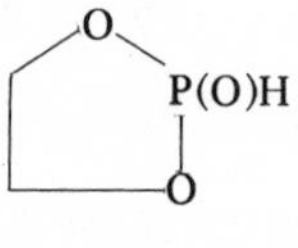

1,3-Dioxa-2-phosphacyclopentane 2-oxide

HALOGEN DERIVATIVES. If one or more halogen atoms are attached to a phosphorus atom carrying no OH or OR groups, the structure is preferably named as an **acid halide**.

$C_6H_5PCl_2$                    Phenylphosphonous dichloride

$(CH_3CH_2)_2P(O)Br$          Diethylphosphinic bromide

$Cl_2(O)P$—⟨ ⟩—$P(O)Cl_2$      1,4-Phenylenediphosphonic tetrachloride

If, in addition to halogen, a phosphorus atom carries one or more OH and/or OR groups, the structure is preferably named as an **acid** or **ester**, and the halogen is designated by the appropriate infix *fluorid, chlorid,* *bromid,* or *iodid.* An infix beginning with a consonant is preceded by an added *o.*

CH$_3$POH
|
I                                    Methylphosphoniodidous acid

CH$_3$CH$_2$OP(O)Br$_2$                     Ethyl phosphorodibromidate

In a like manner, other infixes may be used: *cyanatid* for OCN, *isocyanatid* for NCO, *nitrid* for N (replacing both O and OH simultaneously), and others as specified in the 1952 Rules (*1*).

NITROGEN DERIVATIVES. Compounds having one or more NH$_2$ groups attached to a phosphorus atom carrying no OH or OR groups are preferably named radicofunctionally as **amides.** Attachment of substituting groups to phosphorus is denoted by the locant *P.*

(C$_6$H$_5$)$_2$PNH$_2$                       *P,P*-Diphenylphosphinous amide

CH$_3$P(O)(NH$_2$)$_2$                       *P*-Methylphosphonic diamide

If OH and/or OR groups are also attached to the phosphorus atom, the structure is preferably named as an **acid** or **ester** by using the appropriate infix *amid* or *diamid.*

C$_6$H$_{11}$P(O)OH
|
NH$_2$                               *P*-Cyclohexylphosphonamidic acid

C$_6$H$_5$OP(NH$_2$)$_2$                      Phenyl phosphorodiamidite

Substituents replacing a hydrogen atom of an NH$_2$ group may be designated by using the locants *N, N'*, etc.

$$C_6H_5P\underset{\textstyle NHCH_2CH_3}{\overset{\textstyle NHCH_3}{<}}$$                *N*-Ethyl-*N'*-methyl-*P*-phenylphosphonous
                                     diamide

Compounds containing the NH group coordinately bound to a phosphorus atom carrying no OH, OR, or NH$_2$ groups are preferably named as derivatives of **phosphine imide** (Table 36.1); if OH and/or OR groups are also present, the structure is named as an **acid** or **ester** by using the infix *imid;* if NH$_2$ is the only other negative group present, the structure is named as an *amide.*

| | |
|---|---|
| $(CH_3CH_2CH_2)_3PNH$ | *P,P,P*-Tripropylphosphine imide |
| $C_6H_5P(NH)(OH)_2$ | *P*-Phenylphosphonimidic acid |
| $(CH_3CH_2O)_3PNH$ | Triethyl phosphorimidate |
| $C_6H_5P(NH)(NH_2)_2$ | *P*-Phenylphosphonimidic diamide |

In assigning locants for groups attached to nitrogen, all $NH_2$ groups should be considered first, before coordinately bound NH.

$$\begin{array}{c} NHCH_2CH_3 \\ | \\ C_6H_5PNCH_3 \\ | \\ NH_2 \end{array}$$
     *N*-Ethyl-*N″*-methyl-*P*-phenylphosphonimidic diamide

SULFUR ANALOGS. Acids and esters in which one or more sulfur atoms are attached to phosphorus are treated as analogs of the corresponding oxygen-containing structures and are preferably named by using the appropriate infix *thio, dithio, trithio,* or *tetrathio.* Where required to avoid ambiguity, the locant *O* or *S* is placed immediately preceding the part of the name to which it applies.

| | |
|---|---|
| $(C_6H_5)_2P(S)OH$ | Diphenylphosphinothioic *O*-acid |
| $CH_3CH_2CH_2CH_2P(S)(SH)(OH)$ | Butylphosphonodithioic *O,S*-acid |
| $CH_3CH_2CH_2CH_2P(O)(SH)_2$ | Butylphosphonodithioic *S,S*-acid |
| $(CH_3O)_2P(O)SCH_2CH_3$ | *S*-Ethyl dimethyl phosphorothioate |
| $(CH_3CH_2S)_3P$ | Triethyl phosphorotrithioite |

Halogen and nitrogen derivatives of phosphorus compounds containing sulfur are treated similarly. When more than one infix is required, these are arranged in alphabetical order.

| | |
|---|---|
| $CH_3CH_2P(S)Br_2$ | Ethylphosphonothioic dibromide |
| $(C_6H_5O)_2P(S)Cl$ | *O,O*-Diphenyl phosphorochloridothioate |
| $(C_6H_5)_2P(S)NH_2$ | *P,P*-Diphenylphosphinothioic amide |
| $CH_3CH_2OP(O)(SH)NH_2$ | Ethyl *S*-hydrogen phosphoramidothioate |

Compounds in which coordinately bound S is the only atom other than C or H attached to phosphorus are preferably named as derivatives of phosphine sulfide (Table 36.1).

$(CH_3CH_2CH_2)_2P(S)H$        Dipropylphosphine sulfide

PHOSPHORUS-CONTAINING SUBSTITUENTS. When a phosphorus-containing organic group must be treated as a substituent, a **prefix name** based on one of the names shown in Table 36.2 should be employed.

$(CH_3)_2P-$        Dimethylphosphino

$CH_3CH_2CH_2PH_3-$        Propylphosphoranyl

$(CH_3CH_2O)_2P(O)-$        Diethoxyphosphinyl

$(CH_3)_3\overset{+}{P}CH_2CH_2-$        2-(Trimethylphosphonio)ethyl

In replacement nomenclature, the phosphonium group is designated by the prefix *phosphonia*.

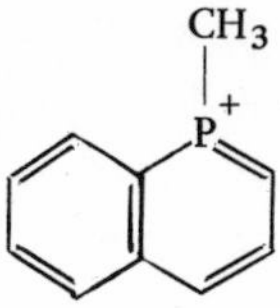

1-Methyl-1-phosphonianaphthalene chloride

### Table 36.2.  Fundamental Phosphorus-Containing Groups

| | | | |
|---|---|---|---|
| $H_2P-$ | Phosphino | $HP\diagdown$ | Phosphoranetetrayl |
| $HP-$ | Phosphinidene | $\diagdown P\diagup$ | Phosphoranepentayl |
| $P$ | Phosphinidyne | $H_3\overset{+}{P}-$ | Phosphonio |
| $H_4P-$ | Phosphoranyl | $H_2(O)P-$ | Phosphinyl |
| $H_3P-$ | Phosphoranediyl | $H(O)P\diagup$ | Phosphinylidene |
| $H_2P-$ | Phosphoranetriyl | $(O)P\diagup$ | Phosphinylidyne |

| | | | |
|---|---|---|---|
| $H_2(S)P$— | Phosphinothioyl | $H_2(HN)P$— | Phosphinimyl |
| $H(S)P<$ | Phosphinothioylidene | $H(HN)P<$ | Phosphinimylidene |
| $(S)P\lessgtr$ | Phosphinothioylidyne | $(HN)P\lessgtr$ | Phosphinimylidyne |

**Polyphosphorus Compounds.** Although various proposals have been made and considered, no comprehensive system for naming organic compounds derived from parent inorganic structures containing more than one phosphorus atom has yet been developed and officially recognized. However, in lieu of official rules, the systematic parent names shown below are recommended for interim use. The corresponding common names are shown for comparison.

| | *Systematic Name* | *Common Name* |
|---|---|---|
| $PH_2PH_2$ | Diphosphane | Diphosphine |
| $PH_2PHPH_2$ | Triphosphane | Triphosphine |
| $(HO)_2(O)P$—$P(O)(OH)_2$ | Diphosphanetetroic acid | Hypophosphoric acid |
| $(HO)_2P$—$O$—$P(OH)_2$ | Diphosphorous acid | Pyrophosphorous acid |
| $(HO)_2(O)P$—$O$—$P(O)(OH)_2$ | Diphosphoric acid | Pyrophosphoric acid |

The systematic approach illustrated can, of course, be extended to form names of analogous parent structures containing more than two phosphorus atoms. Organic derivatives may then be named by applying the usual principles of substitutive and radicofunctional nomenclature.

$CH_3PHPHPHP(CH_3)_2$     1,1,4-Trimethyltetraphosphane

$$(CH_3O)_2(O)P—O—P(O)O—P(O)(OCH_3)_2$$
$$\underset{OCH_3}{|}$$

Pentamethyl triphosphate

## Discussion

The nomenclature recommended in this chapter is essentially the same as that published in 1952 (*1*) and officially adopted shortly thereafter by the American Chemical Society and the Chemical Society of London. It is based on the extension of certain principles well established in organic nomenclature to derivatives of inorganic parent structures, and it has admittedly not received widespread acceptance among inorganic chemists. Nevertheless, it has been used, with a few exceptions, since 1952 by *Chemical Abstracts.*

From the inorganic standpoint, organic phosphorus compounds can be named simply by citing the atoms or groups surrounding the central phosphorus atom without attempting to indicate functionality (*2*). Another approach is to treat all such compounds as derivatives of phosphine ($PH_3$), phosphine oxide ($H_3PO$), or one of the other parent structures shown in Table 36.1. By these two methods the compound $Cl(HO)(H_2N)PO$ would be named amidochlorohydroxooxophosphorus and aminochlorohydroxyphosphine oxide, respectively. Although acceptable, neither method is recommended for organic derivatives of the kinds discussed in this chapter.

Prior to 1952, derivatives of the monophosphorus acids were named by American chemists by prefixing, in alphabetical order, the names of substituent groups to the inorganic parent names phosphoric, phosphinic, phosphonic, phosphorous, phosphinous, and phosphonous acids and their derivatives: $Cl(HO)(H_2N)PO$, amidochlorophosphoric acid; $Cl_2P(S)H$, thiophosphonic dichloride; $CH_3OP(NCH_3)H_2$, methyl *N*-methylimidophosphinate. The substituent groups were thus considered to have replaced hydroxy groups in the parent acid. In contrast, British chemists named the first compound, $Cl(HO)(H_2N)PO$, amidochlorophosphinic acid, the substituent groups being regarded as having replaced hydrogen atoms in the parent acid. This was a significant and confusing inconsistency, but fortunately it and others were eliminated in the 1952 Rules (*1*) that were approved by both British and American representatives. Continued use of the pre-1952 names is not recommended.

In the 1952 Rules the term "thiono" was recommended for the substituting group S=, and "thiolo" and "thiono" were also suggested for use as infixes to specify the positions of sulfur atoms and so avoid ambiguity in names of free acids. Neither of these practices has been otherwise officially recognized (*see also* Chapter 16), and neither is recommended. The preferred prefix name for S= is *thioxo.*

Acid names for hydroxy derivatives of $PH_5$ (*see* Table 36.1) were used by *Chemical Abstracts* in the 1952 and 1953 Indexes but were then dropped in favor of hydroxyphosphorane, dihydroxyphosphorane, etc. It is argued by many chemists that these compounds do not show the characteristic properties of acids.

Some older names for monophosphorus acids having the prefixes *ortho, meta, hypo,* and *holo* continue to appear in print, but their use in naming organic derivatives is no longer considered acceptable.

A few earlier-established prefix names for phosphorus-containing substituting groups, *e.g.*, *phosphono* for $(HO)_2P(O)-$, are still used by *Chemical Abstracts* (*3*). These are currently acceptable, either as alternates to the names recommended in this chapter or as officially recognized names in their own right.

The practice of using infixes rather than prefixes to denote conceptual replacement of oxygen in phosphorus acids by sulfur, halogen, or nitrogen did not exist in phosphorus nomenclature prior to adoption of the 1952 Rules. Thus, names such as potassium *O,O*-dimethyl dithiophosphate for $(CH_3O)_2P(S)SK$, although still acceptable, are now considered to be out of date.

As mentioned at the beginning of this chapter, the methods recommended here for naming organic phosphorus compounds may be extended by analogy to **organic arsenic compounds** but not in all cases to **organic antimony or bismuth compounds**. For the latter two classes, acid-type names are not considered appropriate because of the metallic nature of the central elements involved; coordination names or names denoting substitution of H in parent hydrides, oxides, sulfides, etc. should be used. Arsenic analogs of phosphorane and like-named compounds are named as arsoranes, since arsenane is the name of the six-membered ring containing one arsenic atom; other organic arsenic derivatives take the stem *arsen* or *ars*.

## Table 36.3. Examples of Acceptable Usage

1. $CH_3CH_2CH_2PHCH_3$      Methylpropylphosphine

2. $(C_6H_5)_2P(O)CH_3$      Methyldiphenylphosphine oxide

3. $-CH_2P=O$      (2-Naphthylmethyl)oxophosphine

4. $C_6H_{11}PH_4$      Cyclohexylphosphorane

5. $C_6H_5P(O)(OH)_2$      Phenylphosphonic acid

6. $CH_3CH_2CH_2CH_2\overset{\overset{\displaystyle CH_3}{|}}{P}(O)OH$      Butylmethylphosphinic acid

7. $(CH_3CH_2O)_2POH$      Diethyl hydrogen phosphite

8. $CH_3CH_2CH_2OP(O)(OH)O^-$ $Na^+$      Sodium propyl hydrogen phosphate

9. $C_6H_5\overset{\overset{\displaystyle CH_3}{|}}{\underset{\underset{\displaystyle CH_3}{|}}{\overset{+}{P}}}CH_3$   Br      Trimethylphenylphosphonium bromide

10. $CH_3PHCH_2CH_2PHCH_2CH_2PHCH_3$    2,5,8-Triphosphanonane

11.

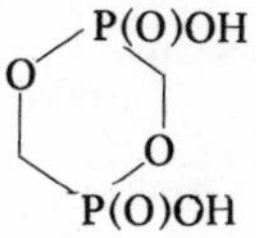

1,4-Dioxa-2,5-diphosphacyclohexane-2,5-diol
2,5-dioxide

12. $CH_3CH_2CH_2PHBr$    Propylphosphinous bromide

13. $CH_3CH_2OPCl_2$    Ethyl phosphorodichloridite

14. $(C_6H_5NH)_2P(O)NHCH_3$    N-Methyl-N',N''-diphenylphosphoric
    triamide

15. $(CH_3CH_2O)_2P(O)NH_2$    Diethyl phosphoramidate

16. $(CH_3)_3P(NH)$    *P,P,P*-Trimethylphosphine imide

17. $C_6H_5P(NCH_3)(OCH_2CH_2CH_3)_2$    Dipropyl N-methyl-
    P-phenylphosphonimidate

18. $CH_3CH_2OP(N)NHCH_3$    Ethyl methylphosphoramidonitridate

19. $(CH_3CH_2O)_2P(O)SH$    Diethyl S-hydrogen phosphorothioate

20. $C_6H_5P(O)(SH)_2$    Phenylphosphonodithioic S,S-acid

21. $HOP(S)(SCH_2CH_3)(OCH_2CH_3)$    O,S-Diethyl O-hydrogen phosphorodithioate

22. $CH_3CH_2CH_2CH_2SP(O)Cl_2$    S-Butyl phosphorodichloridothioate

23. $C_6H_5P(S)(NHCH_3)OH$    N-Methyl-P-phenylphosphonamidothioic
    O-acid

24. $CH_3CH_2NHP(S)Br_2$    Ethylphosphoramidothioic dibromide

25. $(CH_3)_2P(O)CH_2CH_2COOH$    3-(Dimethylphosphinyl)propanoic acid
    3-(Dimethylphosphinyl)propionic acid

26. $(HO)_2P(O)$—⟨benzene ring⟩—$SO_2OH$    4-(Dihydroxyphosphinyl)benzenesulfonic
    acid

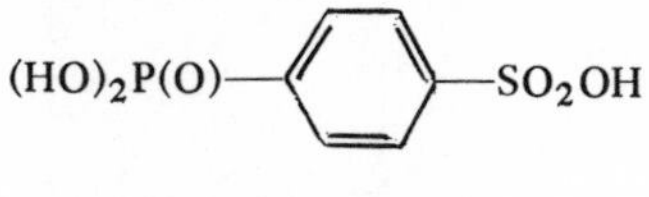

27. $(CH_3)_2PCH_2CH_2N(CH_3)_2$    2-(Dimethylphosphino)-N,N-dimethyl-
    ethanamine

28. $HO$—⟨benzene ring⟩—$PH$—⟨benzene ring⟩—$OH$

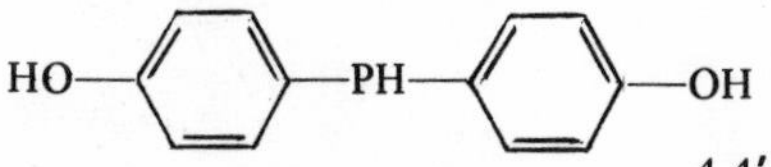

4,4'-Phosphinidenedibenzenol
4,4'-Phosphinidenediphenol

29. $(CH_3CH_2)_2PPH_2$    1,1-Diethyldiphosphane
    1,1-Diethyldiphosphine

| | | |
|---|---|---|
| 30. | $(CH_3O)_2(O)P\text{-}O\text{-}P(O)(OCH_3)_2$ | Tetramethyl diphosphate<br>Tetramethyl pyrophosphate |
| 31. | $(CH_3O)_2As(O)OH$ | Dimethyl hydrogen arsenate |
| 32. | $(C_6H_5)_3As$ | Triphenylarsine |
| 33. | $(CH_3CH_2)_2AsH_3$ | Diethylarsorane |
| 34. | $CH_3As(O)(ONa)_2$ | Disodium methylarsonate |
| 35. | $C_6H_5\overset{+}{As}H_3 \ I^-$ | Phenylarsonium iodide |
| 36. | $CH_3AsHCl$ | Methylarsinous chloride |
| 37. | $CH_3CH_2CH_2AsHAsH_2$ | Propyldiarsane<br>Propyldiarsine |
| 38. | $CH_3SbH_2$ | Methylstibine |

## Literature Cited

1. "Report of ACS Nomenclature, Spelling and Pronunciation Committee for First Half of 1952. Section E," *Chem. Eng. News* (1952) **30**, 4515; *J. Chem. Soc.* **1952**, 5122.
2. IUPAC, "Nomenclature of Inorganic Chemistry. Definitive Rules, 1957," Butterworth and Co., London, 1959.
3. "Index Introductions," *Chem. Abstr.* (1967) **66**, 33-I.

# Polymers

I n addition to the usual chemical structure and functionality para-
meters encountered in naming single organic molecules, the nomenclature of macromolecular substances is beset with problems of including
within a name such factors as physical structure, chemical inhomogeneity,
and even variable composition. As a consequence, the complexity of
names of such materials would be enormously ramified if nomenclature
principles for simple molecules were rigorously applied. A somewhat
empirical approach is therefore preferable, and small variations in
composition (as in the end groups of a chain) and in structure (such as
occasional branches in a chain) are generally ignored.

The term **polymer** implies a multiplicity of **structural repeating units**
in the chain or network of a macromolecule, and most of the substances
considered in this chapter are of this type. These structural repeating units
may be derived from more than one component, as in some condensation polymers. **Copolymers** contain more than one kind of structural
repeating unit, and there are certain classes of copolymers that depend on
the structural relationship of these repeating units to each other. Polymers
in which the steric configuration of the repeating structural units has
assumed some degree of uniformity require yet another type of
nomenclature.

Since no definitive organic polymer nomenclature is in general use at
the present time, the recommendations made in this chapter represent a
codification of current practice with some attempts to channel this
practice along the lines consistent with other chapters.

### Recommended Nomenclature Practice

The names of simple **linear polymers** are based on the names of the
unsaturated monomers from which the polymers were or conceptually
could have been prepared; end groups and structural variations such as
branching or crosslinking are ignored in such names. In general, the term
**poly** is prefixed to the monomer name, with the latter enclosed in
parentheses if it is a complex name or involves more than one word.

$(-CH_2CH-)_n$                  Polystyrene
       $C_6H_5$

$(-CH_2C-\ )_n$             Poly($\alpha$-methylstyrene)
with $CH_3$ above and $C_6H_5$ below

$(-CH_2CH-)_n$               Poly(vinyl alcohol)
       OH

Where the polymer can be viewed as forming conceptually by either an addition or a condensation process, the name of the monomer required for either process may be used.

$(-O\,CH_2CH_2-)_n$          Polyoxirane
Poly(ethylene oxide)
or
Poly(1,2-ethanediol)
Poly(ethylene glycol)

**Condensation polymers** are most frequently derived from more than one reactant. Such polymers may be named by picturing the polymer as having been formed by ring opening of the monomeric product of the reactants.

$(-NH(CH_2)_5CO-)_n$         Poly- $\epsilon$ -caprolactam

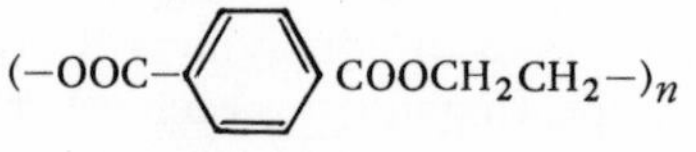

$(-OOC-\langle C_6H_4 \rangle COOCH_2CH_2-)_n$     Poly(ethylene terephthalate)

**Random copolymers** formed from two or more monomers are named by separating the monomer names with the interfix *co:* poly(styrene-*co*-butadiene); poly[(methyl methacrylate)-*co*-styrene].

The official ACS method of naming linear polymers is based on the systematic name of the smallest structural repeating unit (*1*). The name of the repeating polyvalent group is placed within parentheses following the prefix *poly* or, if the degree of polymerization is to be specified, the appropriate multiplying prefix. Polyvalent groups are generally named as indicated elsewhere in this book, and the largest possible group within the repeating unit that has a name of its own is cited.

$(-CH=CHCH_2CH_2-)_n$         Poly(1-butenylene)

Names of polymers whose structural repeating units consist of more than one polyvalent group are formed by citing successively the groups that make up the unit. Since the unit can be written in several ways, the beginning and direction of citation are established by a set of arbitrary rules (*1*). Among bivalent groups, the order of seniority is heterocyclic rings first, followed by chains containing hetero atoms, then carbocyclic rings, and finally chains containing only carbon. Each substituent is named as a prefix using the normal numbering of the ring or chain to which it is attached.

$$\left( \overset{N}{\underset{}{\text{pyridine}}} \text{-} \overset{}{\underset{}{\text{phenylene}}} \text{-CH}_2\overset{O}{\overset{\|}{C}}\text{CH}_2\overset{O}{\overset{\|}{C}}\text{CH}_2\text{CH}_2\text{-} \right)_n$$

Poly[2,4-pyridinediyl-(4-(2-chloroethyl)-
1,3-phenylene)-(2,4-dioxohexamethylene)]

## Discussion

The recommendations of this chapter are admittedly scanty and cover a relatively small area of the field of macromolecular nomenclature. It is realized that here, as in the field of biological chemistry, systematic names that adequately describe substances for which chemical structures are known are bound to be somewhat cumbersome. A logical system for naming linear high polymers on the basis of structure was published by the IUPAC Commission on Macromolecules in 1952 (*2*), but the system has not been widely used: it has since been extended to stereoregular polymers (*3*). The structure-based system described and recommended above represents a modernization of the original IUPAC proposals by the Committee on Nomenclature of the Polymer Division, ACS (*1*). This system has been approved by the ACS Council.

Polymers are ordinarily named on the basis of their sources, as indicated in the previous section, or by employing well-established trivial names. An example of the latter is cellulose triacetate, and such names are acceptable. "Systems" based on newly coined trivial names or on trade names such as nylon-6 or nylon-6,6 should not be used. Names based on source have been used for condensation as well as addition polymers simply by inserting *co* between the names of the reactants involved: poly[(hexamethylenediamine)-*co*-(adipoyl chloride)].

The absence of parentheses can sometimes lead to ambiguous names for polymers derived from complex monomers or those having multiple-word names. An example is that of the polymer prepared from a mixture of monochlorinated styrenes. The correct name of the polymer is poly(chlorostyrene) or, even better, poly(*x*-chlorostyrene); in contrast, the name polychlorostyrene describes monomeric styrene substituted at unknown positions with several chlorine atoms.

Copolymers present a problem since, in addition to random copolymers, one encounters alternating copolymers, graft copolymers, and block copolymers, all of which should receive unambiguous names. Ceresa (4) has suggested a logical extension of systematic copolymer names by replacing the interfix *co* with *alt* for alternating copolymers, —ABABABAB—; *b* for block copolymers, —AAAA...AABBB...BBAA; and *g* for graft copolymers, —AA(BBBB—)AAA—. The first mentioned monomer is the one present in the greatest proportion. Names such as poly(styrene-*g*-acrylonitrile), poly(ethylene-*alt*-carbon monoxide), and poly[(methyl methacrylate)-*b*-(styrene-*co*-butadiene)] result.

Such problems in polymer nomenclature can be side-stepped, as they were by *Chemical Abstracts* prior to the Vol. 66 indexes, by the use of a periphrase: "Styrene, block polymer with methyl methacrylate." Polymers were almost always indexed under the appropriate monomers. *Chemical Abstracts* now utilizes the ACS-approved structure-based system for additional entries (5).

Abbreviations abound in profusion in the polymer field; most, if not all, should be avoided in scientific writing. At the very least, abbreviations should be well defined whenever they are used to designate specific polymers; they should never be used for monomers. Even such established abbreviations as PS, PMMA, and PMA for polystyrene, poly(methyl methacrylate), and poly(methyl acrylate), respectively, should be used with caution. Under the heading PMA, for example, *Chemical Abstracts* refers one to mercury, (acetato)phenyl-; aside from this ambiguity, the question can always arise whether PMA refers to poly(methyl acrylate) or to poly(methacrylic acid).

*Literature Cited*

1. "A Structure-Based Nomenclature for Linear Polymers," Report of the Committee on Nomenclature, Division of Polymer Chemistry, ACS, *Macromolecules* (1968) **1**, 193.
2. "Report on Nomenclature in the Field of Macromolecules," *J. Polymer Sci.* (1952) **8**, 257.
3. Huggins, M. L. *et al., Pure Appl. Chem.* (1966) **12**, 645.
4. Ceresa, R. J., "Block and Graft Copolymers," pp. 8-10, Butterworths, Washington, D. C., 1962.
5. Loening, K. L., W. Metanomski, W. H. Powell, "Indexing of Polymers in *Chemical Abstracts,*" *J. Chem. Doc.* (1969) **9**, 248.

# Silicon Compounds

**B** roadly speaking, molecular structures containing both carbon and silicon are known as **organic silicon compounds.** When at least one organic group is attached directly to silicon through a carbon atom, the more specific term **organosilicon compound** is applicable. As might be expected, the nomenclature of silicon hydrides closely parallels that of hydrocarbons, and in general the naming of functional and nonfunctional derivatives is analogous to that of the corresponding organic structures.

### Recommended Nomenclature Practice

The compound $SiH_4$ is called silane; its acyclic homologs, $H_3 Si-(SiH_2)_n SiH_3$, are called disilane, trisilane, tetrasilane, etc., according to the number of silicon atoms present. As with hydrocarbons, cyclic nonaromatic structures are designated by the prefix *cyclo.* The class name *silanes* includes both open-chain and cyclic systems.

$CH_3 SiH_2 SiH_2 SiH_2 CH_2 CH_3$  1-Ethyl-3-methyltrisilane

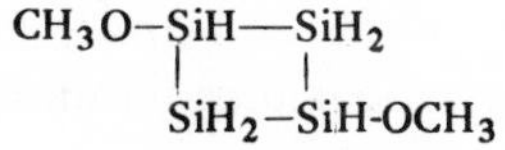

$CH_3 O-SiH-SiH_2$
$\quad\quad | \quad\quad |$
$SiH_2 -SiH-OCH_3$  1,3-Dimethoxycyclotetrasilane

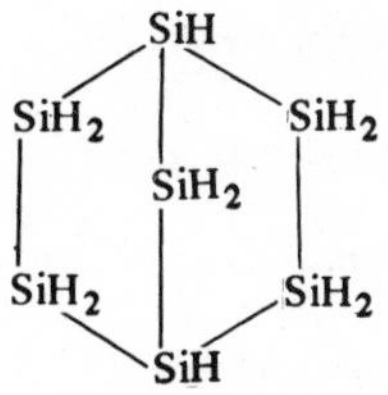

Bicyclo[2.2.1]heptasilane

When, as often occurs, a chain or ring system is composed entirely of alternating silicon and oxygen atoms, the stem name *siloxane* is used, with a multiplying prefix denoting the number of silicon atoms present.

$H_3 SiOSiH_3$  Disiloxane

$H_3SiOSiH_2OSiH_2OSiH_3$                    Tetrasiloxane

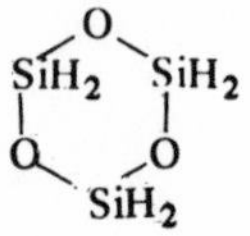

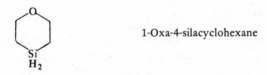

Cyclotrisiloxane

Analogous structures containing sulfur and nitrogen are named by using the stem names *silathiane* and *silazane*, respectively. Examples are included in Table 38.1.

Organic heterocyclic and hetero-acyclic systems containing silicon are named by using the prefixes *sila, disila, trisila*, etc. to designate replacement of carbon, as described in Chapters 6 and 7.

1-Oxa-4-silacyclohexane

$CH_3CH_2SiH_2CH_2CH_2SiH_2CH_2CH_2SiH_2CH_2CH_3$
3,6,9-Trisilaundecane

Simple open-chain structures that do not qualify for replacement nomenclature (*see* Chapter 7) are named as silane derivatives by the principles recommended in other chapters for various organic functional classes.

$H_3SiSiH_2OCH_2CH_2CH_3$              Propoxydisilane

$CH_3SSiH_2SiH_2SiH_2SCH_3$            1,3-Bis(methylthio)trisilane

$H_3SiNHCH_2C_6H_5$                   *N*-(Phenylmethyl)silanamine

$CH_3SiH_2SiH_2COOH$                  2-Methyldisilanecarboxylic acid

$C_6H_5CH_2CH_2SiH(OH)_2$             (2-Phenylethyl)silanediol

Functional derivatives of siloxanes, silathianes, and silazanes are treated in the same manner.

$HSSiH_2OSiH_2SH$                     1,3-Disiloxanedithiol

$Cl_3SiNHSiH_2NHSiH_3$               1,1,1-Trichlorotrisilazane

Compounds having the structures $(RO)_4Si$, $(RO)_3SiOH$, $(RO)_2Si(OH)_2$ and $ROSi(OH)_3$ are named as esters of **orthosilicic acid**, $Si(OH)_4$.

| | |
|---|---|
| $(CH_3CH_2O)_4Si$ | Tetraethyl orthosilicate |
| $(C_6H_5O)_2Si(OH)_2$ | Diphenyl dihydrogen orthosilicate |

**Prefix names** for silicon-containing groups are formed analogously to those for the corresponding carbon-containing groups by using the endings *yl, ylene, ylidene, ylidyne, triyl, tetrayl*, etc., as described in Chapter 2, or, preferably, by simply attaching the suffixes *yl, diyl, triyl*, etc. to the name of the appropriate silicon chain or ring system. For $H_3Si-$, the shortened form *silyl*, rather than *silanyl*, is used (*see* Discussion section).

|  |  |
|---|---|
| $Br_3Si-\langle\!\bigcirc\!\rangle-CONH_2$ | 4-(Tribromosilyl)cyclohexane-carboxamide |

$CH_3SiH_2SiH_2OCH_2CH_2OH$  2-[(2-Methyldisilanyl)oxy]ethanol

$H_3SiOSiH_2-\langle\!\bigcirc\!\rangle-CH_2COCH_2CH_3$

1-(4-Disiloxanylphenyl)-2-butanone

## Discussion

The recommendations made here follow closely those developed by the IUPAC Organic Commission in 1949 and subsequently officially adopted by the American Chemical Society (*1*), taking into account certain changes expected to be made by IUPAC in an updated version of its silicon rules. As can be seen, the nomenclature of organic silicon compounds is well systematized.

The stem name *silathiane* is now preferred to the previously recommended *silthiane* for compounds having the generic structure $H_3Si(SSiH_2)_nSSiH_3$.

Names for chain and ring systems incorporating the stems *siloxane, silathiane*, and *silazane* are used in the same way as silane, disilane, etc. in parent names of functional and nonfunctional derivatives. When there is a choice between two parent polysilicon chains having the same number of silicon atoms, the preference is in the decreasing order listed above, with the silanes coming last.

By using the shortened form *silyl* for the substituting group $H_3Si-$ rather than *silanyl*, the need for *bis, tris,* etc. as multiplying prefixes is eliminated. Thus, *disilanyl* denotes the group $H_3SiSiH_2-$, and *disilyl* denotes two $H_3Si-$ groups; otherwise, this distinction would have to be made by using bis(silanyl) or bissilanyl. An additional simplification is achieved by using *siloxy* for $H_3SiO-$ rather than *silyloxy;* the latter, though properly formed, is a two-part prefix name requiring *bis, tris,* etc. whereas with *siloxy, di, tri,* etc. can be used. Names of other silicon-containing substituting groups connected through oxygen should not be shortened in this manner although such abbreviations were permitted by the 1949 IUPAC Rules.

The officially approved name for $Si(OH)_4$ is orthosilicic acid, but in common practice the prefix *ortho* is often omitted. Since the 1970 IUPAC Inorganic Rules do not recognize the term silicic acid, retention of the prefix is recommended here. Structures in which halogen or nitrogen may be regarded as having replaced one or more OH groups in orthosilicic acid are named as silane derivatives—*e.g.,* chlorosilanetriol for $ClSi(OH)_3$, *not* "chlorosilicic" acid.

## Table 38.1.  Examples of Acceptable Usage

1.
$$\begin{array}{cc} CH_3 & SiH_3 \\ | & | \\ H_3SiSiHSiH_2 & SiHSiH_3 \end{array}$$
2-Methyl-4-silylpentasilane

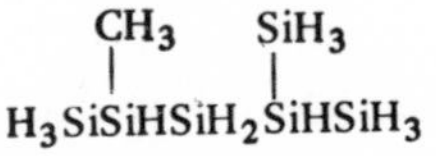

2.
1,3,5-Trichlorocyclohexasilane

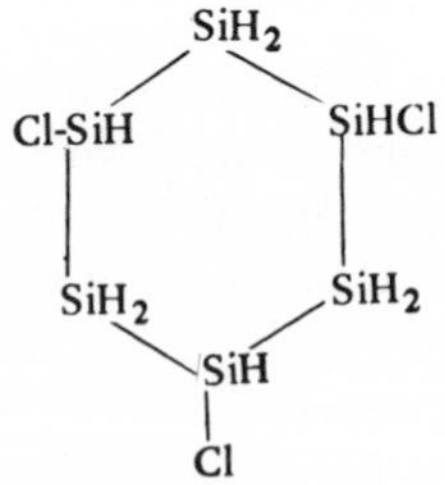

3.
$$\begin{array}{c} H_3SiOSiHOSiH_3 \\ | \\ OCH_3 \end{array}$$
3-Methoxytrisiloxane

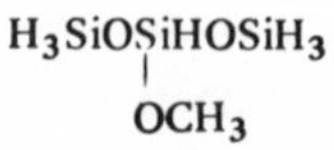

4.
$$\begin{array}{c} H_3SiSSiH_2SSiHSSiH_3 \\ | \\ OSiH_3 \end{array}$$
3-Siloxytetrasilathiane

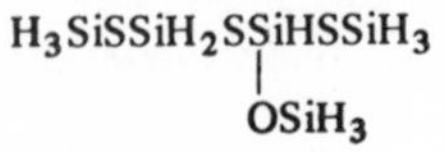

5.　$H_3SiSiH_2OSiH_2SiH_3$　1,3-Disilyldisiloxane
Bis(disilanyl) ether

6. 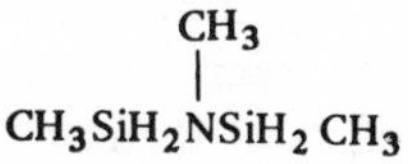
$$CH_3$$
$$|$$
$$CH_3SiH_2NSiH_2CH_3$$
1,2,3-Trimethyldisilazane

7. 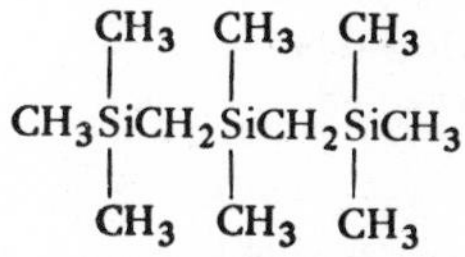
$$CH_3 \quad CH_3 \quad CH_3$$
$$| \qquad | \qquad |$$
$$CH_3SiCH_2SiCH_2SiCH_3$$
$$| \qquad | \qquad |$$
$$CH_3 \quad CH_3 \quad CH_3$$
2,2,4,4,6,6-Hexamethyl-2,4,6-
trisilaheptane

8. 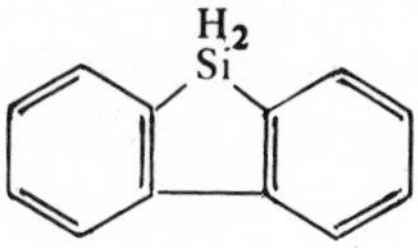
9H-9-Silafluorene

9. 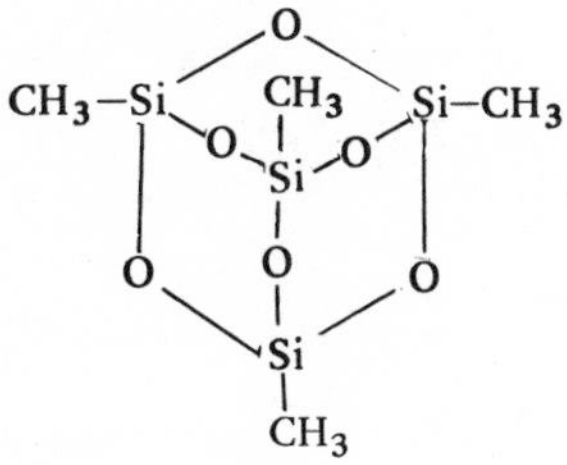
Tetramethyltricyclo[$3.3.1.1^{3,7}$]-
tetrasiloxane

10. 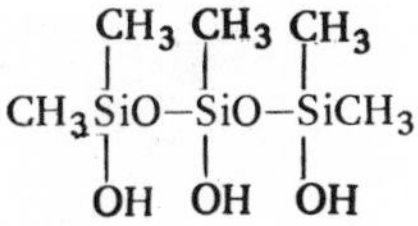
$$CH_3 \quad CH_3 \quad CH_3$$
$$| \qquad | \qquad |$$
$$CH_3SiO-SiO-SiCH_3$$
$$| \qquad | \qquad |$$
$$OH \quad OH \quad OH$$
1,1,3,5,5-Pentamethyl-1,3,5-trisiloxane-
triol

11.　$H_3SiN(CH_3)_2$　*N,N*-Dimethylsilanamine

12.　$C_6H_5NHSiH_2SiH_2NHC_6H_5$　*N,N'*-Diphenyl-1,2-disilanediamine

13.　$(CH_3CH_2CH_2)_2Si(COOH)_2$　Dipropylsilanedicarboxylic acid

14.　$(ClCH_2CH_2O)_3SiCl$　Chlorotris(2-chloroethoxy)silane

15.　$(C_6H_5CH_2O)_4Si$　Tetrakis(phenylmethyl) orthosilicate
Tetrabenzyl orthosilicate

16.　$CH_3CH_2CH_2CH_2OSi(OH)_3$　Butyl trihydrogen orthosilicate

17.  $(CH_3CH{=}CHCOO)_3SiNH_2$        Aminosilanetriyl tri-2-butenoate
                                         Tris(2-butenoyloxy)silanamine

18.  $(ClCH_2COO)_4Si$                  Chloroethanoic orthosilicic tetraanhydride
                                         Chloroacetic orthosilicic tetraanhydride

19.

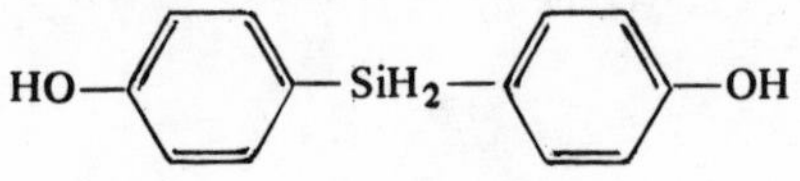

                                 4,4'-Silanediyldibenzenol
                                 4,4'-Silylenediphenol

20.  $(C_6H_5)_3SiLi$                  (Triphenylsilyl)lithium

21.  $(CH_3CH_2CH_2)_3SiSNa$          Sodium tripropylsilanethiolate

## Literature Cited

1.  "Silicon Compounds", *Chem. Eng. News* (1952) **30**, 4517.
2.  "Nomenclature of Inorganic Chemistry. Definitive Rules," IUPAC, 1970, Butterworths, London, 1971; *Pure Appl. Chem.* (1971) **28**, 1-110.

# Sulfides, Sulfoxides, and Sulfones

In this chapter, organic sulfur compounds corresponding to the structures RSR′, RS(O)R′, and $RSO_2R′$, where R and R′ are the same or different, are discussed. These three functional classes, which are known as sulfides, sulfoxides, and sulfones, respectively, also include cyclic systems in which one or more sulfur atoms are members of a ring. However, compounds of the latter type are usually named as heterocycles (*see* Chapter 6).

Sulfides are the sulfur analogs of ethers and are best named similarly to oxygen compounds of corresponding structure (*see* Chapter 26). The same kind of nomenclature is applicable to sulfoxides and sulfones.

Also considered here are disulfides, which have the structure RSSR, and **polysulfides** having more than two sulfur atoms arranged in a straight chain. (The term polysulfide also denotes compounds containing more than one monosulfide group.)

**Thioacetals**, a subclass of sulfides derived from aldehydes and ketones, are treated in Chapter 20. Compounds in which a sulfur atom carries a **positive charge** are discussed in Chapter 24.

*Recommended Nomenclature Practice*

**Sulfides.** **Acyclic sulfides** containing no more than two —S— linkages within a single straight chain otherwise composed of carbon atoms are preferably named substitutively as derivatives of one of the all-carbon chains present in the molecule. Principles to be followed in selecting the parent chain are given in Chapter 26. Names for SR groups in which R is cyclic are formed by combining the prefix name of the R group with *thio*.

| | |
|---|---|
| $CH_3SCH_2CH_2CH_3$ | 1-(Methylthio)propane |
| $CH_3CH_2SCH_2CH_2SCH_2CH_3$ | 1,2-Bis(ethylthio)ethane |
| $\overset{\displaystyle SC_6H_5}{\underset{\displaystyle C_6H_5SCH_2CHCH_2SC_6H_5}{\mid}}$ | 1,2,3-Tris(phenylthio)propane |

Cyclic sulfides having one or more —S— linkages connecting two ring systems are preferably named substitutively as derivatives of the senior ring system carrying the most SR groups. Criteria for use in selecting the senior ring system are given in Chapter 26. Names of SR groups are formed as described above for acyclic groups.

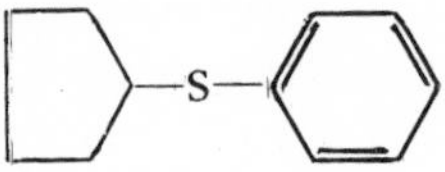

(Cyclopentylthio)benzene

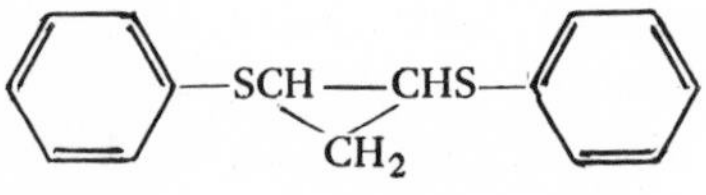

1,2-Bis(phenylthio)cyclopropane

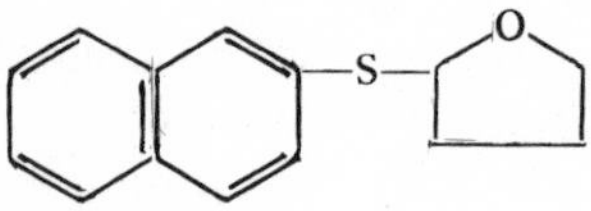

2-(2-Naphthylthio)oxacyclopentane

For both acyclic and cyclic **symmetrical monosulfides**, radicofunctional names ending with the word *sulfide* are acceptable alternatives to the preferred substitutive names.

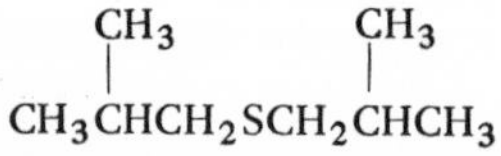

1-(Isobutylthio)-2-methylpropane
Diisobutyl sulfide

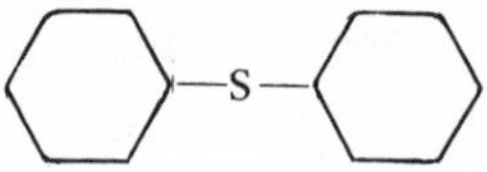

(Cyclohexylthio)cyclohexane
Dicyclohexyl sulfide

**Cyclic-acyclic monosulfides** are preferably named substitutively as derivatives of the senior chain or ring present. When there is a choice, all heterocyclic systems are preferred to all cyclic and acyclic hydrocarbon systems; otherwise, the largest hydrocarbon is selected as the parent.

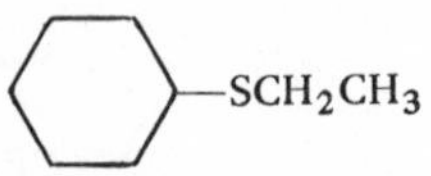

(Ethylthio)cyclohexane

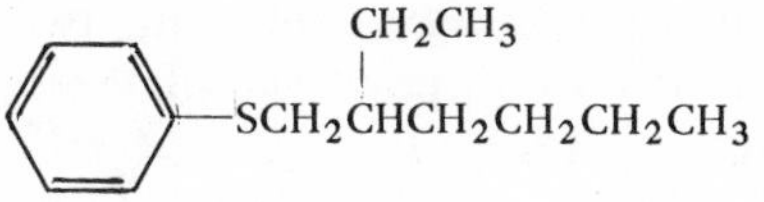

3-[(Phenylthio)methyl] heptane

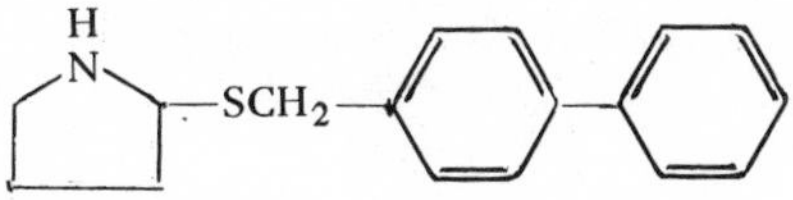

2-[(4-biphenylylmethyl)thio] -
azacyclopentane

**Symmetrical** structures may also be named radicofunctionally.

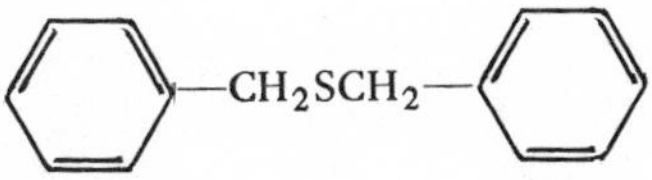

[[(Phenylmethyl)thio] methyl] -
benzene
Dibenzyl sulfide

**Cyclic-acyclic polysulfides** are treated similarly to the monosulfides, except that in selecting the parent chain or ring system consideration is first given to the number of attached SR groups.

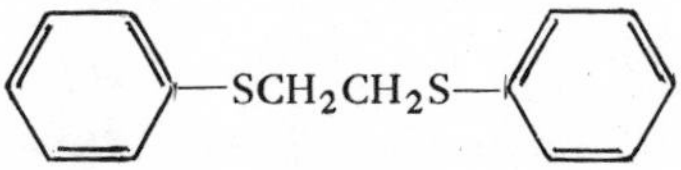

1,2-Bis(phenylthio)ethane

When more than two —S— linkages occur within a single straight chain, replacement nomenclature is applicable (*see* Chapter 7).

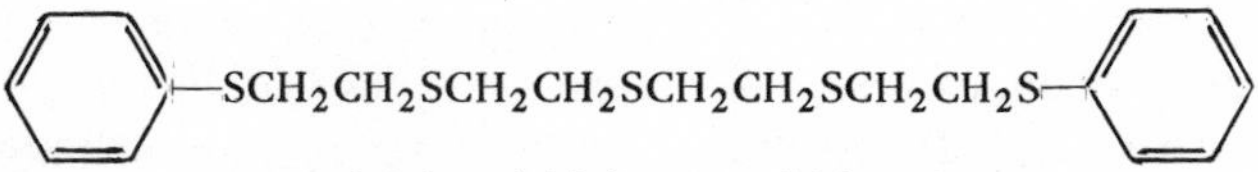

1,11-Bis(phenylthio)-3,6,9-trithiaundecane

Structures that contain one or more —S— linkages in a ring system are preferably named as heterocyclic compounds (*see* Chapter 6).

1,4-Dithiacyclohexane

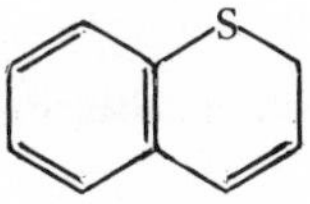

2*H*-1-Thianaphthalene

As an alternative to regularly formed heterocycle names for structures in which S is attached to two adjacent carbon atoms of a chain or ring, the substitutive prefix *epithio* may be used.

CH<sub>3</sub>CHCHOH
\ /
S

           3-Methylthiacyclopropan-2-ol
           1,2-Epithio-1-propanol

The term *epithio* is also used as a nondetachable prefix (*see* Chapter 6, Discussion section) in naming bridged polycyclic aromatic structures.

           1,4-Dihydro-1,4-epithionaphthalene

    The prefix name *thio*, in addition to its meaning as used above, denotes the bivalent group —S— in names of symmetrical functional compounds (*see* Chapter 11).

$$\text{HOOCCH}_2\text{CH}_2\text{SCH}_2\text{CH}_2\text{COOH} \qquad \text{3,3'-Thiodipropanoic acid}$$

    **Polysulfides.** Compounds containing the groups —SS—, —SSS—, etc. (or —$S_2$—, —$S_3$—, etc.) are preferably named substitutively by applying the principles given above for sulfides and using *dithio, trithio,* etc. in place of *thio*. In heterocycle names, *epidithio, epitrithio,* etc. replace *epithio,* and in radicofunctional names *disulfide, trisulfide,* etc. replace *sulfide.* Examples are provided in the list at the end of the chapter.

    In cases where it is desired to emphasize or specify the presence of a **straight chain** of sulfur atoms, polysulfides may be named as derivatives of disulfane (HSSH), trisulfane (HSSSH), tetrasulfane (HSSSSH), etc.

$$\text{CH}_3\text{SSSCH}_2\text{CH}_2\text{CH}_3 \qquad\qquad \text{Methylpropyltrisulfane}$$

    **Sulfoxides and Sulfones.**    With a few exceptions, as described in later paragraphs, **sulfoxides** and **sulfones** are named according to the principles given above for sulfides. Corresponding to the prefix *thio* are *sulfinyl* and *sulfonyl,* respectively.

$$\text{CH}_3\text{S(O)CH}_2\text{CH}_3 \qquad\qquad \text{(Methylsulfinyl)ethane}$$

$$\text{—SO}_2\text{CH}_2\text{CH}_3 \qquad\qquad \text{(Ethylsulfonyl)benzene}$$

In radicofunctional names when used as acceptable alternatives to substitutive names for **symmetrical** structures, the terminal words *sulfoxide* and *sulfone* are used.

$ClCH_2CH_2S(O)CH_2CH_2Cl$

1-Chloro-2-(2-chloroethylsulfinyl)ethane
Bis(2-chloroethyl) sulfoxide

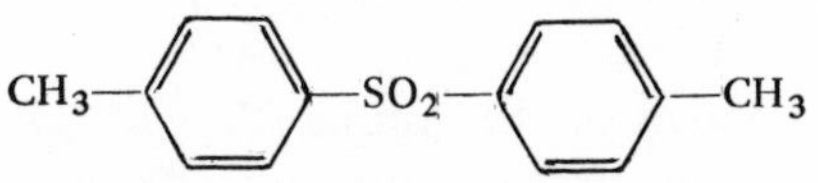

$CH_3$—⟨ ⟩—$SO_2$—⟨ ⟩—$CH_3$

1-Methyl-4-(4-methylphenylsulfonyl)-benzene
Di-*p*-tolyl sulfone

In names of symmetrical functional compounds (*see* Chapter 11) the prefixes *sulfinyl* and *sulfonyl* are used to denote the bivalent groups —S(O)— and —$SO_2$—.

$HOCH_2CH_2SO_2CH_2CH_2OH$           2,2'-Sulfonylbisethanol

In replacement nomenclature there are no prefix names for the sulfoxide and sulfone groups corresponding to *thia* for the sulfide group. However, replacement names may be formed by adding the word *oxide* or *dioxide* (with appropriate locants) to the name of the corresponding sulfide structure.

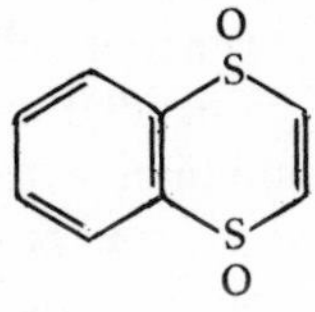

1,4-Dithianaphthalene 1,4-dioxide

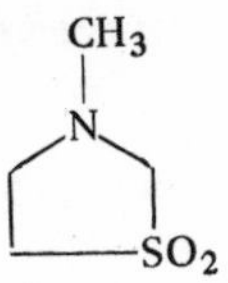

3-Methyl-1-thia-3-azacyclopentane 1,1-dioxide

$CH_3SCH_2CH_2SO_2CH_2CH_2SCH_3$

2,5,8-Trithianonane 5,5-dioxide
Bis[2-(methylthio)ethyl] sulfone

## Discussion

Although the class name "thioether" should not be used in place of *sulfide* in radicofunctional names, it is properly formed and unambiguous when used (usually in the plural) to refer to organic sulfides as a class. For the latter purpose, strictly speaking, the term thioethers is more specific than the term sulfides, which also denotes inorganic salts containing anionic sulfur.

Like the ether function, the functional groups sulfide, sulfoxide, and sulfone are not considered in selecting the principal function when naming compounds of mixed function substitutively. This principle, along with others that also apply to the sulfur analogs of ethers, is discussed in Chapter 26.

As is pointed out in Chapter 18, names of the groups RS—, RS(O)—, and RSO$_2$— take different forms in substitutive and radicofunctional nomenclature. In substitutive names the R portion is simply cited as a substituting group—e.g., phenylthio, phenylsulfinyl, and phenylsulfonyl—while in radicofunctional nomenclature the R portion is named as a hydrocarbon or heterocycle to produce an acyl-type name—e.g., benzene-sulfenyl, benzenesulfinyl, and benzenesulfonyl.

The prefix *thio*, which properly denotes a bivalent sulfur atom named as a substituting group, should no longer be used in names of specific compounds to indicate replacement of oxygen by sulfur (*see* Chapter 9, Discussion section), except for some very common trivial names such as thioacetone, thioacetic acid, and a few others.

The nomenclature of nitrogen analogs of sulfoxides and sulfones having the structures RS(NH)R' and RS(O)(NH)R' has been clarified in the 1965 IUPAC Rules wherein substitutive names based on the hypothetical parent compound H$_2$S(NH), designated sulfimide, and H$_2$S(O)(NH), designated sulfoximide, are recommended. Although the Rules do not so state, the parent name sulfonodiimide for H$_2$S(NH)$_2$ would follow logically. In the past *Chemical Abstracts* has indexed compounds of the type R$_2$S(NR') under the heading Sulfilimine (*see* Table 39.1, Example 14), but the IUPAC names are preferable because the ending *imine* regularly denotes doubly bonded nitrogen (*see* Chapter 23).

## Table 39.1. Examples of Acceptable Usage

| | | |
|---|---|---|
| 1. | CH$_3$CH$_2$SCH$_2$CH$_3$ | (Ethylthio)ethane<br>Diethyl sulfide<br>*not* Ethyl sulfide |
| 2. | CH$_3$S(O)CH$_2$CH$_2$Cl | 1-Chloro-2-(methylsulfinyl)ethane |
| 3. | CH$_3$SO$_2$CH$_2$CH$_2$CH$_2$SO$_2$CH$_3$ | 1,3-Bis(methylsulfonyl)propane |
| 4. | CH$_3$SCH$_2$SCH$_2$SCH$_2$SCH$_3$ | 2,4,6,8-Tetrathianonane |

5.

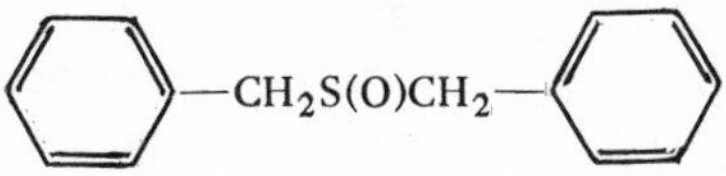

(Phenylmethylsulfinylmethyl)benzene
Dibenzyl sulfoxide

$-CH_2S(O)CH_2-$

6.

$CH_2CHCH_2S-$

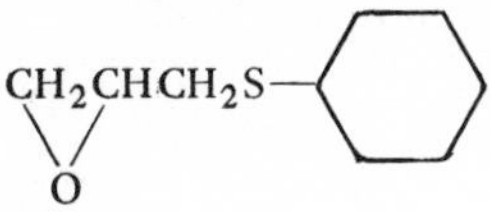

[(Cyclohexylthio)methyl] oxacyclopropane

O

7.

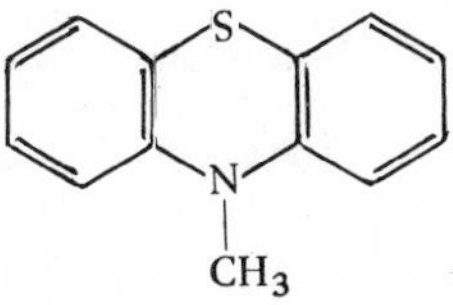

10-Methyl-10$H$-9-thia-10-azaanthracene
10-Methylphenothiazine

$CH_3$

8.

HOOC—⟨ ⟩—$SO_2$—⟨ ⟩—COOH

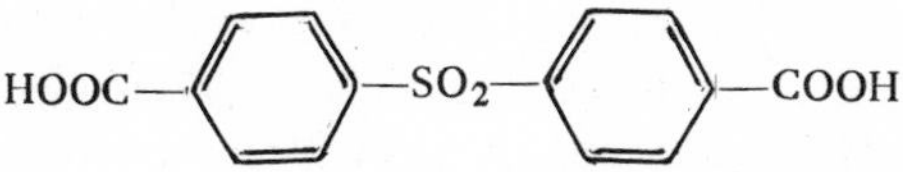

4,4′-Sulfonyldibenzenecarboxylic acid
4,4′-Sulfonyldibenzoic acid

9.   $CH_3CH_2SSCH_2CH_3$

(Ethyldithio)ethane
Diethyl disulfide
Diethyldisulfane

10.

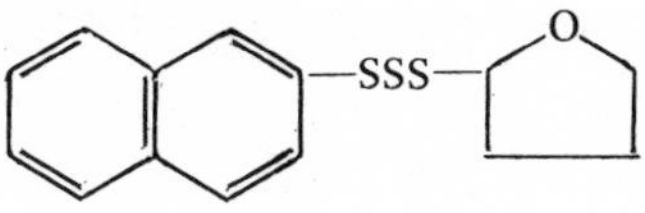

—SSS—

2-(2-Naphthyltrithio)oxacyclopentane
Tetrahydro-2-(2-naphthyltrithio)furan
2-Naphthyloxacyclopent-2-yltrisulfane

11.   $CH_3—S_4—CH_2CH_2CH_3$

1-(Methyltetrathio)propane

12.

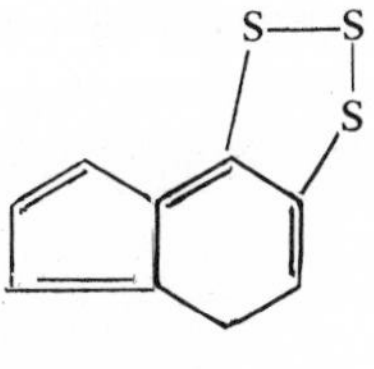

S—S

S

5$H$-1,2,3-Trithia-$as$-indacene
6,7-Epitrithio-4$H$-indene

13.

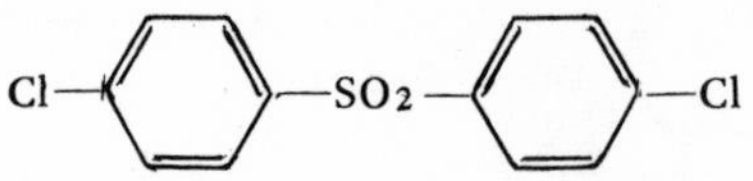

1-Chloro-4-(4-chlorophenylsulfonyl)benzene
Bis(4-chlorophenyl) sulfone

14.

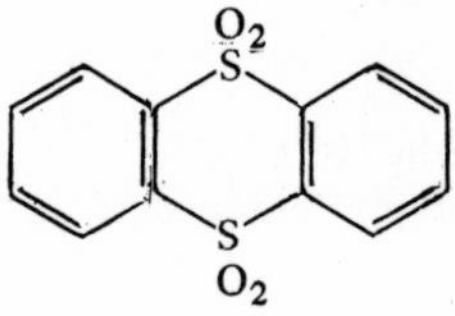

9,10-Dithiaanthracene 9,9,10,10-tetraoxide

15.          CH$_3$
             |
      CH$_3$CH$_2$S(NSO$_2$C$_6$H$_5$)

*S*-Ethyl-*S*-methyl-*N*-(phenylsulfonyl)sulfimide
*S*-Ethyl-*S*-methyl-*N*-(phenylsulfonyl)-
    sulfilimine

# Sulfur Compounds (Miscellaneous)

O ther chapters have been devoted to the nomenclature of organosulfur acids and their functional derivatives (Chapter 18) and the most common nonacidic carbon-sulfur compounds (Chapter 39). In the present chapter most of the remaining classes of **organic sulfur compounds** are considered. The majority of these are functional derivatives of inorganic sulfur acids and oxides wherein the central sulfur atom is not directly attached to carbon.

A few of the organic derivatives treated in this chapter are named on the basis of parent inorganic structures that have been assigned names not officially recognized in the 1970 IUPAC Inorganic Rules. This practice does not imply a recommendation that these parent names be used generally for the inorganic compounds themselves.

Organic thiocyano compounds (RSCN) and isothiocyano compounds (RNCS) are discussed in Chapter 32.

*Recommended Nomenclature Practice*

**Esters of inorganic sulfur acids** are named in the same way as esters of organosulfur acids and carboxylic acids. The first word of the name is that of a cation (other than hydrogen) if one is present; this is followed first by the name of the esterifying group (using the locants *O—* or *S—* when necessary to avoid ambiguity) and then by the name of the anion of the parent acid. If more than one cation or esterifying group is present, their names are cited in alphabetical order. In names of partial esters the word hydrogen immediately precedes the anion name. As with other types of esters and salts, each component of the name is a separate word.

| | |
|---|---|
| $CH_3OSO_2OCH_3$ | Dimethyl sulfate |
| $C_6H_5SS(O)ONa$ | Sodium *S*-phenyl thiosulfite |
| $CH_3CH_2OSOH$ | Ethyl hydrogen sulfoxylate |
| $CH_3CH_2CH_2OSO_2SO_2OCH_3$ | Methyl propyl dithionate |

The sulfur amido acids $H_2NS(O)OH$ and $H_2NSO_2OH$ are preferably given the parent names **sulfinamidic acid** and **sulfamidic acid**, respectively, in forming names of their organic derivatives. Names of substituting groups (except *amino*) attached to the nitrogen atom are cited as prefixes.

| | |
|---|---|
| $(CH_3CH_2)_2NS(O)OH$ | Diethylsulfinamidic acid |
| $C_6H_5NHSO_2OCH_3$ | Methyl phenylsulfamidate |
| $HONHSO_2OCH_2CH_2CH_3$ | Propyl hydroxysulfamidate |

*N*-Amino derivatives are preferably named analogously to the organosulfur amides (*see* Chapter 18) by attaching the appropriate functional suffix to the name of the nitrogen chain system.

| | |
|---|---|
| $H_2NNHS(O)OCH_2CH_3$ | Ethyl diazanesulfinate |

**Organic amides of inorganic sulfur acids** are best named by using regularly formed inorganic names for the parent compounds. The locants *N* and *N'* are used when ambiguity would otherwise result.

| | |
|---|---|
| $CH_3NHSNH_2$ | *N*-Methylsulfoxylic diamide |
| $C_6H_5NHSO_2NHCH_2CH_3$ | *N*-Ethyl-*N'*-phenylsulfuric diamide |
| $CH_3NHS(O)NHOH$ | *N*-Hydroxy-*N'*-methylsulfurous diamide |

As recommended above for acids and esters, *N*-amino derivatives are best treated as nitrogen chains carrying functional groups (*see* Chapter 32).

| | |
|---|---|
| $C_6H_5NHNHSO_2NHCH_3$ | *N*-Methyl-2-phenyldiazanesulfonamide |

Organic derivatives of HN=S(O) and HN=SO$_2$ are preferably named as substituted inorganic imides.

| | |
|---|---|
| $CH_3CH_2CH_2N{=}S(O)$ | *N*-Propylsulfinyl imide |
| $C_6H_5N{=}SO_2$ | *N*-Phenylsulfonyl imide |

Compounds having the generic structure $R_mSX_n$, where X is halogen or some other "negative" group and $m + n$ is greater than 2, are named as organosulfur compounds The organic substituents, R, are cited as prefixes attached to the word *sulfur;* this is followed by a space and the names of the X groups in anionic form.

| | |
|---|---|
| $C_6H_5SCl_3$ | Phenylsulfur trichloride |
| $(CH_3)_5SCN$ | Pentamethylsulfur cyanide |

## Discussion

Most of the compounds considered in this chapter are relatively uncommon and represent classes properly regarded as organic derivatives of inorganic parent structures; they have often been dealt with and named by inorganic principles, and few attempts have been made to systematize their nomenclature. Organic chemists have usually been content to apply organic naming methods to the carbon-containing portion of the molecule, leaving the traditional inorganic parent name intact. In some cases, however, officially recognized inorganic names for parent compounds are not available, usually because these compounds are not known in the free state; here the best solution is to coin a name by analogy or extrapolation. For example, either sulfinamidic acid or sulfinamic acid is a logical name for $H_2NS(O)OH$, since both sulfamidic acid and sulfamic acid have been officially approved for $H_2NSO_2OH$. Such factors have been taken into consideration in developing the recommendations presented here.

In naming neutral esters of dibasic inorganic sulfur acids, the multiplying prefix *di* should not be omitted as in "methyl sulfate" for $CH_3OSO_2OCH_3$. Although names of this type have been used in the past by *Chemical Abstracts* as Subject Index entries for a number of simple esters of common acids, they are not recommended for general use. Both the 1970 IUPAC Inorganic Rules and the 1965 IUPAC Organic Rules specify dimethyl sulfate as the name of the neutral ester; for the partial ester, $CH_3OSO_2OH$, the respective recommendations differ only in the form of name—*i.e.*, methyl hydrogensulfate (inorganic version) and methyl hydrogen sulfate (organic version).

Replacement of OH in inorganic polybasic acids by other functional groups such as Cl or $NH_2$ leads to structures that have been named inconsistently. By inorganic principles the compound $ClSO_2OH$ is called chlorosulfuric acid. Organic chemists, applying the practice of substitution for H, have favored "chlorosulfonic acid," thus implying the parent name "sulfonic acid" for $HSO_2OH$, the hypothetical tautomer of sulfurous acid. However, *sulfonic acid* has become firmly established in organic nomenclature as a functional suffix denoting the group $-SO_2OH$, as in benzenesulfonic acid. The analogous case involving replacement of OH by $NH_2$ is even more complicated. While "aminosulfonic acid" for $H_2NSO_2-OH$ has been avoided by inorganic and organic chemists alike, both amidosulfuric acid and sulfamidic acid are recognized in the 1970 IUPAC Inorganic Rules, and sulfamic acid is added as an acceptable alternative in the 1965 IUPAC Organic Rules. Under these circumstances, and taking note of a general trend toward wider acceptance of *-amic acid* nomenclature, the present recommendation is to use approved inorganic names as far as possible.

The compound $HONHSO_2OH$ has often been assigned the parent name hydroxylamine-$N$-sulfonic acid in naming its organic derivatives. The more systematic name $N$-hydroxysulfamidic acid is much preferred.

Treatment of symmetrically substituted sulfur amides as substituted amines—e.g., $N,N'$-sulfonylbis(methylamine) for $CH_3NHSO_2NHCH_3$—is recognized as an acceptable alternative in the 1965 IUPAC Organic Rules, as is the contracted form sulfamide for the parent compound $H_2NSO_2$-$NH_2$. However, use of the preferred inorganic amide names is recommended here. A similar approach is best taken with inorganic imide derivatives such as $RN=SO_2$, although here again past organic practice has favored the sulfonylamine type of name.

The names sulfinyl and sulfonyl are now officially preferred to the older "thionyl" and "sulfuryl" for the groups $S(O)$ and $SO_2$. The latter forms should not be used in naming organic compounds.

### Table 40.1. Examples of Acceptable Usage

| | | |
|---|---|---|
| 1. | $CH_3CH_2OSO_2OH$ | Ethyl hydrogen sulfate |
| 2. | $ClCH_2CH_2OS(O)Cl$ | 2-Chloroethyl chlorosulfite |
| 3. | $CH_3CH_2CH_2SS(O)(S)OCH_3$ | $O$-Methyl $S$-propyl dithiosulfate |
| 4. | $C_6H_5OSO_2OSO_2OC_6H_5$ | Diphenyl disulfate |
| 5. | $CH_3OS(O)S(O)OCH_3$ | Dimethyl dithionite |
| 6. | $C_6H_{11}NHSO_2OH$ | Cyclohexylsulfamidic acid<br>Cyclohexylsulfamic acid |
| 7. | $(C_6H_5)_2NS(O)Cl$ | Diphenylsulfinamidoyl chloride |
| 8. | $C_6H_5NHSO_2NHC_6H_5$ | $N,N'$-Diphenylsulfuric diamide<br>$N,N'$-Diphenylsulfamide |
| 9. | $(CH_3)_2NNHSO_2OCH_3$ | Methyl 2,2-dimethyldiazanesulfonate<br>Methyl 2,2-dimethylhydrazinesulfonate |
| 10. | $CH_3CH_2OSO_2NHSO_2OCH_3$ | Ethyl methyl imidodisulfate |
| 11. | $\overset{\displaystyle CH_3}{\underset{\displaystyle C_6H_5NHSO_2NSO_2NHC_6H_5}{\vert}}$ | $N,N'$-Diphenyl(methylimido)disulfuric diamide |
| 12. | $C_6H_5CH_2N=SO_2$ | $N$-(Phenylmethyl)sulfonyl imide<br>$N$-Benzylsulfonyl imide |
| 13. | $F_3CSF_5$ | (Trifluoromethyl)sulfur pentafluoride |
| 14. | $C_6H_5N=SF_2$ | (Phenylimino)sulfur difluoride |
| 15. | $(CH_3CO)_2SI_4$ | Diethanoylsulfur tetraiodide<br>Diacetylsulfur tetraiodide |

# Appendix A

*Generic Group Names*

Given below are the principal simple and selected composite generic group names, together with their structure types and the chapters in which they or closely related groups are discussed. In the structures, R denotes an aliphatic hydrocarbon residue and Ar denotes an aromatic ring. Most of the composite names beginning with alk-, alkyl-, and alkan(e)- have aromatic analogs beginning with ar-, aryl-, and aren(e)-, respectively.

| Group | Structure | Chapter |
|---|---|---|
| Alkyl | R— | 2 |
| Alkylene | —R— | 2 |
| Alkanediyl | R= or R< | 2 |
| Alkanetriyl | R≡ or —R< | 2 |
| Alkylidene | RCH= or RCH< | 2 |
| Alkylidyne | RC≡ or RC< | 2 |
| | | |
| Aryl | Ar— | 3 |
| Arenyl | Ar— | 4 |
| Arylene | —Ar— | 3,4 |
| Arenylene | —Ar— | 3,4 |
| Arenetriyl | Ar≡ | 3,4 |
| Biarylyl | Ar—Ar— | 5 |
| Alkylaryl | R—Ar— | 3 |
| Arylalkyl | Ar—R— | 3,4 |
| | | |
| Alkanoyl | RCO— | 17 |
| Alkylcarbonyl | RCO— | 17 |
| Aroyl | ArCO— | 17 |
| Alkanedioyl | —CORCO— | 17 |
| Alkanethioyl | RCS— | 17 |
| Alkanimidoyl | RC(:NH)— | 17 |
| Alkoxycarbonyl | ROCO— | 17 |
| | | |
| Alkoxy | RO— | 26 |
| Alkyloxy | RO— | 26 |
| Alkyldioxy | ROO— | 34 |
| Alkylenedioxy | —ORO— | 26 |
| Oxydialkylene | —ROR— | 26 |
| | | |
| Alkylamino | RNH— | 23 |
| Alkylimino | RN= or RN< | 23 |
| | | |
| Alkanamido | RCONH— | 21 |
| Alkylcarboxamido | RCONH— | 21 |
| Alkanoylamino | RCONH— | 21 |
| Alkanoylimino | RCON= or RCON< | 21 |

311

| Alkylthio | RS— | 18,39 |
| Alkanesulfenyl | RS— | 18,39 |
| Alkylsulfinyl | RSO— | 18,39 |
| Alkanesulfinyl | RSO— | 18,39 |
| Alkylsulfonyl | RSO$_2$— | 18,39 |
| Alkanesulfonyl | RSO$_2$— | 18,39 |
| Alkanesulfonamido | RSO$_2$NH— | 18 |

# Appendix B

## *Group Name Index*

This list includes most of the group names cited in the text, plus some related names, with the following exceptions: (1) only a few of the most common trivial names have been listed; (2) no heterocyclic group names have been cited (*see* Chapter 6); and (3) other than a few prototypes, names that are denoted by the generic names cited in Appendix A have been omitted. Examples of the latter category are such names as Toluenesulfonyl (*see* Alkanesulfonyl, Chapter 18) or Butylidene (*see* Alkylidene, Chapter 2). Fairly exhaustive lists of group names may be found in the introductions to the subject indexes of *Chemical Abstracts*, particularly that of Vol. 66, and in the IUPAC publications.

| Group Name | Structure | Chapter |
|---|---|---|
| Acetamido | $CH_3CONH-$ | 21 |
| Acetimidoyl | $CH_3C(:NH)-$ | 16 |
| Acetyl | $CH_3CO-$ | 16 |
| Acetylamino | $CH_3CONH-$ | 21 |
| Acetylimino | $CH_3CON=$ or $CH_3CON{<}$ | 21 |
| Acryloyl | $CH_2=CHCO-$ | 16 |
| Adipoyl | $-CO(CH_2)_4CO-$ | 11,16 |
| Allyl | $CH_2=CHCH_2-$ | 2 |
| Amidino | $H_2NC(:NH)-$ | 22 |
| Amino | $H_2N-$ | 23 |
| Aminomethanamido | $H_2NCONH-$ | 21 |
| Aminooxy | $H_2NO-$ | 32 |
| Aminothio | $H_2NS-$ | 32 |
| Ammonio | $H_3N^+-$ (ion) | 24 |
| Arsino | $H_2As-$ | 33 |
| Azanyl | $H_2N-$ | 32 |
| Azido | $N_3-$ | 27 |
| Azino | $=N-N=$ or $>N-N{<}$ | 11,32 |
| Azo | $-N=N-$ | 11,32 |
| Azyl | $H_2N-$ | 32 |
| Azyloxy | $H_2NO-$ | 32 |
| Azylthio | $H_2NS-$ | 32 |
| | | |
| Benzamido | $C_6H_5CONH-$ | 21 |
| Benzenedicarbonyl | $-COC_6H_4CO-$ | 17 |
| Benzhydryl | $(C_6H_5)_2CH-$ | 3 |
| Benzoyl | $C_6H_5CO-$ | 16 |
| Benzoylamino | $C_6H_5CONH-$ | 21 |
| Benzoylimino | $C_6H_5CON=$ or $C_6H_5CON{<}$ | 21 |
| Benzoyloxy | $C_6H_5COO-$ | 16 |
| Benzyl | $C_6H_5CH_2-$ | 3 |
| Benzylidene | $C_6H_5CH=$ | 3 |

| | | |
|---|---|---|
| Benzylidyne | $C_6H_5C\equiv$ | 3 |
| Biphenylyl | $C_6H_5-C_6H_4-$ | 5 |
| Bismuthino | $H_2Bi-$ | 33 |
| Boryl | $H_2B-$ | 25 |
| Borylene | $HB<$ | 25 |
| Borylidyne | $B\lessgtr$ | 25 |
| Bromo | $Br-$ | 27 |
| 2-Butenyl | $CH_3CH=CHCH_2-$ | 2 |
| 2-Butenylene | $-CH_2CH=CHCH_2-$ | 2,11 |
| | | |
| Carbamimidoyl | $H_2NC(:NH)-$ | 22 |
| Carbamimidoylamino | $H_2NC(:NH)NH-$ | 22 |
| Carbamoyl | $H_2NCO-$ | 21 |
| Carbamoylimino | $H_2NCON=$ or $H_2NCON<$ | 21 |
| Carbonyl | $OC=$ or $OC<$ | 11,20 |
| Carbonylamino | $OC=N-$ | 32 |
| Carbonyldioxy | $-OCOO-$ | 11 |
| Carboxy | $HOOC-$ | 16 |
| Carbylamino | $CN-$ | 32 |
| Chloro | $Cl-$ | 27 |
| Chlorocarbonyl | $ClCO-$ | 17 |
| Chloroformyl | $ClCO-$ | 17 |
| Chloromethanoyl | $ClCO-$ | 17 |
| Chlorosyl | $OCl-$ | 8 |
| Chloryl | $O_2Cl-$ | 8 |
| Cyanato | $NCO-$ | 32 |
| Cyano | $NC-$ | 30 |
| Cyanoamino | $NCNH-$ | 32 |
| Cyanoimino | $NCN=$ or $NCN<$ | 32 |
| 1,2-Cyclobutanediylidene | $=\overline{CCH_2CH_2C}=$ | 2 |
| Cyclohexanecarboxamido | $C_6H_{11}CONH-$ | 21 |
| Cyclohexyl | $C_6H_{11}-$ | 2 |
| Cyclohexylcarbonyl | $C_6H_{11}CO-$ | 17 |
| | | |
| Diacetylamino | $(CH_3CO)_2N-$ | 21 |
| Diazanyl | $H_2NNH-$ | 32 |
| 1,1-Diazanediyl | $H_2NN-$ | 32 |
| 1,2-Diazanediyl | $-NHNH-$ | 11,32 |
| Diazenediyl | $-N=N-$ | 11,32 |
| Diazanetetrayl | $=NN=$ or $>NN<$ | 11,32 |
| Diazenyl | $HN=N-$ | 32 |
| Diazo | $N_2=$ | 32 |
| Diazoamino | $-NHN=N-$ | 32 |
| Diboran(4)yl | $H_3B_2-$ | 25 |
| Diboran(6)yl | $H_5B_2-$ | 25 |
| (Dihydroxyiodo) | $(HO)_2I-$ | 8 |
| Diimino | $-NHNH$ | 32 |
| Dioxy | $-OO-$ | 34 |
| Disilanyl | $H_3SiSiH_2-$ | 38 |
| Disiloxanyl | $H_3SiOSiH_2-$ | 38 |
| Dithio | $-SS-$ | 11 |
| Dithiocarboxy | $HSSC-$ | 16 |

| Epidioxy (as a bridge) | $-OO-$ | 34 |
|---|---|---|
| Epidithio (as a bridge) | $-SS-$ | 39 |
| Epimino (as a bridge) | $-NH-$ | 23 |
| Epithio (as a bridge) | $-S-$ | 39 |
| Epoxy (as a bridge) | $-O-$ | 26 |
| Epoxythio (as a bridge) | $-OS-$ | 39 |
| Ethanediylidene | $=CHCH=$ | 2,11 |
| Ethanetetrayl | $=CHCH=$ or $>CHCH<$ | 2 |
| Ethanethioyl | $CH_3CS-$ | 16 |
| Ethanimidoyl | $CH_3C(:NH)-$ | 16 |
| Ethanoyl | $CH_3CO-$ | 16 |
| Ethanoylimino | $CH_3CON=$ or $CH_3CON<$ | 21 |
| 1-Ethanyl-2-ylidene | $-CH_2CH=$ | 2 |
| Ethenyl | $CH_2=CH-$ | 2 |
| Ethenylene | $-CH=CH-$ | 2,11 |
| Ethyl | $CH_3CH_2-$ | 2 |
| Ethylene | $-CH_2CH_2-$ | 2,11 |
| Ethylenedioxy | $-OCH_2CH_2O-$ | 2,11 |
| Ethylidene | $CH_3CH=$ | 2,11 |
| Ethylidyne | $CH_3C\equiv$ or $CH_3C\leqq$ | 2 |
| Ethynyl | $HC\equiv C-$ | 2 |
| Ethynylene | $-C\equiv C-$ | 2,11 |
| | | |
| Fluoro | $F-$ | 27 |
| Formamido | $HCONH-$ | 21 |
| Formimidoyl | $HC(:NH)-$ | 16 |
| Formyl | $HCO-$ | 16 |
| Formyloxy | $HCOO-$ | 16 |
| | | |
| Germyl | $H_3Ge-$ | 33 |
| Guanidino | $H_2NC(:NH)NH-$ | 22 |
| Guanyl | $H_2NC(:NH)-$ | 22 |
| | | |
| Hydrazino | $H_2NNH-$ | 32 |
| Hydrazo | $-NHNH-$ | 11,32 |
| Hydroperoxy | $HOO-$ | 34 |
| Hydroxy | $HO-$ | 19,35 |
| Hydroxyamino | $HONH-$ | 32 |
| Hydroxyazyl | $HONH-$ | 32 |
| Hydroxyazylene | $HON=$ or $HON<$ | 32 |
| Hydroxyimino | $HON=HON<$ | 32 |
| Hydroxy(thiocarbonyl) | $HOSC-$ | 16 |
| | | |
| Imidocarboxy | $HOC(:NH)-$ | 16 |
| Imino | $HN=$ or $HN<$ | 11,21,23 |
| Iminodicarbonyl | $-CONHCO-$ | 11,21 |
| Iodo | $I-$ | 27 |
| Iodosyl | $OI-$ | 27 |
| Iodoxy | $O_2I-$ | 27 |
| Isocyanato | $OCN-$ | 8,32 |
| Isocyano | $CN-$ | 8 |

| | | |
|---|---|---|
| Isopropenyl | $CH_2=C(CH_3)-$ | 2 |
| Isothiocyanato | SCN$-$ | 8,32 |
| | | |
| Mercapto | HS$-$ | 19,35 |
| Mercaptoamino | HSNH$-$ | 32 |
| Mercaptoazyl | HSNH$-$ | 32 |
| Mercaptocarbonyl | HSOC$-$ | 16 |
| Methanetetrayldinitrilo | $-N=C=N-$ | 11 |
| Methanethioyl | HCS$-$ | 20 |
| Methanimidoyl | HC(:NH)$-$ | 17 |
| Methanoyl | HCO$-$ | 16,20 |
| Methyl | $H_3C-$ | 2,11 |
| Methylene | $H_2C=$ or $H_2C<$ | 2,11 |
| Methylidyne | $HC\equiv$ or $HC\leqq$ | 2,11 |
| | | |
| Nitrilo | $N\equiv$ or $-N<$ | 11,23,30 |
| Nitro | $O_2N-$ | 27 |
| *aci*-Nitro | HO(O)N= or HO(O)N$<$ | 8 |
| Nitroso | ON$-$ | 27 |
| | | |
| Oxido | $^{-}$O$-$ (ion) | 19,35 |
| Oxidodiazenediyl | $-N=N(O)-$ | 32 |
| Oxo | O= | 20,29 |
| Oxoethylene | $-COCH_2-$ | 20,29 |
| Oxonio | $H_2C^+-$ (ion) | 24 |
| Oxy | $-O-$ | 11,26 |
| Oxydicarbonyl | $-COOCO-$ | 17 |
| Oxydimethylene | $-CH_2OCH_2-$ | 11 |
| | | |
| Phenyl | $C_6H_5-$ | 3 |
| Phosphinidene | HP= or HP$<$ | 36 |
| Phosphinidyne | P$\equiv$ or $-P<$ | 36 |
| Phosphinimyl | $H_2P(:NH)-$ | 36 |
| Phosphino | $H_2P-$ | 36 |
| Phosphinothioyl | $H_2(S)P-$ | 36 |
| Phosphinyl | $H_2(O)P-$ | 36 |
| Phosphinylidene | H(O)P= or H(O)P$<$ | 36 |
| Phosphinylidyne | (O)P$\equiv$ or (O)P$\leqq$ | 36 |
| Phosphonio | $H_3P^+-$ (ion) | 36 |
| Phosphono | $(HO)_2(O)P-$ | 36 |
| Phosphoranyl | $H_4P-$ | 36 |
| Plumbyl | $H_3Pb-$ | 33 |
| 1-Propenyl | $CH_3CH=CH-$ | 2 |
| 2-Propenyl | $CH_2=CHCH_2-$ | 2 |
| Propenylene | $-CH_2CH=CH_2-$ | 2,11 |
| Propylene | $-CH(CH_3)CH_2-$ | 2 |
| 2-Propynyl | $HC\equiv CCH_2-$ | 2 |
| | | |
| Siloxy | $H_3SiO-$ | 38 |
| Silyl | $H_3Si-$ | 38 |
| Stannyl | $H_3Sn-$ | 33 |
| Stibino | $H_2Sb-$ | 33 |

| | | |
|---|---|---|
| Sulfamoyl | $H_2NSO_2-$ | 17 |
| Sulfenamoyl | $H_2NS-$ | 17 |
| Sulfeno | $HOS-$ | 17 |
| Sulfido | $^-S-$ (ion) | 19,35 |
| Sulfinamoyl | $H_2NSO-$ | 17 |
| Sulfino | $HOSO-$ | 17 |
| Sulfinyl | $-S(O)-$ | 39 |
| Sulfo | $HOSO_2^-$ | 17 |
| Sulfonio | $H_2S^+-$ (ion) | 19,35 |
| Sulfonyl | $-SO_2-$ | 39 |
| Sulfonyldioxy | $-OSO_2O-$ | 11 |
| | | |
| Thio | $-S-$ | 39 |
| Thiocarbamoyl | $H_2NCS-$ | 21 |
| Thiocarbonyl | $-CS-$ | 29 |
| Thiocarbonylamino | $SCN-$ | 32 |
| Thiocarboxy | $HSOC-, HOSC-$ | 16 |
| Thiocyanato | $NCS-$ | 32 |
| Thioformyl | $HCS-$ | 20 |
| Thioxo | $S=$ | 29 |
| 1-Triazanyl | $H_2NNHNH-$ | 32 |
| 1,3-Triazenediyl | $-N=NNH-$ | 32 |
| 2-Triazen-1-yl | $HN=NNH-$ | 32 |
| Trimethylene | $-CH_2CH_2CH_2-$ | 2 |
| | | |
| Ureido | $H_2NCONH-$ | 21 |
| Ureylene | $-NHCONH-$ | 11,21 |
| | | |
| Vinyl | $CH_2=CH-$ | 2 |
| Vinylidene | $CH_2=C=$ or $CH_2=C<$ | 2 |

# INDEX

The text of this book is set in 11 point Journal Roman with two points of leading. The chapter numerals are set in 30 point Garamond; the chapter titles are set in 18 point Times Roman Bold. All pages are reduced to 93% of typeset size.

The book is printed offset on Danforth 550 Machine Blue White text, 50-pound. The cover is Joanna Book Binding blue linen.

Jacket design by Norman Favin.
Editing and production by Mary C. Westerfeld

The book was composed by William B. and Adrienne Lodder, Rockville, Md., printed and bound by The Maple Press Co., York, Pa.